TK 453 ROB

UNIVERSITY OF BRISTOL
ENGINEERING

UNIVERSITY OF BRISTOL
QUEEN'S BUILDING LIBRARY
UNIVERSITY WALK
BRISTOL BS8 1TR

Date due for return
UNLESS RECALLED EARLIER

07 NOV 2000

21 NOV 2000

31/7/2002

22 JUL 2003

29 JAN 2008

08 MAY 2009

WITHDRAWN
University of Bristol

Electrical and Magnetic Properties of Materials

The Artech House Materials Science Library

DiLorenzo, James V., and Deen D. Khandelwal, **GaAs FET Principles and Technology**

Gilmour, A.S., Jr., **Microwave Tubes**

Granatstein, Victor L., and Igor Alexeff, eds., **High-Power Microwave Sources**

Robert, Philippe, **Electrical and Magnetic Properties of Materials**

Soares, Robert A., ed., **GaAs MESFET Circuit Design**

Williams, Ralph E., **Gallium Arsenide Processing Techniques**

Electrical and Magnetic Properties of Materials

Philippe Robert

Artech House

Library of Congress Cataloging-in-Publication Data

Robert, P. (Philippe)

Electrical and magnetic properties of materials.

Translation of: Matériaux de l'électrotechnique.

Bibliography: p.

Includes index.

1. Electric engineering—Materials. 2. Materials—Electric properties.
3. Materials, Magnetic properties. I. Title

TK453.R5613 1988 620.1'1297 88-6357

ISBN 0-89006-262-5

Copyright © 1988

ARTECH HOUSE, INC.
685 Canton Street
Norwood, MA 02062

All rights reserved. Printed and bound in the United States of America. No part of this book may be reproduced or utilized in any form or by any means, electronic or mechanical, including photocopying, recording, or by any information storage and retrieval system, without permission in writing from the publisher.

International Standard Book Number: 0-89006-262-5
Library of Congress Catalog Card Number: 88-6357

Translation of *Matériaux de l'électrotechnique*, originally published in French as Volume II of the *Traité d'Électricité* by the Presses Polytechniques Romandes, Lausanne, Switzerland. © 1979.

10 9 8 7 6 5 4 3 2 1

WITHDRAWN
University of Bristol

UNIVERSITY
OF BRISTOL
LIBRARY

ENGINEERING

Contents

Introduction

Engineering science essentially deals with the concrete world. Engineers therefore need materials in order to carry out their work.

The nature of the materials available to them is so important for humanity that it has been used to characterize significant epochs in the past: the Stone Age, the Bronze Age, the Iron Age. Today, it is no exaggeration to say that a highly significant part of technological development relies directly on the advances in the science of materials arising from basic disciplines such as physics and chemistry. We will take three of the most spectacular cases to serve as examples. Pocket calculators and powerful computers owe their existence to the methods of purification and doping of semiconductor materials. The production and transport of electricity would be inconceivable on the scale that we know today without the magnetic materials used for the construction of large-scale generators and transformers. Finally, the rise of the air transport industry results directly from the development of the metallurgy of high-resistance light alloys.

In the calculations of the electrical engineer, materials most often appear in the form of a simple letter in an equation, ρ for the resistivity, ε for the permittivity, and μ for the permeability, in Maxwell's equations. In addition, other symbols represent, for example, a loss tangent, thermal conductivity, and elastic limit.

This book presents the relationships, in quantitative terms when this is possible, connecting these parameters to the corresponding mechanisms that take place inside matter. The book is aimed at making the reader capable of judiciously choosing and using materials as a function of the applications, and of understanding their role and their limitations in components and systems.

GENERAL ORGANIZATION OF THE BOOK

Although the electrical engineer can be faced with problems relating to all the physicochemical properties of materials, it is obvious that the electric and magnetic properties are of particular interest to him or her. This is the reason that Chapter 2 (conducting properties), Chapter 3 (magnetic properties) and Chapter 4 (dielectric properties) are by far the largest. These chapters actually form the heart of the work. Chapter 1 provides the indispensible elements of the physics of the solid state to which a large number of references are subsequently made. Chapter 5 deals with thermodynamic transformations the importance of which, in the manufacture of industrial materials, has been emphasized at appropriate points. Chapter 6 gives a summary of mechanical properties, whereas Chapter 7, consisting of various appendices, provides the proof for the main statistical distributions used and certain additional information relating to the preceding chapters. Several tables summarize the essential electric and magnetic properties of the most important industrial materials.

An adequate knowledge of the basic principles of solid state physics makes it possible to read Chapters 2–6 in any order, referral to the indispensable knowledge of the preceding sections being provided either by the set of references or by the subject index.

The approach adopted throughout the book differs from the one used in physics, on the one hand, and the one used in the science of materials, on the other hand, by the fact that the selection of the models, their presentation, as well as the choice of material in general, has been made in the light of the benefit to the electrical engineer.

CONVENTIONS

The *Treatise on Electricity* consists of volumes (Vol.) referred to by a Roman numeral (Vol. V). Each volume is divided into chapters (Ch.) referenced by an Arabic number (Ch. 2). A complete list of all volumes in the series, including those translated by Artech House, is given in the Select Bibliography at the end of this book.

Each chapter is divided into sections referenced by two Arabic numbers separated by a decimal point (Section 2.3). Each section is divided into subsections referenced by three Arabic numbers separated by two decimal points (Section 2.3.11). The internal references specify the volume, the chapter, the section, or the subsection of the Treatise to which reference is being made. In the case of reference to a part of the same volume, the volume number is omitted.

The literature references are numbered consecutively per volume and are referenced by a single Arabic number in square brackets; pages may be given in parentheses: [33] (pp. 12–15).

A term appears in light italics the first time it is defined in the text. An important passage is indicated by italics.

The displayed equations are numbered consecutively by chapter and are referenced by two Arabic numbers in parentheses and separated by a decimal point (3.14). The figures and tables are numbered consecutively in a common sequence by chapter, and are referenced by two Arabic numbers preceded by Fig. (Fig. 4.12) or Table (Table 4.13).

Chapter 1
The Constitution of Matter

1.1 ELEMENTARY PARTICLES AND HYDROGEN ATOMS

1.1.1 History of the Theories of Matter

About 35,000 years ago, Homo Sapiens appeared on earth. Some 30,000 years later, *circa* 3,000 BC, the invention of writing and arithmetic helped solve the commercial problems of the time. Human concerns which would now be called scientific were then geometry, astronomy, and geography. We had to wait another 2,500 years for the first theory of matter to emerge around 450 BC: this involved the four elements, fire, air, water, and earth, propounded by Empedocles. A little later, Leucippus and then Democritus (about 420 BC) developed the first atomic theory in history. What remains of this in modern physics?

The theory of Democritus can be summarized by three propositions:

- all matter is composed of very small indivisible particles, atoms;
- all atoms are formed from the same substance;
- all the differences that are observed between different materials can be explained in terms of the form, order, and position of atoms.

While accepting that the atoms of Democritus are what we ourselves call atoms, the first proposition still has to be modified. However, this is not fundamentally important because what one does today is just to go a little further into the infinitely small world of particles which are called elementary. The second and third propositions can be kept as they are. They clearly foreshadow some of the chapters of modern physics such as the study of crystalline media.

Once propounded, the atomic theory of Democritus was never completely forgotten, but we had to wait until the 18th century and especially the 19th century for chemical experiments to give atoms a more concrete form. The concepts of atomic weight and valence date from the 19th century, and it was in 1869 that Mendeleev published his periodic table of the elements.

1.1.2 Elementary Particles and Atoms: Definition

Any particle that cannot be decomposed into other particles is called an *elementary particle*.

By 1910, the atom was considered to be formed by a positive nucleus with electrons revolving around it. It was then found that the nucleus is formed of neutrons and protons. This was a very satisfactory state of knowledge for the human intellect. All matter was considered to be formed by the combination of just three elementary particles: the proton, the electron, and the neutron.

It was not long before this simplistic picture of reality was destroyed by the subsequent discovery of new particles. Today, the number of known particles numbers several hundred, and it has become very difficult to know which of them deserve to be called elementary.

For most of our needs, it is sufficient to know about four particles, the main characteristics of which are summarized in Table 1.1.

Table 1.1

Particle	*Resting Mass (kg)*	*Charge (C)*	*Magnetic Moment* ($A \cdot m^2$)	*Spin*
Photon	0	0	0	1
Electron	$9.109 \cdot 10^{-31}$	$-1.602 \cdot 10^{-19}$	$9.27 \cdot 10^{-24}$	1/2
Neutron	$1.675 \cdot 10^{-27}$	0	$-9.64 \cdot 10^{-27}$	1/2
Proton	$1.673 \cdot 10^{-27}$	$1.602 \cdot 10^{-19}$	$1.41 \cdot 10^{-26}$	1/2

If we combine neutrons, protons, and electrons, obeying the following rules:

- the number of protons is equal to the number of electrons;
- the light atoms (except hydrogen) have more or less the same number of protons and neutrons, whereas the heavy atoms have more neutrons than protons,

it is theoretically possible to construct the atoms of all the elements. The simplest of them all, the hydrogen atom, is formed by only one proton and one electron.

The theoretical study of the hydrogen atom by quantum mechanics was one of the first successes of this discipline. The quantum model of the hydrogen atom is interesting from several points of view. In particular,

- it gives the most general behavior of this atom by using relatively simple calculations;
- it serves as the basis for constructing more complex models which allow the elements to be arranged logically in the periodic table and give a better understanding of the properties of groups of elements appearing in the table.

1.1.3 Classical Mechanics *versus* Quantum Mechanics

A physical system is correctly described by one of these disciplines depending on the order of magnitude of its action [1], or on other variables such as its energy or temperature. When the action is large compared to Planck's constant $h = 6.626 \cdot 10^{-34}$ J · s, classical mechanics is sufficient, otherwise it is necessary to use quantum mechanics.

1.1.4 Quantum Model of the Hydrogen Atom

The elementary analysis of the hydrogen atom is based on the following two simplifications:

- the spins (Section 1.2.2) of the electron and proton are omitted because their effects can be neglected in the first analysis;
- the proton, approximately 2,000 times heavier than the electron, is considered to be stationary, which justifies the use of the absolute frame of reference in Fig. 1.2.

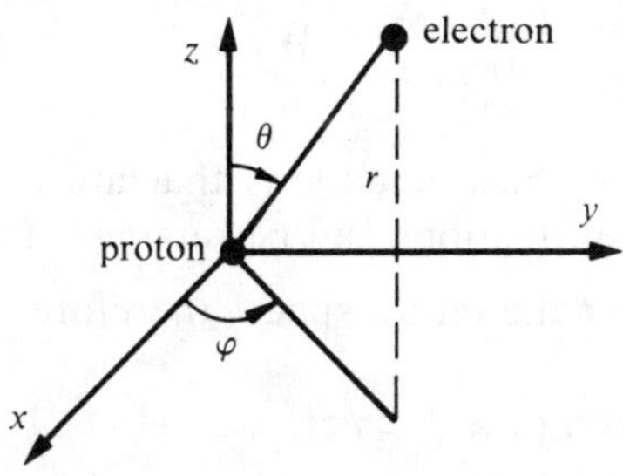

Fig. 1.2

The problem is thus reduced to the analysis of the behavior of a point charge in a potential $V(r)$ (Vol. III, 3.3.1), satisfying the expression:

$$V(r) = \frac{1}{4\pi\varepsilon_0}\frac{e}{r} \tag{1.1}$$

The constants ε_0 and e represent the permittivity in a vacuum and the absolute value of the electronic charge, respectively.

The electron-proton system is a typical object of study by quantum mechanics. Since the problem is independent of time, the Schrodinger equation [1] has the form:

$$\mathrm{H}\Psi = W\Psi \tag{1.2}$$

where Ψ is the wave function of the system. This function depends on the space coordinates and has the property that $|\Psi|^2\,\mathrm{d}\Omega$ expresses the probability of finding the electron inside the volume $\mathrm{d}\Omega$. On the other hand, W represents the total energy of the electron in the state described by Ψ, and H denotes the Hamiltonian operator:

$$\mathrm{H} = -\frac{\hbar^2}{2m_\mathrm{n}}\nabla^2 + W_\mathrm{p} \tag{1.3}$$

In this expression, m_n denotes the mass of the electron and W_p is its potential energy. According to standard notation $\hbar = h/2\pi$ and ∇^2 represents the Laplace operator.

Expressed in terms of the frame of reference in Fig. 1.2, the Schrödinger equation (1.2) takes the form:

$$\frac{\hbar^2}{2m_\mathrm{n}}\left[\frac{1}{r^2}\frac{\partial}{\partial r}\left(r^2\frac{\partial\Psi}{\partial r}\right) + \frac{1}{r^2\sin\theta}\frac{\partial}{\partial\theta}\left(\sin\theta\frac{\partial\Psi}{\partial\theta}\right)\right.$$

$$\left. + \frac{1}{r^2\sin^2\theta}\frac{\partial^2\Psi}{\partial\varphi^2}\right] + \left(\frac{1}{4\pi\varepsilon_0}\frac{e^2}{r} + W\right)\Psi = 0 \tag{1.4}$$

Only solutions of these equations that are in agreement with the properties of the wave functions can be retained. In particular,

- Ψ is uniform over the entire space, therefore,

$$\Psi\,(r, \theta, \varphi) = \Psi\,(r, \theta, \varphi + 2n\pi), \tag{1.5}$$

where n is any integer.

- Ψ satisfies the normalization condition:

$$\int |\Psi(r, \theta, \varphi)|^2 \, d\Omega = 1, \tag{1.6}$$

the integral being over the entire space. This condition implies that

$$\lim_{r \to \infty} \Psi(r, \theta, \varphi) = 0 \tag{1.7}$$

- Finally, Ψ must be sufficiently regular at the origin, in particular,

$$\Psi(r = 0, \theta, \varphi) < \infty \tag{1.8}$$

Equation (1.4), subject to conditions (1.5), (1.6), and (1.8), *can only be integrated for certain values of W called eigenvalues*. In general, for each eigenvalue, (1.4) has several solutions which may all be obtained in closed form. The procedure for obtaining these solutions is rather laborious [2] and is abbreviated here for the sake of simplicity.

Following the variable separation method, the wave function is decomposed into a product of three variables R, Θ, and Φ, each of which depends on only one variable:

$$\Psi(r, \theta, \varphi) = R(r)\,\Theta(\theta)\,\Phi(\varphi) \tag{1.9}$$

By substituting (1.9) in (1.4), we obtain an equation in which one side depends only on the variables r and θ, and the other side depends only on φ. Each side of this equation must therefore be equal to a constant. In fact, if we only vary φ, the side of the equation that depends only on r and θ remains constant. As it is equal to the expression that depends only on φ, the latter is also a constant, traditionally denoted by m^2. The same reasoning applies to the variations of r and θ.

The expression depending on φ can be integrated immediately. This gives

$$\Phi(\varphi) = A \exp(jm\varphi) \tag{1.10}$$

where A is an integration constant. The periodicity condition (1.5) means that m must be an integer:

$$m = 0, \pm 1, \pm 2 \ldots \tag{1.11}$$

By arranging the terms of the expression that is a function of r and θ, we obtain two ordinary differential equations, one in terms of θ and the other in terms of r. These equations each have a constant on the right-hand side, where the solution of the equation in θ may be written in the form $(l + 1)$.

In solving for Θ, we start with $m = 0$. For the solution to be regular at the origin and satisfy (1.6), it is necessary that l is either a positive integer or zero. By then taking $m \neq 0$ and by retaining the result that l is a positive integer or zero, it appears that the only regular solutions satisfying (1.6) correspond to cases when $m \leqslant l$.

Finally, R is found by making appropriate changes of variables. The choice of acceptable solutions makes a new constant n appear, which can only take positive integer values, less than or equal to $l + 1$.

1.1.5 Quantum Numbers: Comments

In the language of mathematics, m, l, and n characterize the eigenvalues of the partial differential equations for the functions Φ, Θ, and R, associated with the regularity conditions at the origin, and satisfying (1.5) and (1.6).

Physicists call m, l, and n quantum numbers, each with a special name (Table 1.3).

Table 1.3

$n = 1, 2, 3$	principal quantum number
$l = 0, 1, 2, \ldots n - 1$	azimuthal quantum number
$m = 0, \pm 1, \pm 2, \ldots \pm l$	magnetic quantum number

Since the form of the wave function depends on the three quantum numbers, Ψ will have the values of these three numbers as subscripts when necessary in the order n, l, m. The same applies to functions calculated from Ψ.

The above calculations produce a probabilistic picture of the hydrogen atom. It is a map giving the probability for each triplet (n, l, m) of observing the electron at a given point in space. More exactly, the probability of finding the electron in the volume $d\Omega$ surrounding the point with coordinates (r, θ, φ) is equal to

$$|\Psi_{nlm}(r, \theta, \varphi)|^2 \, d\Omega \tag{1.12}$$

The function (1.12) is shown in Sections 1.1.7 and 1.1.10 for a few simple cases.

1.1.6 Definition of an Orbit

Each solution Ψ_{nlm} is called an *orbit*. An orbit gives a complete description of the electron in the state corresponding to the values specified by the three quantum numbers. This means that it contains all the information that it is possible to know about the electron, such as its position, kinetic energy, and potential energy.

1.1.7 Orbit of the Hydrogen Atom in the Ground State

By definition, the ground state is that of minimum energy. It corresponds to the case $n = 1$, $l = 0$, $m = 0$. The wave function reduces to

$$\Psi_{100}(r, \theta, \varphi) = \frac{1}{\sqrt{\pi}} \left(\frac{1}{r_0}\right)^{3/2} \exp\left(-r/r_0\right) \tag{1.13}$$

where

$$r_0 = \frac{4\pi h^2 \,\varepsilon_0}{\varepsilon^2 \, m_{\mathrm{n}}} \tag{1.14}$$

The probability of observing the electron at a distance lying between r and $r + \mathrm{d}r$ from the nucleus is

$$P_{100}(r)\,\mathrm{d}r = |\Psi_{100}(r)|^2 \, 4\pi r^2 \,\mathrm{d}r \tag{1.15}$$

The most probable distance from the electron to the nucleus is obtained by finding the value of r that makes $P(r)$ a maximum:

$$\frac{\mathrm{d}P(r)}{\mathrm{d}r} = 0 = \frac{8}{r_0^3}\left(r - \frac{r^2}{r_0}\right) \exp\left(-2r/r_0\right) \tag{1.16}$$

from which we have

$$r = r_0 \tag{1.17}$$

We call r_0 the *radius of the first Bohr orbit*. According to (1.14), $r_0 = 0.0529$ nm.

This concept of an orbit must not be interpreted in a deterministic and rigid manner. The orbit is simply the geometric locus of points where the probability of finding the electron is a maximum (Fig. 1.4).

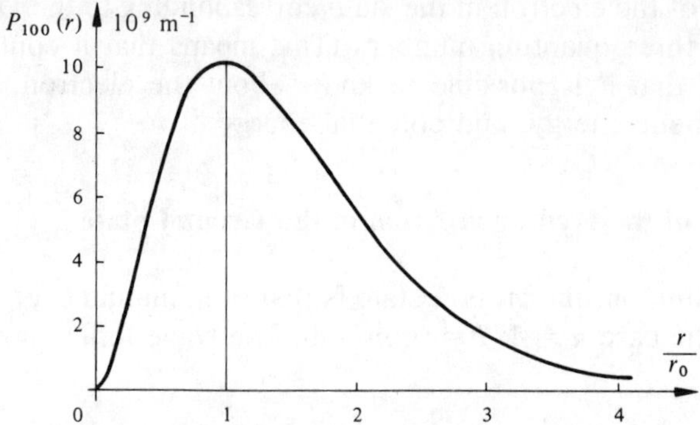

Fig. 1.4 Probability Density for Finding the Electron at a Distance *r* from the Nucleus (1.15).

1.1.8 Radius of the Hydrogen Atom in the Ground State

It can be assumed that the radius of the atom is equal to the distance r from the nucleus to the electron. An observer would find that, at each observation, the radius of the atom takes a different value, but most often it remains close to r_0. The atomic radius is a random variable with its most probable value r_0.

In order to illustrate the fluctuations in r, it is worth calculating the probability of observing an atom with radius $0.1r_0$, and then $10r_0$. By referring these probabilities to those of observing an atom with radius r_0, we have according to (1.15):

$$\frac{P_{100}(r)}{P_{100}(r_0)} = \frac{r^2}{r_0^2} \exp\left[2(1 - r/r_0)\right] \tag{1.18}$$

from which

$$\frac{P_{100}(10r_0)}{P_{100}(r_0)} = 1.52 \cdot 10^{-6} \quad \text{and} \quad \frac{P_{100}(0,1r_0)}{P_{100}(r_0)} = 6.05 \cdot 10^{-2} \tag{1.19}$$

In round numbers, we therefore have one chance out of 17 to observe an atom with radius $r = 0.1r_0$ and one chance out of 658,000 to observe an atom with radius $r = 10r_0$.

1.1.9 The Energy *W* of the Hydrogen Atom in the Ground State

This energy appears as an eigenvalue of (1.4) (Section 1.1.4), when this equation is integrated, leading to the wave function Ψ_{100}. The details of these calculations are not given, but it is now worth finding this energy by introducing the solution Ψ_{100} in (1.4). In this case, where the wave function does not depend on θ or φ, (1.4) reduces to

$$\frac{\hbar^2}{2m_n}\frac{1}{r^2}\frac{\partial}{\partial r}\left(r^2\frac{\partial\Psi}{\partial r}\right) + \left(\frac{1}{4\pi\varepsilon_0}\frac{e^2}{r} + W\right)\Psi = 0 \tag{1.20}$$

By setting $\Psi = \Psi_{100}$ in this equation and, for the sake of simplification, writing

$$A = \frac{1}{\sqrt{\pi}}\left(\frac{1}{r_0}\right)^{3/2} \tag{1.21}$$

we obtain

$$\frac{\hbar^2}{2m_n}\left(-\frac{2A}{r\,r_0}\exp(-r/r_0) + \frac{A}{r_0^2}\exp(-r/r_0)\right) + \left(\frac{1}{4\pi\varepsilon_0}\frac{e^2}{r} + W\right)A\exp(-r/r_0) = 0 \tag{1.22}$$

This equation must be satisfied for every positive value of r other than zero. It follows that the coefficients of $1/r\,\exp(-r/r_0)$, on the one hand, and of $\exp(-r/r_0)$, on the other hand, must be zero. This leads to the following relationships, respectively:

$$\frac{\hbar}{m_n r_0} = \frac{e^2}{4\pi\varepsilon_0} \tag{1.23}$$

and

$$W = -\frac{\hbar^2}{2m_n r_0^2} = -\frac{m_n e^4}{8\hbar^2\varepsilon_0^2} \tag{1.24}$$

The first equation does not contribute anything new because it once again gives the value of r_0 given by (1.14). The second can be used to calculate

$$W = -2{,}18 \cdot 10^{-18}\mathrm{J} \mathrel{\hat{=}} -13{,}6\ \mathrm{eV} \tag{1.25}$$

Experiments in spectroscopy showed well before 1928 (Schrödinger's equation) that the lowest energy level in the hydrogen atom is at exactly −13.6 eV. It was a considerable success for the infant discipline of quantum mechanics to find this value by a purely theoretical approach.

1.1.10 Orbits of the Hydrogen Atom in an Excited State: Particular Cases

We consider here cases for any n, with $l = 0$ and $m = 0$. The wave function therefore keeps a spherical symmetry and takes the form:

$$\Psi_{n00}(r, \theta, \varphi) = A_n \exp\left(-\frac{r}{nr_0}\right) \mathrm{L}_n^1\left(\frac{2r}{nr_0}\right) \tag{1.26}$$

where L_n^1 is a Laguerre associated polynomial of the first order and degree $n - 1$ (Section 7.1). The probability density for finding the electron at a distance r from the nucleus : $P_{n00}(r)$ is shown in Fig. 1.5. It can be seen that this function has a number of maxima equal to n, and the most probable radius increases with n.

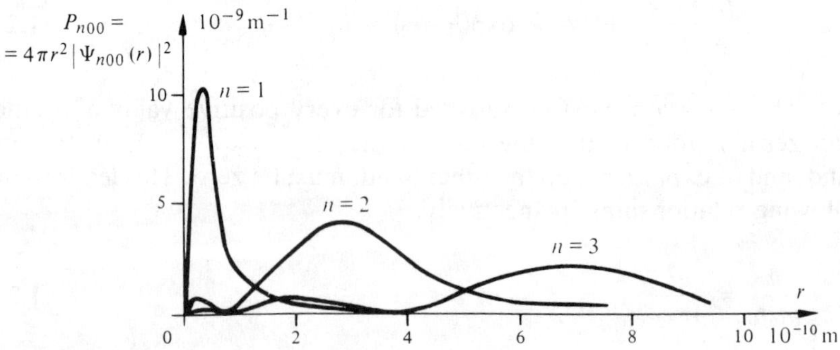

Fig. 1.5

The energy W_n of the atom is given by

$$W_n = -\frac{1}{n^2}\frac{\hbar^2}{2m_n r^2} \tag{1.27}$$

It can be seen that this expression coincides with (1.24) if $n = 1$. On the other hand, it can be shown that (1.27) remains valid in the general case when n, l, and m can take any value.

The interpretation of this is simple. When n increases, the negative energy of the electron approaches zero. Now, expression (1.1), used to describe the potential $V(r)$, implies that a free electron at rest has zero energy. Therefore, the larger the n, the less the electron is tied to the nucleus and the greater is the most probable radius of the atom.

1.1.11 Orbits of the Hydrogen Atom: General Case

As soon as $l \geqslant 1$, the orbits and the probability density $P_{nlm} = |\Psi_{nlm}|^2$ have an angular dependence. However, the form of Φ, given by (1.10), shows that P_{nlm} remains independent of φ, and therefore keeps a symmetry of rotation about the z-axis. This is why it is customary to represent P_{nlm} in a polar diagram (Fig. 1.6), plotted in a plane containing the z-axis.

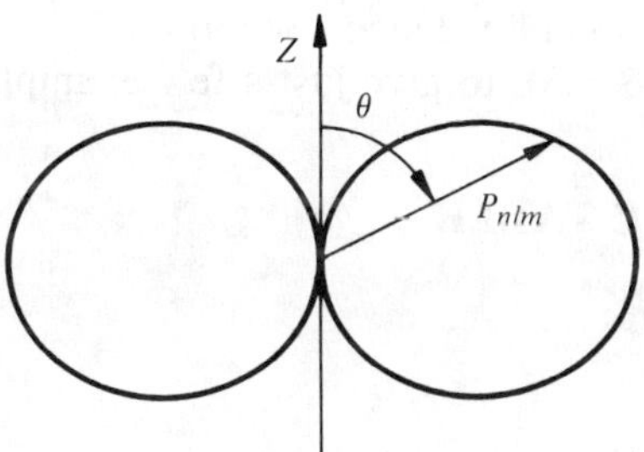

Fig. 1.6

The shape of P_{nlm} quickly becomes complicated when l increases, as Fig. 1.7 shows, where all the forms of this function have been shown schematically for $l = 3$.

1.1.12 Comments

The probability density in atoms heavier than hydrogen has lobes similar to those of Fig. 1.7, although with an even more complex form.

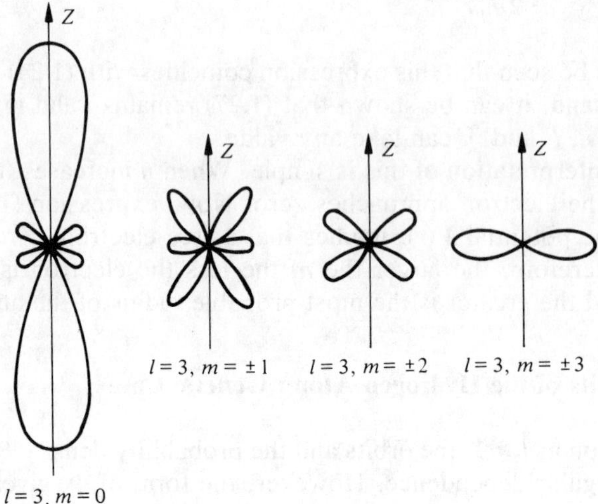

Fig. 1.7

The existence of these lobes plays a very important part in a large number of areas such as those of valence (Section 1.3) or the superexchange interaction (Section 3.5.3), to give just a few examples.

1.2 THE PERIODIC TABLE

1.2.1 Introduction

After having obtained remarkable results by quantum mechanics as described above for the hydrogen atom, it is natural to use the same tool for studying more complex atoms. To arrive finally at the periodic table of Mendeleev by specifying all the electron states for all the elements and, consequently, their properties to a certain extent, would be a highly valuable operation.

Unfortunately, the complexity of the calculations increases very quickly when the atom has more than one electron so that the use of approximation methods is unavoidable. The simplest of these, Hartree's method, will be briefly described.

First of all, it is necessary to recall two imporant concepts: the spin and Pauli's principle. Although both of these concepts are derived from relatively complex theories, they can be interpreted in a simple manner.

1.2.2 Electron Spin: Definitions

Around 1925, Uhlenbech and Goudsmit showed that several experimental results which could not be predicted by the theories of the time could be explained, provided that it was assumed that the electron had an intrinsic kinetic moment. This intrinsic kinetic moment is called the *spin* and denoted $\boldsymbol{S}$.

From a classical point of view, it can be imagined that the spin is due to a rotation of the electron about itself. Since it is charged, it can be easily seen that this rotation gives the electron its own magnetic moment called the *spin magnetic moment*.

Although plausible, the picture of the electron rotating about itself must be regarded with care because it is only an analogy to represent phenomena that take place in an infinitely small dimension with the help of models borrowed from the real world.

There is a quantum number s associated with the projection S_z of $\boldsymbol{S}$ on a z-axis, defined, for example, by an external magnetic field. This number can only take the values $+1/2$ and $-1/2$, and it is related to S_z by the expression:

$$S_z = s\hbar \tag{1.28}$$

The spin is a concept that is totally compatible with the nonrelativistic quantum mechanics of Schrödinger, based on (1.2) and (1.3). However, this theory does not predict the existence of spin, which must be introduced in an independent manner. Starting from the same postulates as Schrödinger, Dirac (1929) developed a relativistic quantum theory, which *proved* that the electron had a spin and $s = \pm 1/2$.

Taking the spin into account, it is therefore necessary to have four quantum numbers n, l, m, and s for completely defining the state of the electron in the hydrogen atom.

1.2.3 Occupation of Energy Levels

The study of atoms with several electrons, as in the case of the hydrogen atom, reveals the existence of a theoretically infinite number of *possible* energy levels for the electrons. Hartree's model (Section 1.2.7) also makes it possible to associate a set of four quantum numbers n, l, m, and s with each energy level.

In the ground state, which is the state corresponding to the minimum energy, all the electrons should occupy the lowest energy level. Assuming that (1.27) is still valid, this level would correspond to

$$n = 1, \quad l = 0, \quad m = 0, \quad s = \pm 1/2 \tag{1.29}$$

However, experience (observation of spectra) absolutely contradicts this point of view. In the ground state, *the system of electrons certainly has the minimum possible energy, but each electron does not have the minimum possible energy for an electron*. In reality, the number of available places for electrons is restricted in each level. This is a consequence of Pauli's exclusion principle.

1.2.4 Statement of Pauli's Exclusion Principle

This principle can be expressed in a more or less general manner. The simplest statement that can be used directly in Hartree's model is as follows. In an atom with more than one electron, two electrons cannot have the same set of quantum numbers n, l, m, and s.

1.2.5 Electronic States of the Helium Atom

Although still very simple, the helium atom already leads to appreciably more complex mathematical calculations than does the hydrogen atom. In an absolute frame of reference centered on the nucleus, three distances, r_1, r_2, and r_{12}, must be taken into consideration for the same reasons as previously.

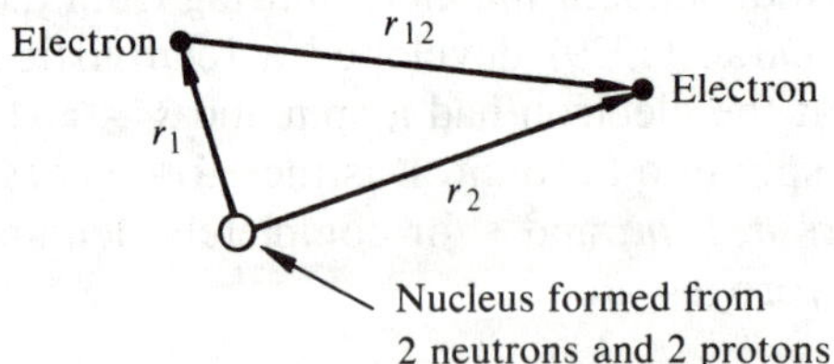

Fig. 1.8 Helium Atom

The potential energy of this system contains three terms: the first two represent the potential energy of each electron in the field of the nucleus, and the third is the potential energy of the subsystem formed by the two electrons. The subscripts 1 and 2 are used to distinguish the two electrons. By neglecting the spin, Schrödinger's equation takes the form:

$$-\frac{\hbar^2}{2m_n}(\nabla_1^2\Psi + \nabla_2^2\Psi) + \frac{1}{4\pi\varepsilon_0}\left(-\frac{2e^2}{r_1} - \frac{2e^2}{r_2} + \frac{e^2}{r_{12}}\right)\Psi = W\Psi \tag{1.30}$$

where Ψ depends on six spatial variables, which are the three coordinates of each electron.

This equation cannot be solved in closed form. The separation of the variables is made impossible by the presence of the term e^2/r_{12}, representing the electron-electron interaction. It is therefore necessary to resort to numerical methods and, among these, the variation theorem is the most useful.

In simple terms, the procedure is as follows for the ground state. A first reasonable estimate for Ψ is established, based on the solutions obtained in the case of the hydrogen atom, for example. The energy of the system can then be calculated due to the orthogonal properties of the wave functions. The quality of this first estimate must then be assessed with respect to a second estimate, which, in principle, should be better. The assessment criterion is given by the variation theorem, which states here that the exact solution for Ψ minimizes the energy. Algorithms exist to help find a better function Ψ at each step. In this way, it is possible to find a wave function as exact as necessary by iteration.

1.2.6 Electronic States of Atoms Where $Z > 2$

Let us consider, for example, an atom of gallium, of considerable practical interest in electro-optics for the manufacture of light-emitting diodes, or in certain high-power systems where particular alloys of gallium are used as superconductors.

Although gallium, which has 31 electrons, is still a relatively light element, the approach used for helium would lead to an excessively complex Schrödinger equation:

- the function Ψ would depend on 93 spatial variables;
- the potential energy would contain 496 terms, or 465 electron-electron interaction terms plus 31 electron-nucleus interaction terms.

It is not possible to solve such an equation because its existence has more of a philosophical than practical interest! Under these conditions, it is necessary to simplify the atomic model, and so Hartree's model [3] forms the basis for a large number of calculations carried out these days in this area.

1.2.7 Hartree's Atomic Model

The main feature of this model is a simplified method for calculating the electron-electron interaction energy. Instead of individually and successively considering all possible electron pairs, Hartree made the assumption that all the orbits were spherical. Each electron therefore evolves in a spherical potential, which is the sum of the nuclear potential and that created by the $N - 1$ other electrons. In this situation, Schrödinger's equation for N electrons can be decomposed into N equations for one electron. These equations are of the same type as used for the hydrogen atom, except that the potential $V(r)$ is no longer in terms of $1/r$, but has a more complex form. An iteration procedure makes it possible to find all the exact orbits according to the model, starting from arbitrarily chosen initial orbits.

These orbits, solutions of Schrödinger equations with a spherical potential, can still be identified by four quantum numbers n, l, m, and s.

All the electrons with the same n form an atomic shell designated by a capital letter.

Table 1.9
Equivalence between the quantum number *n* and atomic shell.

n	1	2	3	4	5	6	7
Shell	K	L	M	N	O	P	Q

Each of these shells is divided into subshells corresponding to the values of l, and designated by lower-case letters.

Table 1.10
Equivalence between the quantum number *l* and the atomic subshell.

l	0	1	2	3	4	5	6	7
Subshell	s	p	d	f	g	h	i	j

The electronic configuration of an atom is given by the sequence of occupied subshells. Each subshell is described in the notation xa^y, where a is the identification of the subshell by a letter from Table 1.10, x is the value of n, and y is the number of electrons in the subshell. For example, the configuration of aluminum ($Z = 13$) is written as

$$1s^2 \quad 2s^2 \quad 2p^6 \quad 3s^2 \quad 3p^1$$

1.2.8 Energy and Quantum Numbers

The observation of atomic emission spectra provides a precise and powerful means of comparing theoretically determined energy levels with actual levels.

Despite the relatively risky simplifying hypothesis of a potential with spherical symmetry, Hartree's model predicts rather exactly the energies of electrons belonging to one atom. The four quantum numbers n, l, m, and s therefore often serve to reference these energies. The variations in energy associated with the values of m and s, being much less than those associated with n and l, will be neglected here.

Figure 1.11 shows the energy W of the electrons in all the subshells forming shells K to Q as a function of the charge Z of the nucleus, according to Hartree's model. Although the energies are represented as continuous functions, it is obvious that only values corresponding to integer Z have any physical meaning.

For small values of Z, for example $Z \leqslant 18$ (argon), the energy of all the subshells of the same shell tend toward the same value when Z decreases. In fact, the reduction in the number of electrons causes a reduction in the relative importance of electron-electron interactions compared to electron-nucleus interactions. The result is an increasing resemblance of the atom in question to the hydrogen atom, the limiting case being when only the electron-nucleus interaction exists.

For high values of Z with small n, (i.e., in the deeper shells), another convergence is observed for the energies of the subshells belonging to the same shell. This is due to the increase in the electron-nucleus interaction, resulting from the increase in the charge of the nucleus. The energies of the outer subshells, where the electron-electron interaction is less influenced by the presence of the nucleus, remain quite different from each other.

For intermediate values of Z, the situation is more complex, but for a given shell, the energy always increases when l increases.

This rule is inadequate for classifying all the subshells in terms of increasing energy. If we consider the M shell, for example, the energy increases when we go from the 3s to the 3p subshells, and then to the 3d subshells, *but,* as Fig. 1.11 shows, the energy of 4s is less than that of 3d in a range of Z from 6 to 23, approximately. The subshells 1s, 2s, 2p, 3s, and 3p contain a total of 18 electrons so that the position of the level 4s between 3p and 3d can only affect the electronic configuration of the ground state of atoms where $Z > 18$. The position of the 4s level is responsible for the particular structure of the first period of transition metals (Section 1.2.14).

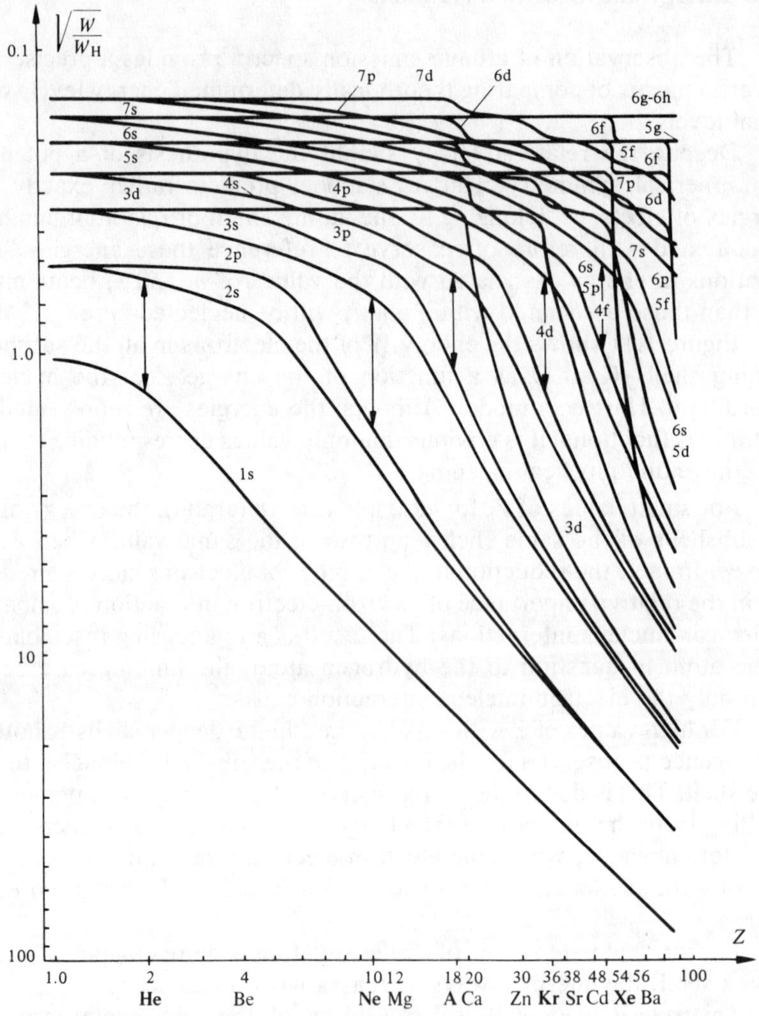

Fig. 1.11 W_H Represents the Energy of the Atom H in the Ground State (According to [4].)

The overlapping of the levels 3p, 4s, and 3d is related to the fact that the energy of 3d remains constant over a large interval of values of Z ($1 < Z \lesssim 18$).

While the subshells outside 3d are unoccupied, an electron finding itself alone in 3d is practically subjected to the potential of a virtual nucleus, carrying an invariable charge e, regardless of the occupation of the subshells inside 3d.

Examination of Fig. 1.11 shows that this is not an exceptional case. In each shell, starting from shell L, and in an ever more marked manner when n increases, the subshells corresponding to the highest values of l have a constant energy over an ever increasing range of Z. The overlapping between levels of different shells that results from this is responsible for a large number of irregularities in the periodic table, some of which form families: the transition metals already mentioned; the lanthanide elements; the actinide elements.

1.2.9 Quantum Interpretation of the Periodic Table

Several chemists in the second half of the 19th century observed that ranking the elements in increasing order of their atomic weight revealed a certain periodicity of their chemical and physical properties. In 1869, D. I. Mendeleev produced his famous table, the significance of which has not diminished ever since. Although the intuitive development of this table was one of the most significant milestones in the history of chemistry, its interpretation by quantum physics was also an extremely important step. During the course of this, an equation of modest appearance, Schrödinger's equation, revealed the solution of a large number of problems that had haunted chemists for centuries!

The theoretical derivation of the periodic table from the results of the Hartree model is very simple in principle. It is sufficient to list the possible orbits by combining the values of n, l, m, and s, following the rules of Table 1.3, and then to fill them with electrons obeying Pauli's exclusion principle and to choose the configuration corresponding to the minimum total energy, regardless of the number of electrons. The result of this operation is shown in Table 1.12. The entries in this table give the number of electrons in the subshells. These numbers are printed against a grey background when the subshells are complete.

The chemical properties of an element are determined by the behavior of the electrons that it has in the outer subshell or subshells. In order to form groups of chemically similar elements, it is sufficient to group those having an outer subshell that corresponds to the same value of l and has the same number of electrons. In the periodic table (Table 1.13), these groups appear in columns with the exception of the lanthanide elements and the actinide elements, which appear in rows. These groups are traditionally identified by letters and roman numerals.

Table 1.12 (first part)

Shell		K	L		M			N				O					P						Q	
Subshell		1s	2s	2p	3s	3p	3d	4s	4p	4d	4f	5s	5p	5d	5f	5g	6s	6p	6d	6f	6g	6h	7s	
Row 1*	1 H	1																						
	2 He	2																						
	3 Li	2	1																					
	4 Be	2	2																					
	5 B	2	2	1																				
Row 2*	6 C	2	2	2																				
	7 N	2	2	3																				
	8 O	2	2	4																				
	9 F	2	2	5																				
	10 Ne	2	2	6																				
	11 Na	2	2	6	1																			
	12 Mg	2	2	6	2																			
	13 Al	2	2	6	2	1																		
Row 3*	14 Si	2	2	6	2	2																		
	15 P	2	2	6	2	3																		
	16 S	2	2	6	2	4																		
	17 Cl	2	2	6	2	5																		
	18 A	2	2	6	2	6																		
	19 K	2	2	6	2	6		1																
	20 Ca	2	2	6	2	6		2																
	21 Sc	2	2	6	2	6	1	2																
	22 Ti	2	2	6	2	6	2	2																
	23 V	2	2	6	2	6	3	2																
	24 Cr	2	2	6	2	6	5	1																First period of transition metals
	25 Mn	2	2	6	2	6	5	2																
	26 Fe	2	2	6	2	6	6	2																
	27 Co	2	2	6	2	6	7	2																
Row 4*	28 Ni	2	2	6	2	6	8	2																
	29 Cu	2	2	6	2	6	10	1																
	30 Zn	2	2	6	2	6	10	2																
	31 Ga	2	2	6	2	6	10	2	1															
	32 Ge	2	2	6	2	6	10	2	2															
	33 As	2	2	6	2	6	10	2	3															
	34 Se	2	2	6	2	6	10	2	4															
	35 Br	2	2	6	2	6	10	2	5															
	36 Kr	2	2	6	2	6	10	2	6															
	37 Rb	2	2	6	2	6	10	2	6			1												
	38 Sr	2	2	6	2	6	10	2	6			2												
	39 Y	2	2	6	2	6	10	2	6	1		2												
	40 Zr	2	2	6	2	6	10	2	6	2		2												
	41 Nb	2	2	6	2	6	10	2	6	4		1												
	42 Mo	2	2	6	2	6	10	2	6	5		1												Second period of transition metals
	43 Tc	2	2	6	2	6	10	2	6	5		2												
	44 Ru	2	2	6	2	6	10	2	6	7		1												
	45 Rh	2	2	6	2	6	10	2	6	8		1												
Row 5*	46 Pd	2	2	6	2	6	10	2	6	10														
	47 Ag	2	2	6	2	6	10	2	6	10		1												
	48 Cd	2	2	6	2	6	10	2	6	10		2												
	49 In	2	2	6	2	6	10	2	6	10		2	1											
	50 Sn	2	2	6	2	6	10	2	6	10		2	2											
	51 Sb	2	2	6	2	6	10	2	6	10		2	3											
	52 Te	2	2	6	2	6	10	2	6	10		2	4											
	53 I	2	2	6	2	6	10	2	6	10		2	5											
	54 Xe	2	2	6	2	6	10	2	6	10		2	6											

* *See* Table 1.13 for the rows of the periodic table. (*Continued*)

Table 1.12 (continued)

Shell		K	L		M			N				O					P						Q	
Subshell		1s	2s	2p	3s	3p	3d	4s	4p	4d	4f	5s	5p	5d	5f	5g	6s	6p	6d	6f	6g	6h	7s	
Row 6*	55 Cs	2	2	6	2	6	10	2	6	10		2	6				1							
	56 Ba	2	2	6	2	6	10	2	6	10		2	6				2							
	57 La	2	2	6	2	6	10	2	6	10		2	6	1			2							Lanthanide elements
	58 Ce	2	2	6	2	6	10	2	6	10	2	2	6				2							
	59 Pr	2	2	6	2	6	10	2	6	10	3	2	6				2							
	60 Nd	2	2	6	2	6	10	2	6	10	4	2	6				2							
	61 Pm	2	2	6	2	6	10	2	6	10	5	2	6				2							
	62 Sm	2	2	6	2	6	10	2	6	10	6	2	6				2							
	63 Eu	2	2	6	2	6	10	2	6	10	7	2	6				2							
	64 Gd	2	2	6	2	6	10	2	6	10	7	2	6	1			2							
	65 Tb	2	2	6	2	6	10	2	6	10	9	2	6				2							
	66 Dy	2	2	6	2	6	10	2	6	10	10	2	6				2							
	67 Ho	2	2	6	2	6	10	2	6	10	11	2	6				2							
	68 Er	2	2	6	2	6	10	2	6	10	12	2	6				2							
	69 Tm	2	2	6	2	6	10	2	6	10	13	2	6				2							
	70 Yb	2	2	6	2	6	10	2	6	10	14	2	6				2							
	71 Lu	2	2	6	2	6	10	2	6	10	14	2	6	1			2							Third period of transition elements
	72 Hf	2	2	6	2	6	10	2	6	10	14	2	6	2			2							
	73 Ta	2	2	6	2	6	10	2	6	10	14	2	6	3			2							
	74 W	2	2	6	2	6	10	2	6	10	14	2	6	4			2							
	75 Re	2	2	6	2	6	10	2	6	10	14	2	6	5			2							
	76 Os	2	2	6	2	6	10	2	6	10	14	2	6	6			2							
	77 Ir	2	2	6	2	6	10	2	6	10	14	2	6	7			2							
	78 Pt	2	2	6	2	6	10	2	6	10	14	2	6	9			1							
	79 Au	2	2	6	2	6	10	2	6	10	14	2	6	10			1							
	80 Hg	2	2	6	2	6	10	2	6	10	14	2	6	10			2							
	81 Tl	2	2	6	2	6	10	2	6	10	14	2	6	10			2	1						
	82 Pb	2	2	6	2	6	10	2	6	10	14	2	6	10			2	2						
	83 Bi	2	2	6	2	6	10	2	6	10	14	2	6	10			2	3						
	84 Po	2	2	6	2	6	10	2	6	10	14	2	6	10			2	4						
	85 At	2	2	6	2	6	10	2	6	10	14	2	5	10			2	5						
	86 Rn	2	2	6	2	6	10	2	6	10	14	2	6	10			2	6						
Row 7*	87 Fr	2	2	6	2	6	10	2	6	10	14	2	6	10			2	6					1	
	88 Ra	2	2	6	2	6	10	2	6	10	14	2	6	10			2	6					2	
	89 Ac	2	2	6	2	6	10	2	6	10	14	2	6	10			2	6	1				2	Actinide elements
	90 Th	2	2	6	2	6	10	2	6	10	14	2	6	10			2	6	2				2	
	91 Pa	2	2	6	2	6	10	2	6	10	14	2	6	10	2		2	6	1				2	
	92 U	2	2	6	2	6	10	2	6	10	14	2	6	10	3		2	6	1				2	
	93 Np	2	2	6	2	6	10	2	6	10	14	2	6	10	4		2	6	1				2	
	94 Pu	2	2	6	2	6	10	2	6	10	14	2	6	10	5		2	6	1				2	
	95 Am	2	2	6	2	6	10	2	6	10	14	2	6	10	6		2	6	1				2	
	96 Cm	2	2	6	2	6	10	2	6	10	14	2	6	10	7		2	6	1				2	
	97 Bk	2	2	6	2	6	10	2	6	10	14	2	6	10	8		2	6	1				2	
	98 Cf	2	2	6	2	6	10	2	6	10	14	2	6	10	10		2	6					2	
	99 E	2	2	6	2	6	10	2	6	10	14	2	6	10	11		2	6					2	
	100 Fm	2	2	6	2	6	10	2	6	10	14	2	6	10	12		2	6					2	
	101 Md	2	2	6	2	6	10	2	6	10	14	2	6	10	13		2	6					2	
	102 No	2	2	6	2	6	10	2	6	10	14	2	6	10	14		2	6					2	
	103 Lw	2	2	6	2	6	10	2	6	10	14	2	6	10	14		2	6	1				2	

* *See* Table 1.13 for the rows of the periodic table.

1.2.10 The Noble Gases

The elements occupying column 0 on the right-hand side of the periodic table are called the noble gases.

The noble gases, with the exception of helium, owe their properties to the fact that their outer subshell is complete and of the p type.

Figure 1.11 shows that, for values of Z corresponding to the noble gases, the energy separating the subshell p from the next subshell s is very high. However, if one or two units are added to the value of Z for a noble gas, the electrons required to make the new atoms neutral will invariably be accommodated in the next s subshell, even if subshells d, f, g, or h of the same shell as p are available (Table 1.12).

This fact is important because if we tried to excite the electrons of a noble gas, they would follow the same route. The noble gases are therefore very difficult to excite. Because they do not create external fields (at least in terms of an average value over time) due to the symmetry of their electronic configuration, they are particularly incapable of forming chemical bonds.

Although helium does not have a p subshell, it is also a noble gas. Helium has the two specific characteristics of a noble gas: the next subshell capable of being occupied is of the s type, and it is situated at a very high level with respect to the last normally occupied subshell. Simply, this is also of the s type, whereas in all the other noble gases, it is of the p type.

The noble gases provide a useful reference point for describing the other elements.

1.2.11 Alkali Metals

The elements of column IA of Table 1.13, with the exception of hydrogen, are called the alkali metals.

The electronic configuration of an alkali metal is that of a noble gas to which one electron has been added. As we have just seen, this electron occupies the outer s subshell. It has a relatively very high energy and so is weakly bound. This is why the alkali metals are chemically highly active (Section 1.3.7).

1.2.12 The Alkaline Earth Metals

The elements of column IIA of Table 1.13 are called the alkaline earth metals. Their electronic configuration is that of an alkali metal to which one electron has been added. Their outer s subshell is therefore

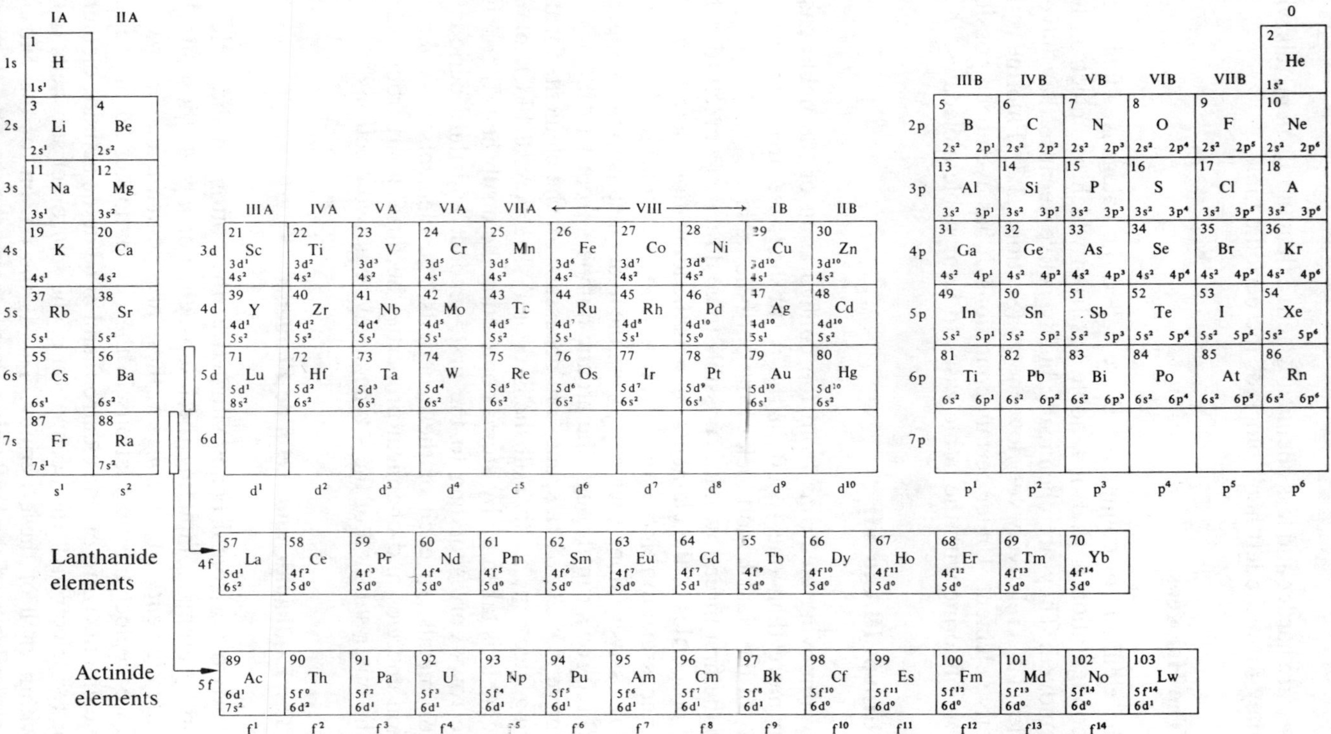

	IA	IIA
1s	1 H $1s^1$	
2s	3 Li $2s^1$	4 Be $2s^2$
3s	11 Na $3s^1$	12 Mg $3s^2$
4s	19 K $4s^1$	20 Ca $4s^2$
5s	37 Rb $5s^1$	38 Sr $5s^2$
6s	55 Cs $6s^1$	56 Ba $6s^2$
7s	87 Fr $7s^1$	88 Ra $7s^2$
	s^1	s^2

	IIIA	IVA	VA	VIA	VIIA	← VIII	VIII	VIII →	IB	IIB
3d	21 Sc $3d^1$ $4s^2$	22 Ti $3d^2$ $4s^2$	23 V $3d^3$ $4s^2$	24 Cr $3d^5$ $4s^1$	25 Mn $3d^5$ $4s^2$	26 Fe $3d^6$ $4s^2$	27 Co $3d^7$ $4s^2$	28 Ni $3d^8$ $4s^2$	29 Cu $3d^{10}$ $4s^1$	30 Zn $3d^{10}$ $4s^2$
4d	39 Y $4d^1$ $5s^2$	40 Zr $4d^2$ $5s^2$	41 Nb $4d^4$ $5s^1$	42 Mo $4d^5$ $5s^1$	43 Tc $4d^5$ $5s^2$	44 Ru $4d^7$ $5s^1$	45 Rh $4d^8$ $5s^1$	46 Pd $4d^{10}$ $5s^0$	47 Ag $4d^{10}$ $5s^1$	48 Cd $4d^{10}$ $5s^2$
5d	71 Lu $5d^1$ $8s^2$	72 Hf $5d^2$ $6s^2$	73 Ta $5d^3$ $6s^2$	74 W $5d^4$ $6s^2$	75 Re $5d^5$ $6s^2$	76 Os $5d^6$ $6s^2$	77 Ir $5d^7$ $6s^2$	78 Pt $5d^9$ $6s^1$	79 Au $5d^{10}$ $6s^1$	80 Hg $5d^{10}$ $6s^2$
6d										
	d^1	d^2	d^3	d^4	d^5	d^6	d^7	d^8	d^9	d^{10}

	IIIB	IVB	VB	VIB	VIIB	0
1s						2 He $1s^2$
2p	5 B $2s^2$ $2p^1$	6 C $2s^2$ $2p^2$	7 N $2s^2$ $2p^3$	8 O $2s^2$ $2p^4$	9 F $2s^2$ $2p^5$	10 Ne $2s^2$ $2p^6$
3p	13 Al $3s^2$ $3p^1$	14 Si $3s^2$ $3p^2$	15 P $3s^2$ $3p^3$	16 S $3s^2$ $3p^4$	17 Cl $3s^2$ $3p^5$	18 A $3s^2$ $3p^6$
4p	31 Ga $4s^2$ $4p^1$	32 Ge $4s^2$ $4p^2$	33 As $4s^2$ $4p^3$	34 Se $4s^2$ $4p^4$	35 Br $4s^2$ $4p^5$	36 Kr $4s^2$ $4p^6$
5p	49 In $5s^2$ $5p^1$	50 Sn $5s^2$ $5p^2$	51 Sb $5s^2$ $5p^3$	52 Te $5s^2$ $5p^4$	53 I $5s^2$ $5p^5$	54 Xe $5s^2$ $5p^6$
6p	81 Ti $6s^2$ $6p^1$	82 Pb $6s^2$ $6p^2$	83 Bi $6s^2$ $6p^3$	84 Po $6s^2$ $6p^4$	85 At $6s^2$ $6p^5$	86 Rn $6s^2$ $6p^6$
7p						
	p^1	p^2	p^3	p^4	p^5	p^6

Lanthanide elements

4f														
57 La $5d^1$ $6s^2$	58 Ce $4f^2$ $5d^0$	59 Pr $4f^3$ $5d^0$	60 Nd $4f^4$ $5d^0$	61 Pm $4f^5$ $5d^0$	62 Sm $4f^6$ $5d^0$	63 Eu $4f^7$ $5d^0$	64 Gd $4f^7$ $5d^1$	65 Tb $4f^9$ $5d^0$	66 Dy $4f^{10}$ $5d^0$	67 Ho $4f^{11}$ $5d^0$	68 Er $4f^{12}$ $5d^0$	69 Tm $4f^{13}$ $5d^0$	70 Yb $4f^{14}$ $5d^0$	

Actinide elements

5f														
89 Ac $6d^1$ $7s^2$	90 Th $5f^0$ $6d^2$	91 Pa $5f^2$ $6d^1$	92 U $5f^3$ $6d^1$	93 Np $5f^4$ $6d^1$	94 Pu $5f^5$ $6d^1$	95 Am $5f^6$ $6d^1$	96 Cm $5f^7$ $6d^1$	97 Bk $5f^8$ $6d^1$	98 Cf $5f^{10}$ $6d^0$	99 Es $5f^{11}$ $6d^0$	100 Fm $5f^{12}$ $6d^0$	101 Md $5f^{13}$ $6d^0$	102 No $5f^{14}$ $6d^0$	103 Lw $5f^{14}$ $6d^1$
f^1	f^2	f^3	f^4	f^5	f^6	f^7	f^8	f^9	f^{10}	f^{11}	f^{12}	f^{13}	f^{14}	

Table 1.13 The periodic table.

full. They are harder and less chemically active than the alkali metals with which they form the left-hand block of the periodic table.

1.2.13 The Halogens

The elements of column VIIB are called the halogens. Their electronic configuration is that of a noble gas from which one electron has been removed. They are chemically highly active because they have a vacant level at a relatively very low energy (Section 1.3.7). Fluorine is the most active element. Under certain conditions, it has even been possible to combine fluorine with the noble gases to form stable molecules.

1.2.14 Groups IIIA to VIB

By removing electrons from the configuration of the noble gases until no more than one is left in the p subshell, we arrive at the six columns of the right-hand block of the periodic table.

In the first three rows, it is possible to go directly from left to right. Several properties of the elements such as the ionization energy (i.e., the energy necessary to remove one electron from the atom) or the valence vary progressively along each row, from the alkali metals to the halogens. These three rows run through the atomic numbers from 1 to 18.

Above $Z = 20$, extra spaces are necessary for the elements formed by adding electrons in the subshells d or f. The central block of the periodic table is taken up by atoms completing the d subshell. It consists of three rows corresponding to the first, second, and third periods of transition metals. In each of these periods, the elements have similar ionization energies and chemical properties because, apart from exceptions, the outer subshell of the s type always has two electrons.

1.2.15 The Lanthanide and Actinide Elements

The two horizontal bands situated at the bottom of the table are taken up by atoms completing the f subshell. For $n = 4$, these are the *lanthanide elements,* and for $n = 5$, these are the *actinide elements*.

If we neglect the fluctuations in the occupation of 5d in the lanthanide elements (zero or one electron) and of 6d in the actinide elements (zero or two electrons), the occupation of the *four* outer subshells is the same within each of these periods. This is the reason for the extreme chemical similarity of the lanthanide elements and the actinide elements.

1.3 BONDS

1.3.1 Introduction

Two atoms separated by a very large distance with respect to their diameters form two independent systems. If they are brought sufficiently close together, their orbits will overlap, and this modifies the form of the potential governing the electrons. This modification produces a change in the orbits themselves and, consequently, a variation in the energy of the electrons. At this stage, the two atoms form a system.

These atoms exert a repulsive force on each other if the energy of the system increases when they approach each other. Otherwise, the force is attractive.

An attractive force is a frequent occurrence, as is demonstrated by the extremely large number of different molecules and substances present in the condensed form, i.e., solid or liquid.

1.3.2 Definition: Valence

The bonds that join atoms and molecules are called chemical bonds. The *valence of an element* is the number of hydrogen atoms with which it may combine or which it may replace.

1.3.3 Definition: Cohesion Energy

The minimum energy necessary to dissociate the atoms that make up a molecule and remove them sufficiently far from each other so that there is no longer an interaction is called the *cohesion energy* of a molecule.

The cohesion energy is usually expressed in eV/molecule or in kJ/mole (1 kJ/mole $\hat{=}$ $1.037 \cdot 10^{-2}$ eV/atom).

This concept also applies to crystals (Section 1.4), which may be considered as macromolecules.

1.3.4 Types of Chemical Bonds

Depending on their positions in the periodic table, atoms form different types of bonds between each other. It can be easily understood that the mechanism that joins atoms of the same type, carbon (C) in diamond, for example, is quite far removed from that which joins chlorine (Cl^-) and sodium (Na^+) in table salt! It is also possible that atoms of the same type can form different types of bonds between each other. Carbon, for example, can occur in the form of graphite or diamond. The very large difference between these two substances (in particular, one is a conductor and the other is an insulator) gives an idea of the influence of the bonding mechanisms on the properties of the materials in the most general sense of the term.

The study of chemical bonds from the fundamental equations of quantum physics has until now only been carried out with success in crystalline media and simple molecules.

Three types of *strong bonds* are distinguished. They are so-called because of the high cohesion energy associated with them. These are

- the ionic bond
- the covalent (or unipolar) bond
- the metallic bond

In parallel, there are *weak bonds*:

- the Van der Waals bond
- the dipolar bond
- the hydrogen bond

It is rare that a given bond strictly belongs to one of the above-mentioned types. Most of the time, it presents characteristics of two or even three types at the same time with the possible predominance of one of them.

For example, in the gallium arsenide (GaAs) crystal, the covalent character is approximately twice as apparent as the ionic character, whereas in NaCl the ionic character dominates over the covalent character by 94%. Silicon is purely covalent.

1.3.5 General Properties of Bonds

The existence of a bond between two atoms implies that they exert an attractive force on each other. The fact that this bond maintains the atoms at a certain distance $r = r_0$ may be explained *classically* by the existence of a repulsive force with a smaller range than the attractive force. The repulsive and attractive forces balance each other exactly at $r = r_0$.

The potential energy $W_s(r)$ of the system formed by the two atoms is given by

$$W_s(r) = W_a(r) + W_r(r) \tag{1.31}$$

In this expression, $W_a(r)$ and $W_r(r)$ represent the potential energies due to the attractive force and the repulsive force, respectively. $W_s(r)$ passes through a minimum for $r = r_0$ (Fig. 1.14). We can arbitrarily choose $W_s(r = \infty) = 0$.

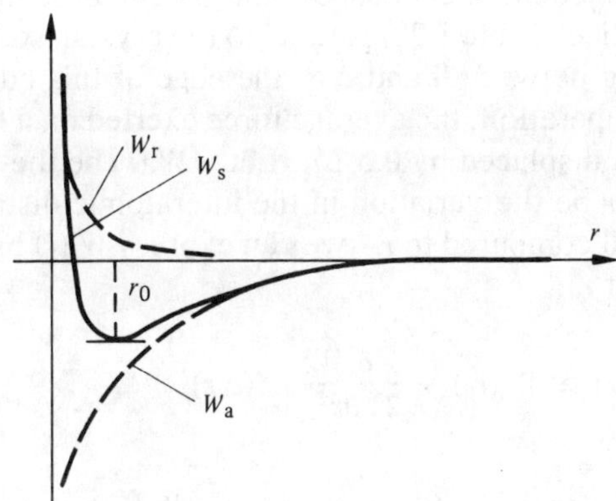

Fig. 1.14 Diagrammatic Representation of W_s

The functions $W_a(r)$ and $W_r(r)$ may be represented, for example, by expressions of the type:

$$W_a(r) = -\frac{B}{r^n} \tag{1.32}$$

$$W_r(r) = \frac{A}{r^m} \tag{1.33}$$

where A and B are positive constants. It is necessary that $n < m$ so that the repulsive force has a shorter range than the attractive force. In the Lennard-Jones model for the Van der Waals bond, we have $n = 6$ and $m = 12$.

It follows from the choice of the value of W_s at infinity that the cohesion energy W_c of the two atoms is simply given by

$$W_c = -W_s(r_0) \tag{1.34}$$

1.3.6 Calculation of the Compressibility

The compressibility κ is defined by the relationship:

$$\kappa = -\frac{1}{\Omega}\frac{\Delta\Omega}{\Delta P} \tag{1.35}$$

where $\Delta\Omega$ is the variation of the volume Ω for a variation in pressure ΔP. It is related in a simple manner to the cohesion energy.

Consider a cubic specimen of side a, cut in a monocrystal with simple cubic lattice (Table 1.28) parallel to the crystal axes. Subjected to a pressure varying between 0 and ΔP, the edge of this cube decreases by Δa. During the operation, the average force exerted on a face is $0.5a^2\,\Delta P$, and this force is displaced by $0.5\,\Delta a$. Let $-W_s(r)$ be the cohesive energy per atom and Δr be the variation in the interatomic distance due to ΔP. Since Δr is small compared to r_0, we can express $W_s(r)$ by a Taylor series expansion about r_0:

$$W_s(r_0 - \Delta r) \cong W_s(r_0) + \frac{1}{2}\frac{\partial^2 W_s}{\partial r^2}\bigg|_{r_0}(\Delta r)^2 \tag{1.36}$$

The first-order derivative is 0 because $W_s(r)$ passes through a minimum at r_0 (Fig. 1.14).

Let N be the total number of atoms. By expressing the equivalence between the variation in the cohesion energy of the crystal and the work performed on it by ΔP, we obtain

$$6a^2\frac{\Delta P}{2}\frac{\Delta a}{2} = \frac{1}{2}\frac{\partial^2 W_s}{\partial r^2}\bigg|_{r_0}\Delta r^2 \cdot Na^3 \tag{1.37}$$

However,

$$\frac{\Delta a}{a} = \frac{\Delta r}{r_0} \tag{1.38}$$

and

$$\frac{\Delta\Omega}{\Omega} = \frac{(a - \Delta a)^3 - a^3}{a^3} \cong -\frac{3\Delta a}{a} \tag{1.39}$$

and

$$N = \frac{1}{r_0^3} \tag{1.40}$$

which makes it possible to derive from (1.37) that

$$\kappa = 9r_0 \Big/ \left. \frac{\partial^2 W_s}{\partial r^2} \right|_{r_0} \tag{1.41}$$

This equation directly relates the cohesion energy to a macroscopic quantity that can be easily measured.

1.3.7 The Ionic Bond

The *ionic bond* is a bond between ions with opposite charges. The alkali metals and the halogens form these ions most easily because both of them acquire the electronic configuration of a noble gas in this way. The classical example is sodium chloride in which Na^+ has the configuration of neon and Cl^- has that of argon.

The distribution of the electronic charge in the noble gases has a spherical symmetry, which is also mainly found in ions having the same electronic configuration. The field created by these ions therefore also has approximate spherical symmetry. It follows that *the ionic bond is nondirectional.* The ions can therefore be considered as spheres with a diameter that varies as a function of the chemical nature of the ion. Under the action of the attractive and repulsive forces mentioned in Section 1.3.5, these spheres combine so that the cohesion energy is maximum. The attractive forces are coulombic in nature. The repulsive forces are due to an increase in the energy of the electrons required by the Pauli principle when the orbits of the ions overlap.

Compounds formed from a halogen and an alkali metal in particular form crystals called ionic crystals. They possess the NaCl or CsCl structure (Fig. 1.15), depending on the ratio of the diameters of the ions involved.

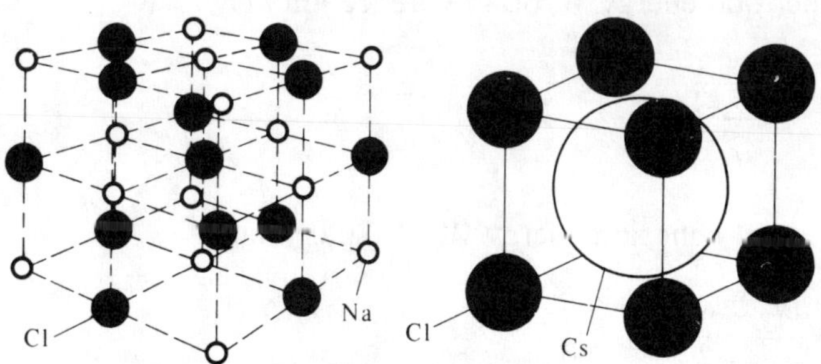

Fig. 1.15 Ionic Crystals

UNIVERSITY
OF BRISTOL
LIBRARY
ENGINEERING

1.3.8 Characteristics of Ionic Crystals

- Ionic crystals are *hard* because their cohesion energy is high, on the order of 750 kJ/mole for halogen–alkali metal compounds. By way of comparison, the typical cohesion energy of the Van der Waals bond is on the order of 10 kJ/mole. For the same reason, their *melting point is high.*
- Their *electronic conductivity is low* because the electrons are rigidly attached to the ions. Ionic conduction appears at high temperatures.
- Their *permittivity is high* and varies weakly with the temperature and frequency (Section 4.3.5).

1.3.9 Cohesion Energy of an Ionic Crystal

Let us consider a crystal formed from $N/2$ ions with charge $+q$ and $N/2$ ions with charge $-q$. We assume that N is sufficiently large so that it is possible to neglect the reduction in cohesion energy of atoms close to the surface.

The potential energy of the subsystem formed by ions i and j, separated by a distance r_{ij}, is

$$W_{ij} = \frac{A}{r_{ij}^{m}} \pm \frac{1}{4\pi\varepsilon_0} \frac{q^2}{r_{ij}} \tag{1.42}$$

The first term represents the energy resulting from the work of the short-range repulsive force (1.33). The second term represents the energy due to the Coulomb force, attractive ($-$) if the ions are of opposite sign and repulsive ($+$) otherwise.

The total energy W_i of a reference ion i is

$$W_i = \sum_{\substack{j=1 \\ j\neq i}}^{N} W_{ij} \tag{1.43}$$

and the total cohesion energy W_c of the crystal is

$$W_c = -\frac{N}{2} W_i \tag{1.44}$$

This expression contains the coefficient $N/2$ and not N because each pair of ions must only be considered once.

The exponent m is large with respect to 1. It is equal to 12 in the Lennard-Jones model (Section 1.3.5). Compared to the Coulomb forces, the short-range force therefore decreases quickly enough that it is sufficient to consider its effect on the closest neighbors of the reference ion. Let z be the number of these neighbors and a be the distance that separates them from the reference ion, then (1.44) takes the form:

$$W_c = -\frac{N}{2}\left[zAa^{-m} - \frac{1}{4_\pi \varepsilon_0}\left(\sum_{\substack{j=1\\ j\neq i}}^{N} \frac{\pm q^2}{r_{ij}}\right)\right] \tag{1.45}$$

There are approximately $5 \cdot 10^{19}$ atoms in one cubic millimetre of NaCl so that, even if the sum in (1.45) converges slowly, no appreciable error is made by taking $N = \infty$. By definition, let p_{ij} be

$$r_{ij} = ap_{ij} \tag{1.46}$$

We obtain

$$W_c = -\frac{N}{2}\left(zAa^{-m} - \frac{1}{4\pi\varepsilon_0}\frac{\alpha q^2}{a}\right) \tag{1.47}$$

where

$$\alpha = \sum_{\substack{j=1\\ j\neq i}}^{\infty} \frac{\pm 1}{p_{ij}} \tag{1.48}$$

is the *Madelung constant*. This constant is dimensionless and only depends on the crystal structure (Section 1.3.10).

The value of a at equilibrium, denoted a_0, is obtained by making the derivative of (1.47) equal to zero:

$$\frac{\mathrm{d}\, W_c}{\mathrm{d}\, a} = 0 = \frac{N}{2}\left(-mzA\; a_0^{-(m+1)} + \frac{\alpha q^2}{4\pi\varepsilon_0}\; a_0^{-2}\right) \tag{1.49}$$

from which we have

$$a_0 = \left(\frac{4\pi\varepsilon_0 mzA}{\alpha q^2}\right)^{\frac{1}{m-1}} \tag{1.50}$$

to give for the total cohesion energy of the crystal:

$$W_c = \frac{N}{2} \frac{\alpha q^2}{4\pi\varepsilon_0 a_0} \left(1 - \frac{1}{m}\right) \tag{1.51}$$

This expression shows that the part of the energy resulting from the Coulomb forces is m times higher than that resulting from the short-range forces.

1.3.10 Calculation of the Madelung Constant

The Madelung constant is very easily calculated in the theoretical case of a row of ions (Fig. 1.16). In this case, we simply have $p_{ij} = 1, 2, 3, \ldots$

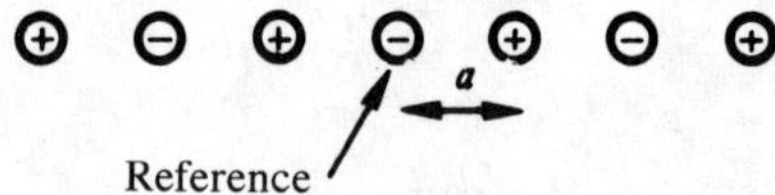

Fig. 1.16

By taking an anion as the reference ion:

$$\alpha = \sum \frac{\pm 1}{p_{ij}} = 2\left(1 - \frac{1}{2} + \frac{1}{3} - \frac{1}{4} + \cdots\right) \tag{1.52}$$

Now, we know that

$$\ln(1 + x) = x - \frac{x^2}{2} + \frac{x^3}{3} - \frac{x^4}{4} \tag{1.53}$$

from which, by setting $x = 1$, we have

$$\alpha = 2 \ln 2 \tag{1.54}$$

In real three-dimensional structures, the calculation of α requires a few stratagems because of the slow convergence of the series. Table 1.17 gives the Madelung constant for three important structures.

Table 1.17

Type of Structure		α
Sodium chloride	NaCl (Fig. 1.15)	1.747565
Cesium chloride	CsCl (Fig. 1.15)	1.762675
Zinc blende	ZnS (Fig. 2.44)	1.6381

1.3.11 The Covalent Bond

As the ionic bond, the *covalent bond* brings the electronic configuration of the atoms involved to that of a noble gas. However, the mechanisms of these two bonds are quite different. In the covalent bond, there is no transfer of electrons from one atom to another, but certain electrons of each atom are pooled.

The chlorine molecule Cl_2 is a typical example of the covalent bond. The chlorine atom needs 1 electron in order to have the configuration of argon. The *M* shell of the chlorine molecule therefore contains seven electrons, i.e., three pairs of electrons with opposite spins (paired electrons) plus a nonpaired electron (also called a sole electron). If their nonpaired electrons have opposite spin, two chlorine atoms may combine in order to pool their sole electrons and thus complete their 3p subshell without violating the Pauli principle (Fig. 1.18). During this process, the energy of each atom is lowered by 1.25 eV, which amounts to saying that the chlorine molecule is formed with a cohesion energy of 2.5 eV.

Fig. 1.18 Diagrammatic Representation of the Molecule Cl_2 : ●, Electrons of the M Shell

The possible candidates for this type of bond can be found in the right-hand part of the periodic table (Table 1.19) with the exception of hydrogen, which has a molecule H_2 that is also formed by the covalent bond.

Table 1.19

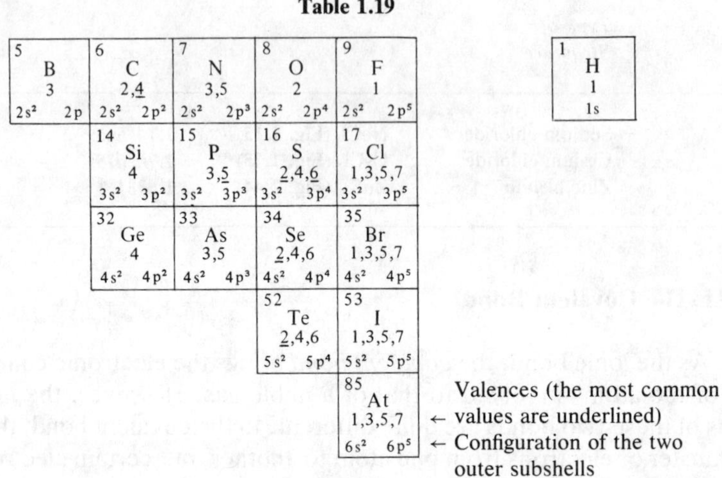

5 B 3 $2s^2$ $2p$	6 C 2,4 $2s^2$ $2p^2$	7 N 3,5 $2s^2$ $2p^3$	8 O 2 $2s^2$ $2p^4$	9 F 1 $2s^2$ $2p^5$	1 H 1 1s
	14 Si 4 $3s^2$ $3p^2$	15 P 3,5 $3s^2$ $3p^3$	16 S 2,4,6 $3s^2$ $3p^4$	17 Cl 1,3,5,7 $3s^2$ $3p^5$	
	32 Ge 4 $4s^2$ $4p^2$	33 As 3,5 $4s^2$ $4p^3$	34 Se 2,4,6 $4s^2$ $4p^4$	35 Br 1,3,5,7 $4s^2$ $4p^5$	
			52 Te 2,4,6 $5s^2$ $5p^4$	53 I 1,3,5,7 $5s^2$ $5p^5$	
				85 At 1,3,5,7 $6s^2$ $6p^5$	

← Valences (the most common values are underlined)

← Configuration of the two outer subshells

Table 1.19 shows that the relationship between the valence of an element and the position which it occupies in the periodic table is not simple in the case of covalence. Fluorine always has a valence equal to 1. It could be expected that the same would apply to chlorine, which belongs to the same column. Now, this element has valences of 1, 3, 5, and even 7, depending on the case. This shows that in addition to the electrons of the subshell p, the electrons of subshell s sometimes participate in the chemical bonds. The modification of orbits associated with this phenomenon is called *hybridization of the orbits*. The hybridization results from the overlapping of the original orbits of the atom in question with those of its neighbors. The very strong anisotropy of the p orbits (Fig. 1.20) and of the orbits formed by hybridization (Fig. 1.21) gives the covalent bond a *marked directional character*.

1.3.12 Hybridization of Carbon

Carbon (C) represents a particularly important case of hybridization. It is well known that C nearly always has a valence of 4, although this element only has two electrons in 2p. It is therefore necessary that one of the electrons in 2s passes to a vacant level in 2p in order to provide the four sole electrons required. The increase in energy associated with this transfer is largely compensated by the formation of four bonds instead of two.

The modification of the probability density functions $P_{nlm} = |\Psi_{nlm}|^2$ by hybridization is shown diagrammatically in Figs. 1.20 and 1.21. In the free atom, P_{nlm} has a rotational symmetry about the z-axis (Section 1.1.11) for all values of n, l, and m. After hybridization, each probability density function has its own symmetry axis directed to the vertices of a regular tetrahedron with the carbon atom occupying the center.

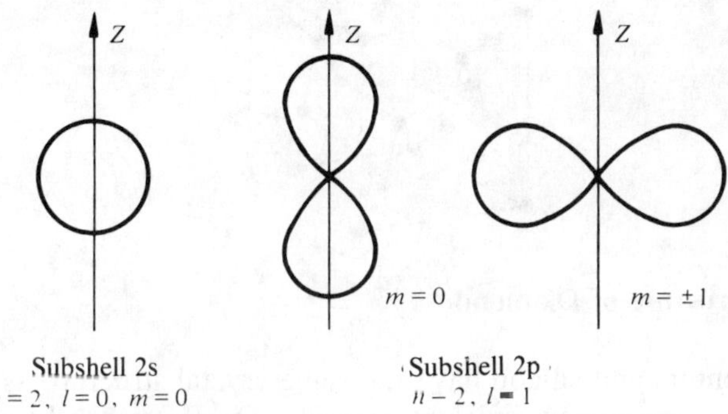

Fig. 1.20 Probability Density Functions $P_{nlm} = |\Psi_{nlm}|^2$ of the Free Atom

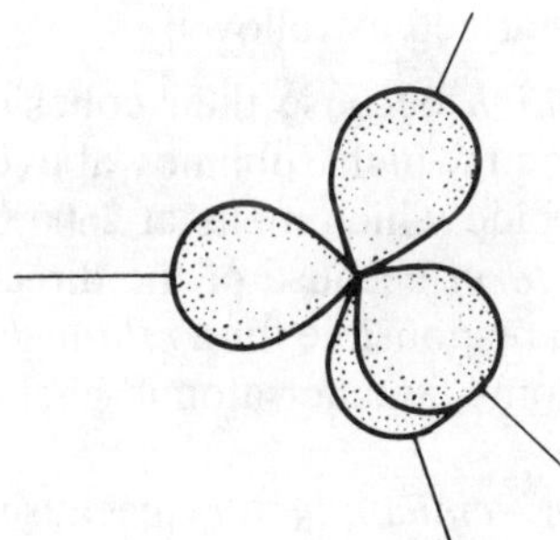

Fig. 1.21 Probability Density Functions after Hybridization

Other modifications of the orbits of the free atom are possible. That of Fig. 1.21 corresponds to *tetrahedral hybridization,* which can be observed in a very large number of carbon compounds, such as polyethylene, polyvinylchloride (Fig. 4.15), or methane CH_4, *et cetera,* and diamond.

1.3.13 Characteristics of Covalent Crystals

Atoms capable of forming three covalent bonds or more may form covalent crystals. The most representative of these is diamond (Fig. 1.22).

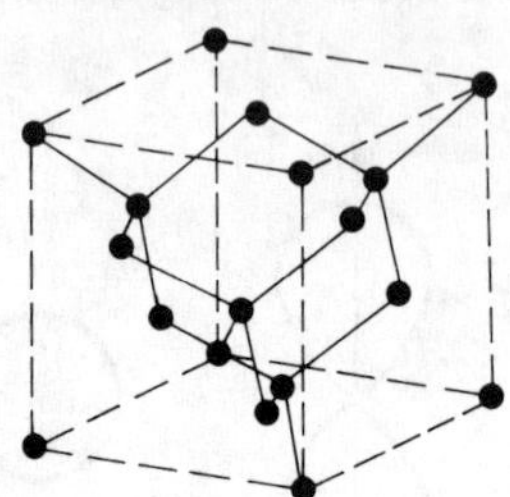

Fig. 1.22 Structure of Diamond

Germanium and silicon have the same crystal structure as diamond and all three are purely covalent crystals. On the other hand, silicon carbide (SiC) is not purely covalent. The bonds in this crystal have a certain ionic character.

The properties of covalent crystals or crystals with a dominant covalent character may be summarized as follows:

- their *melting point is high* because their cohesion energy is large: 714 kJ/mole for diamond which sublimes above 3652°C and 1186 kJ/mole for silicon carbide which melts at 2600°C;
- they are *difficult to deform* because of the directional character of the bond, which is also responsible for a *relatively less dense spatial arrangement* of the atoms (smaller atomic packing factor (Section 1.4.8));
- these are *insulators* or *semiconductors* because the valence electrons are strongly attached to the atoms, and the conductivity due to ions is negligible.

1.3.14 The Metallic Bond

The electrical conductivity of metals as well as their examination by x-ray diffraction (Section 1.6) shows that they are formed by a lattice of positive ions immersed in a gas of highly mobile electrons. The bond responsible for constructing this lattice may be considered as a limiting case of the covalent bond in which the electrons are pooled by all the ions of the crystal. It is called the *metallic bond*.

Metals, which occupy the left-hand part of the periodic table, have few electrons in their outer subshell. Since the energy of these electrons is relatively high, they can be easily torn out to form a bond. Let us consider the simple case of an alkali metal, sodium, for example. This element only has one electron in 3s, and crystallizes in a face-centered cubic lattice (Table 1.28). Sodium therefore must combine with eight neighbors by means of a single electron! This is why it cannot form a covalent bond with any of them, strictly speaking, but it is possible to consider that it forms such a bond with each of them on average for one-eighth of the time. The mobility of the valence electrons involved in this process explains the high conductivity of metals.

The metallic bond involving more than two electrons from 3s does not violate the Pauli principle because each electronic level of the isolated atom broadens into a band formed by as many very close discrete levels as there are atoms in the crystal (Section 2.6.17).

In metals with more complex electronic configurations than Na, there is often hybridization. This is particularly the case for the transition metals.

1.3.15 Characteristics of Metallic Crystals

The metallic bond is characterized by a cohesion energy which on average is lower than that for the covalent or ionic bond. For example, this energy is 311 kJ/mole in aluminum and 386 kJ/mole in iron.

- Metallic crystals are *ductile*. This property follows from the nondirectional nature of the metallic bond.
- By means of certain physico-chemical similarities, different metals may be mixed, sometimes in any proportion, to form *alloys*. This is due to the fact that the exact number of electrons available for forming the bond does not affect this in a significant manner.
- The high *electrical conductivity* of metallic crystals has already been mentioned. It is accompanied by a high *thermal conductivity*, also linked to the mobility of the electrons.

1.3.16 The Van der Waals Bond

Ionic, covalent, and metallic bonds all result from the tendency of the electronic configuration of atoms to approximate that of the noble gases. What therefore is the nature of the bond allowing the noble gases to exist in the liquid or solid form at low temperatures? There should not be any interaction between two noble gas atoms because they do not produce any external electrical field due to the perfect symmetry of their electronic charge.

In reality, this symmetry is only perfect in terms of the average value over time. At any moment, the probability that an atom of a rare gas has a dipole moment $\boldsymbol{p}_1$ different from zero is not zero. The electric field E produced by $\boldsymbol{p}_1$ induces a dipole moment $\boldsymbol{p}_2$ in an atom at distance r, given by (Section 4.3.2):

$$\boldsymbol{p}_2 = \alpha \boldsymbol{E} \tag{1.55}$$

where α is the polarization coefficient of the $\boldsymbol{p}_2$ atom.

The *Van der Waals bond* is due to the attractive force that $\boldsymbol{p}_1$ and $\boldsymbol{p}_2$ exert on each other.

In the case of Fig. 1.23, this force $\boldsymbol{F}$ evaluated for $\boldsymbol{p}_2$ is equal to ([52] p. 461):

$$\boldsymbol{F} = -\frac{3p_1p_2}{2\pi\varepsilon_0 r^4}\frac{\boldsymbol{r}}{r} \tag{1.56}$$

The potential energy W_s of the system formed by the two dipoles is expressed by

$$W_s = \int_\infty^r \boldsymbol{F} \cdot \mathrm{d}\boldsymbol{r}' \tag{1.57}$$

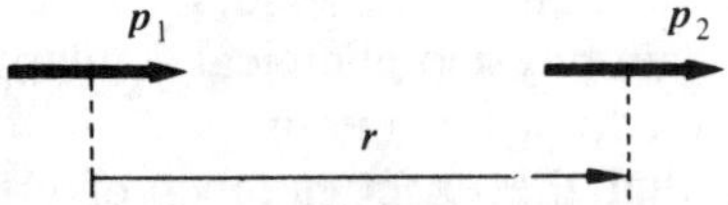

Fig. 1.23

Taking into account (1.55), (1.56), and (4.29), we obtain

$$W_s = \int_\infty^r \frac{3p_1}{2\pi\varepsilon_0 r^4}\frac{\alpha p_1}{2\pi\varepsilon_0 r^3}\,\mathrm{d}r' = -\frac{2\alpha p_1^2}{(4\pi\varepsilon_0)^2}\frac{1}{r^6} \tag{1.58}$$

The $1/r^6$ dependence given for the attraction energy in the Lennard-Jones model (Section 1.3.5) is found again in (1.58).

The higher the atomic number, the larger are the instantaneous dipolar moments. In substances where the Van der Waals bonds dominate, the cohesion energy and consequently the melting and boiling points increase with Z. The example of the noble gases (Table 1.24) is particularly representative.

Table 1.24

Gas	Z	*Melting Point °C*	*Boiling Point °C*
He	2	—	−268.9
Ne	10	−248.7	−245.9
A	18	−189.2	−185.5
Kr	36	−156.6	−152.9
Xe	54	−112	−107.1
Rn	86	−71	−61.8

The Van der Waals bond is present in all materials, but it is most often masked by stronger bonds. It is responsible for the liquefaction and the solidification of gases with symmetric molecules such as H_2, N_2, O_2, CH_4, *et cetera*. It plays an important part in a large number of organic compounds, in particular in thermoplastic polymers where it joins macromolecules with linear structure to each other (Section 4.11). The low cohesion energy of the Van der Waals bond (<40 kJ/mole) is responsible for the softening of these materials on heating.

1.3.17 The Dipole Bond

The *dipole bond* is the name given to the bond which is formed between molecules having a *permanent* dipole moment. Water is a typical example ($p = 6.2 \cdot 10^{-30}$ C · m). This is a directional bond characterized by a cohesion energy less than 20 kJ/mole.

1.3.18 The Hydrogen Bond

Only having one electron, the hydrogen atom cannot form a covalent bond with more than one atom. However, once this bond is formed, with a halogen for example, the molecule has a strong asymmetry because it actually consists of an anion with a proton on its periphery. In this situation, the proton may attract a second atom which is thus bound to the

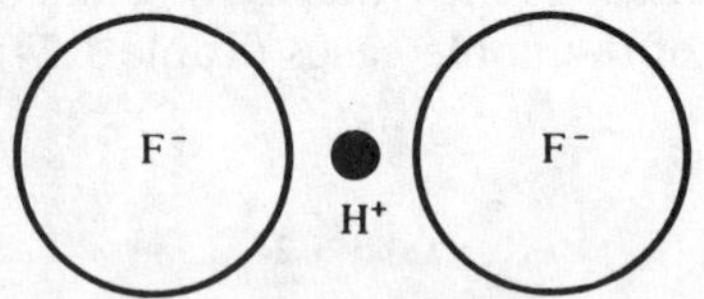

Fig. 1.25 Example of the Hydrogen Bond: the HF_2^- Ion

first by the so-called *hydrogen bond*. The cohesion energy of this bond is comparable to that of the Van der Waals bond.

1.4 CRYSTALS

1.4.1 Basic Concepts, Definitions

The fundamental characteristic of a material in the *crystalline phase* is the periodic arrangement of atoms or molecules.

The study of crystals is the object of a science, *crystallography,* which by means of a specific language, makes it possible to describe the most complex crystal structures in a concise manner. Only the elements of this science necessary for the study of simple crystals will be given here.

The description of a crystal makes use of the concepts of the *space lattice* and the *basis*.

In three-dimensional space, there are 14 different ways of periodically distributing points so that *each point has the same number of first neighbors, situated at the same distances, in the same directions*. The 14 corresponding *space lattices* are the *Bravais lattices*.

Each of the points forming a Bravais lattice is called a *lattice point*.

The entire lattice can be produced by translation operations along three space axes of a unit cell.

A lattice may be defined by its three base vectors $\boldsymbol{a}_1$, $\boldsymbol{a}_2$, and $\boldsymbol{a}_3$ constructed along three edges of the unit cell.

The conventional unit cells of the Bravais lattices are shown in Table 1.28.

We go from the geometric entity represented by a Bravais lattice to a real crystal by limiting the extent of the lattice to a finite volume and associating the same *basis* formed by one or more atoms or molecules with each lattice point.

The Bravais lattice and the basis together define the *crystal structure* (Fig. 1.26) which, supplemented by the knowledge of the base vectors, gives a complete description of the crystal.

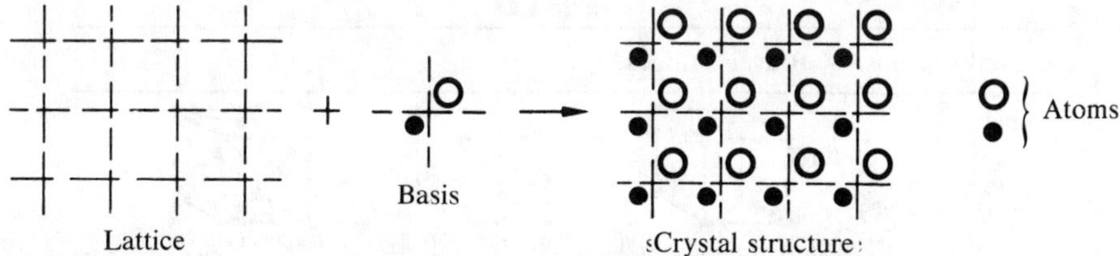

Fig. 1.26 Example of a Two-Dimensional Crystal Structure

The position of the lattice points with respect to the basis may be chosen arbitrarily. Instead of being placed roughly in the middle of the two atoms (Fig. 1.26), the lattice points could, for example, be situated on one or the other of the atoms.

By grouping together Bravais lattices with the same unit cell, we can define seven *crystal systems* (Fig. 1.27 and Table 1.28).

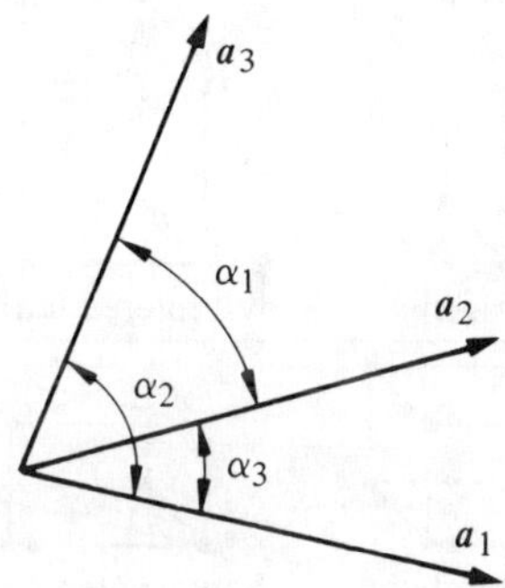

Fig. 1.27 Notation for the Angles between the Base Vectors

A crystal specimen in which the base vectors have the same orientation at all points is called a *monocrystal*. Otherwise, we are dealing with a *polycrystal*. A polycrystal is an assembly of monocrystals which are usually called, in this case, *grains*

1.4.2 Position of a Point in a Crystal

The position of a point is given by its coordinates along three axes x_1, x_2, x_3, the extension of the base vectors $\boldsymbol{a}_1$, $\boldsymbol{a}_2$, $\boldsymbol{a}_3$, respectively. On each axis, the modulus of the base vector is taken as unity. It follows from the periodicity of the crystal and the definition of the unit cell that the

Table 1.28

Crystal systems	Bravais lattices		
Cubic $a_1 = a_2 = a_3$ $\alpha_1 = \alpha_2 = \alpha_3 = 90°$	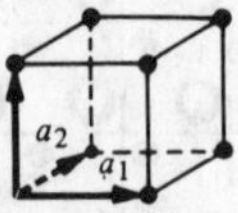 Simple cubic (SC)	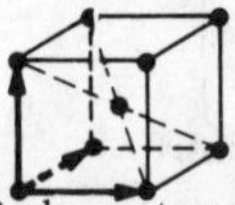 Body-centered cubic (BCC)	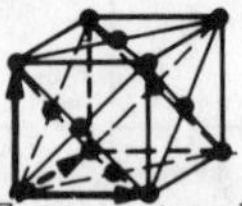 Face-centered cubic (FCC)
Hexagonal $a_1 = a_2 \neq a_3$ $\alpha_1 = \alpha_2 = 90°$ $\alpha_3 = 120°$	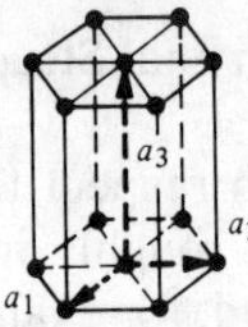 (HC)		
Rhombohedral $a_1 = a_2 = a_3$ $\alpha_1 = \alpha_2 = \alpha_3 \neq 90°$	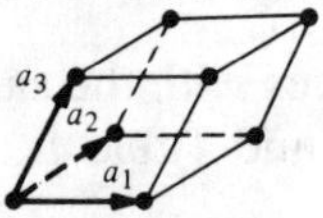		
Tetragonal or quadratic $a_1 = a_2 \neq a_3$ $\alpha_1 = \alpha_2 = \alpha_3 = 90°$	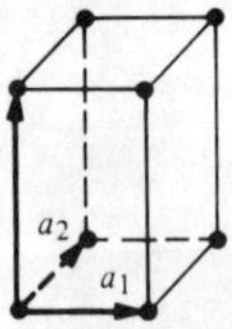 Simple tetragonal	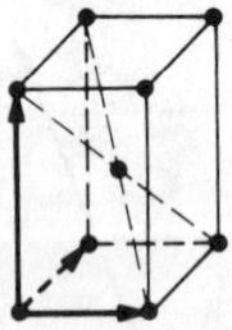 Body-centered tetragonal	
Orthorhombic $a_1 \neq a_2 \neq a_3$ $\alpha_1 = \alpha_2 = \alpha_3 = 90°$	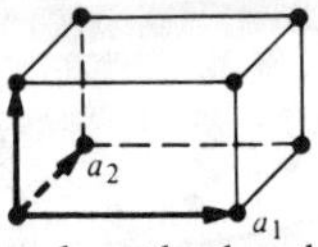 Simple orthorhombic	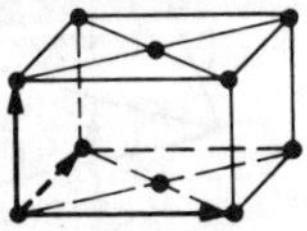 Body-centered orthorhombic	
	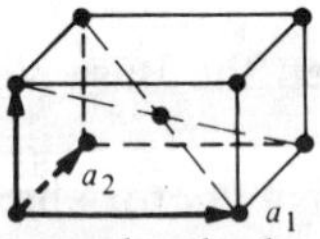 End-centered orthorhombic	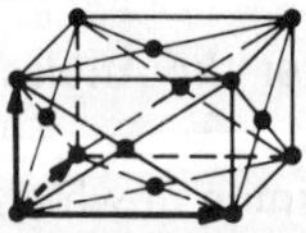 Face-centered orthorhombic	
Monoclinic $a_1 \neq a_2 \neq a_3$ $\alpha_1 = \alpha_2 = 90° \neq \alpha_3$	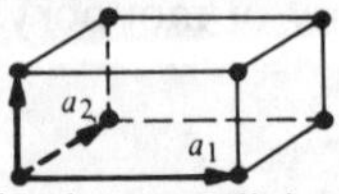 Simple monoclinic	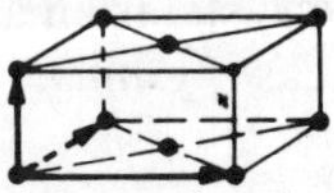 End-centered monoclinic	
Triclinic $a_1 \neq a_2 \neq a_3$ $\alpha \neq \alpha_2 \neq \alpha_3$.	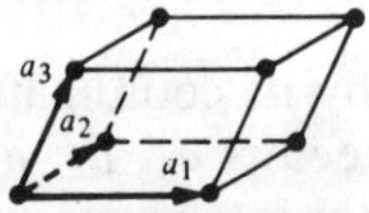		

coordinates of a point may always be expressed as three numbers less than one. In NaCl, for example, (FCC lattice, Fig. 1.15), if the origin of the coordinates is chosen on Na, the basis is formed by Na at (0,0,0) and by Cl at (½,0,0).

1.4.3 Directions in a Crystal

A direction is given by its direction cosines on x_1, x_2, x_3, multiplied by the smallest constant making these three integer numbers. They are written between square brackets and have a bar over them if they are negative. A set of equivalent directions is placed between angle brackets.

In a simple cubic crystal, for example, the directions parallel to the axes are written as [100] [010] [001] $[\bar{1}00]$ $[0\bar{1}0]$ $[00\bar{1}]$. All these equivalent directions are designated collectively as *directions of the form* {100}.

1.4.4 Planes in a Crystal Described by Miller Indices

The equation of a plane intersecting x_1, x_2, x_3 at the abscissas u_1, u_2, u_3 has the form:

$$\frac{x_1}{u_1} + \frac{x_2}{u_2} + \frac{x_3}{u_3} = 1 \tag{1.59}$$

This plane is described by the Miller indices (h,k,l) always written between parentheses and defined by their relationships

$$h = A\frac{1}{u_1} \qquad k = A\frac{1}{u_2} \qquad l = A\frac{1}{u_3} \tag{1.60}$$

where the constant A makes h, k, and l three integers as small as possible.

A set of equivalent planes is placed between braces. For example, the planes parallel to the coordinate planes in a simple cubic system (Fig. 1.29) have the Miller indices (100), (010), (001), $(\bar{1}00)$, $(0\bar{1}0)$, $(00\bar{1})$. They are equivalent to each other. These are *planes of the form* {100}.

1.4.5 Comments

The crystal structure of a material depends above all on the relative dimensions of the atoms and the nature of the valence bonds. A good

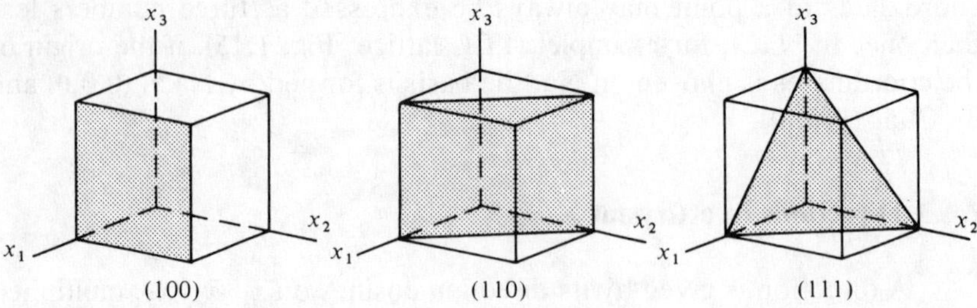

Fig. 1.29 Miller Indices and Corresponding Planes

number of materials with great practical importance have relatively simple crystal structures. Before describing the most common ones, it is useful to define three parameters frequently employed to characterize them.

1.4.6 Atomic Radius: Definition

The *atomic radius* of an element is half the distance separating the center of an atom from the center of its nearest neighbor in the crystal of this element.

1.4.7 Coordination Number: Definitions

The *coordination number of an atom* is the number of its nearest neighbors.

When all the atoms of a crystal have the same coordination number, this is called the *coordination number of the crystal.*

1.4.8 Atomic Packing Factor and the Hard-Ball Model: Definitions

The *hard-ball model* is the diagrammatic representation of a crystal as an arrangement of spheres in contact with each other in certain directions. In this model, the *atomic packing factor* is the quotient of the volume of the spheres in the unit cell by the volume of the unit cell itself.

1.4.9 Crystals With a Single Type of Atoms

In crystals where all the atoms are identical, the structure only depends on the degree of isotropy of the orbits involved in the valence bonds.

When these orbits are spherical, the atoms are arranged (hard-ball model) so that the minimum volume is occupied. The distance between the atoms, which is also a minimum, makes the cohesion energy a maximum.

Two different structures called *close-packed structures* satisfy the condition of minimum volume. These are:

- The *hexagonal close-packed structure,* abbreviated HCP
- The *face-centered cubic structure,* abbreviated FCC

Table 1.30 shows how these structures are obtained by arranging layers of spheres in a hexagonal manner.

Table 1.30
Close-Packed Structures. For the Sake of Clarity, the Diameter of the Spheres Has Been Reduced Without Modifying the Unit Cell

Hexagonal close-packed structure 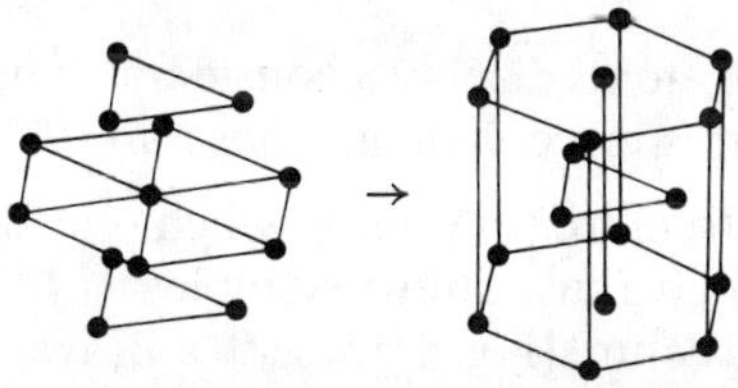	Example of HCP crystals Be, Mg, Ti, Co, Zn, Y, Zr, Ru, Hf, Re, Os, Tl. The majority of the lanthanide elements
Face-centered cubic structure 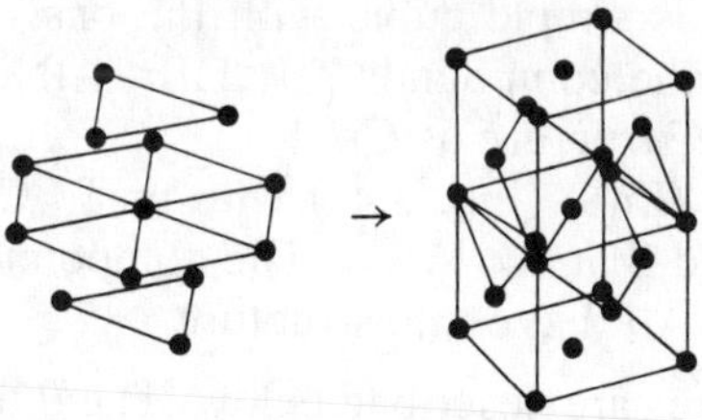	Example of FCC crystals Al, Ca, Ni, Cu, Sr, Rh, Pd, Ag, Ir, Pt, Au, Pb, Th. The noble gases

The close-packed structures have the highest possible coordination number: 12, and the highest possible atomic packing factor: 0.740 (Section 1.8.3). Thirty-two metal elements have a close-packed structure at room temperature.

When the valence bonds have an anisotropic character, the *centered cubic* (cc) structure is most commonly observed followed by the diamond structure (Fig. 1.22 and Table 1.31) and then the hexagonal, rhombohedral, and other structures.

Table 1.31

Structure	*Coordination Number*	*Atomic Packing Factor*	*Examples (20°C)*
Centered cubic	**8**	$\pi\sqrt{3}/8$	**V, Cr, Fe, Mo, W**
Diamond	**4**	$\pi\sqrt{3}/16$	**C, Si, Ge**

The *simple cubic* structure used as an example in a large number of qualitative descriptions is very rare. Of the elements, only polonium has this structure at room temperature.

1.4.10 Crystals with Several Types of Atoms

Crystals with several types of atoms display a considerable diversity of structure. Three simple structures are common. These are

- the *sodium chloride* (NaCl) structure (Fig. 1.15) which is also shared by MgO, MnS, PbS, FeO to give just a few examples.
- the *cesium chloride* (CsCl) structure (Fig. 1.15) with a Bravais lattice that is not centered cubic! In fact, the atom at the center of the cube does not have the same neighbors as an atom situated at one corner. The Bravais lattice of CsCl is simple cubic with the basis Cs^+ at (000) and Cl^- at (½ ½ ½). The compounds TlI, TlBr, AlNi, BeCu among others have the same structure as CsCl.
- the structure of *zinc blende* (ZnS) (Fig. 2.44) with an FCC lattice with the basis Zn at (000) and S at (¼ ¼ ¼). The compounds AgI, ZnSe, InAs, InSb, SiC, and AlP have this structure.

Other more complex structures are described below. The β tungsten structure is described in connection with superconductors (Fig. 2.79), the spinel structure in connection with the ferrites (Fig. 3.33), and the Perovskite structure in connection with the ferroelectrics (Fig. 4.40).

1.4.11 Allotropic Varieties and Polymorphism

Some elements can occur in several crystalline structures forming *allotropic varieties* of the element. This property, called *polymorphism,* was mentioned above (Section 1.3.4) in connection with carbon. Certain chemical compounds, silica, for example, also have the property of polymorphism, which can occur in the form of quartz, tridimite, or crystobalite.

1.5 CRYSTAL IMPERFECTIONS

1.5.1 Introduction

The perfect crystal formed from a Bravais lattice and an associated basis is an abstraction. It serves as a reference and makes it possible to study certain properties such as density, elastic behavior, or specific heat.

In reality, all crystalline media have imperfections which do not have to be considered as harmful *a priori*. On the contrary, perfect crystals would most often not have any interesting properties at all. It is not possible to manufacture transistors with perfect silicon (Section 2.7); it is not possible to obtain permanent magnets using perfect iron (Section 3.8.6), *et cetera*. The list of these examples can be extended almost indefinitely covering all the fields of science and engineering.

It is therefore hardly an exaggeration to say that crystalline materials are of value by virtue of their imperfections above all. These imperfections are classified into four categories which will be covered in the following sections.

The presence of each type of imperfection may have an effect on several properties. This effect varies from one material to another which is why only information of general application will be given in this section. More specific information will be provided in the following chapters when specific materials are examined.

1.5.2 Thermal Vibrations, Phonons

The thermal energy stored in a solid essentially takes the following forms:

- oscillating motion of the atoms
- rotational motion of the molecules
- displacement and excitation of the electrons

Thus, the atoms do not occupy the positions that are reserved for them in a crystal in a static manner and this constitutes the first imperfection. At room temperature, the amplitude of their vibrational movement is on the order of 5–10% of an interatomic distance. An overall analysis of

UNIVERSITY OF BRISTOL LIBRARY
ENGINEERING

this movement is possible based on the representation of the energy involved in a bond given in Section 1.3.5.

The repulsive force F_r acting on the atom, at a distance Δr from its mean position r_0 is according to (1.36):

$$F_r = -\frac{\mathrm{d}W_s(r)}{\mathrm{d}r} + -\left.\frac{\partial^2 W}{\partial r^2}\right|_{r_0} \Delta r \tag{1.61}$$

Let α be the return constant:

$$\alpha = \left.\frac{\partial^2 W}{\partial r^2}\right|_{r_0} \tag{1.62}$$

The equation for the movement of the atom can be written

$$m \frac{\mathrm{d}^2 \Delta r}{\mathrm{d}t^2} = -\alpha \Delta r \tag{1.63}$$

where m is the mass of the atom. A solution of (1.63) is of the type:

$$\Delta r = A \cos \sqrt{\frac{\alpha}{m}}\, t \tag{1.64}$$

The frequency f of the movement is therefore equal to

$$f = \frac{1}{2\pi} \sqrt{\frac{\alpha}{m}} \tag{1.65}$$

The constant α is related to the compressibility (1.41) and therefore can be experimentally determined. For the crystals of elements with medium atomic weight (Fe, Cu, Ge), we obtain $f \approx 10^{13}$ Hz at room temperature using this model.

When we do not just consider one atom but an *ensemble* of atoms connected together, it appears that the vibrational state of this lattice is quantized. This state may be represented by a sum of acoustic waves with different frequencies. The energy quantum associated with these waves is called the *phonon,* by analogy with the energy quantum associated with an electromagnetic wave, the photon. The study of the spectrum (i.e., the energy distribution) of the phonons in a crystal forms one of the chapters in solid-state physics [5].

1.5.3 Point Defects: Definitions

Point defects are the defects occurring at isolated points in the crystal. They are of several types, the most important of which are shown diagrammatically in Fig. 1.32.

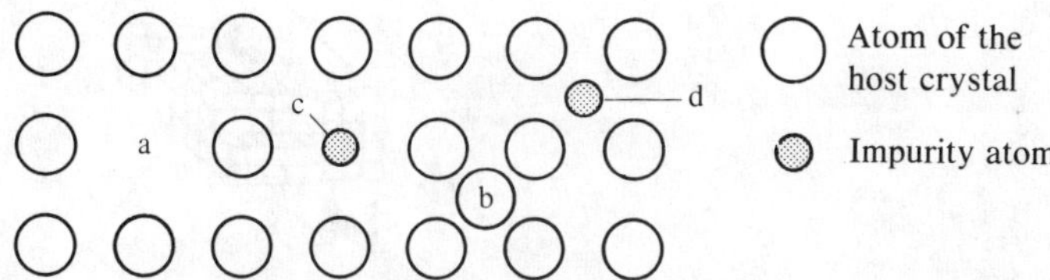

Fig. 1.32 Point Defects: (a) vacancy; (b) self-interstitial; (c) substitutional impurity atom; (d) interstitial impurity atom.

When there is no atom at a site that should normally be occupied by one, the defect is called a *vacancy*. All crystals contain vacancies constantly produced and annihilated by thermal motion. The total number n of vacancies varies as a function of the temperature according to the relationship:

$$n = N \exp\left(-\frac{W_\ell}{k_B T}\right) \tag{1.66}$$

where n is the total number of sites that should contain atoms and W_ℓ is a so-called vacancy activation energy. In aluminum, for example, $W_\ell =$ 0.75 eV.

When an atom of a crystal occupies a position that is normally empty, it forms a fault which is called *self-interstitial*. The density of self-interstitial defects is appreciably lower than that of the vacancies.

The presence of *impurity atoms,* in a *substitution* position or in an *interstitial* position, forms a defect whose effect on conductivity is used in semiconductor technology (Sections 2.7.2 and 2.7.3).

Other types of point defects exist in ionic crystals.

1.5.4 Dislocations: Definitions

Dislocations are line-shaped defects in the arrangement of the atoms. Their behavior determines the mechanical properties of metals to a very large extent. There are two types of dislocations, edge dislocation and screw dislocation.

Dislocations appear and move in crystals (Fig. 1.33) when stresses applied to them produce a plastic deformation (Section 6.3).

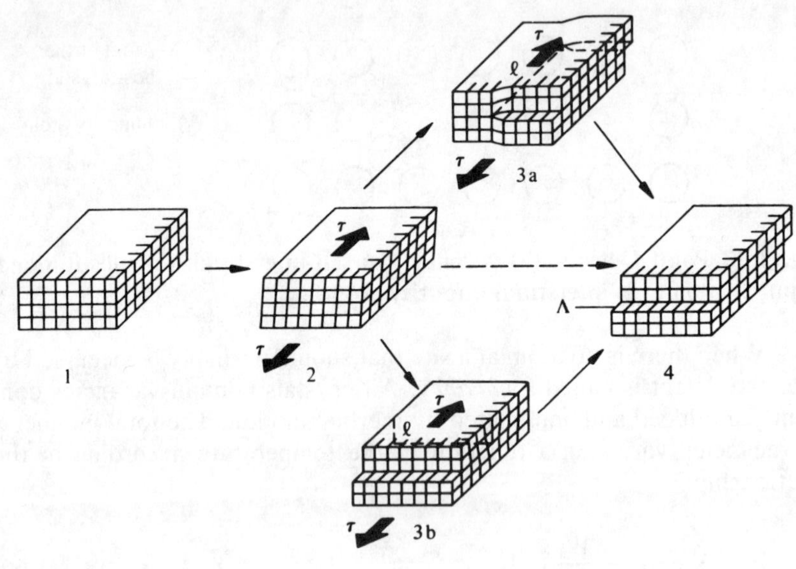

Fig. 1.33

At stage 1, the crystal is free of any stress. Subjected progressively to a shear stress τ, it first of all presents an elastic deformation (reversible) corresponding to stage 2. If τ is further increased, the crystal will finally be subjected to the permanent deformation shown in stage 4. This results in a slipping of the atoms in a plane Λ.

It is possible to imagine that this slipping occurs abruptly, all the valence bonds simultaneously breaking up and immediately reforming after a translation of the upper part of the crystal with respect to the lower part. In fact, the minimum shear stress τ_0 necessary to obtain a permanent deformation by this process would be much larger (Section 1.5.5) than that found by experiment. This shows that this mechanism does not occur in reality.

On the contrary, permanent deformation is produced *progressively* passing through the intermediate stages 3a or 3b. In this way, the mechanical energy involved is much less because only a small region of the crystal is deformed at any one time.

The dislocation is identified by the line ℓ, drawn in the most deformed area of the crystal. Fig. 1.33 shows that ℓ may take two extreme orientations.

The dislocation appearing at stage 3a is called a *screw dislocation*. It has the characteristic of being oriented parallel to τ and of moving perpendicular to τ.

The dislocation appearing at stage 3b is called an *edge dislocation*. It has the characteristic of being oriented perpendicular to τ and of moving parallel to τ.

It can be seen that both types of dislocation lead to the same final deformation.

From one point to another, a dislocation line may progressively go from the edge type to the screw type and vice versa (Fig. 1.34).

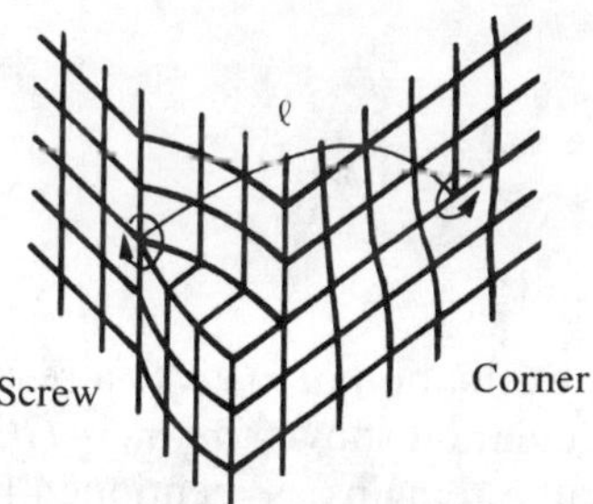

Fig. 1.34

1.5.5 Calculation of τ_0

In order to simplify the calculation, it is assumed that the deformation is concentrated between two adjacent lattice planes (Fig. 1.35).

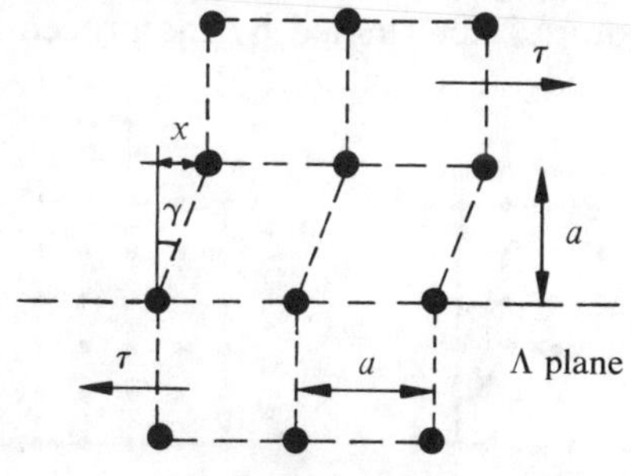

Fig. 1.35

The necessary energy W per unit of area in Λ, to obtain a displacement x, is expressed as a first approximation by a relationship of the type:

$$W = W_0 \left(1 - \cos\left(\frac{2\pi x}{a}\right)\right) \tag{1.67}$$

where W_0 is a constant. Now,

$$\tau = \frac{\mathrm{d}W}{\mathrm{d}x} = \frac{2\pi W_0}{a} \sin\left(\frac{2\pi x}{a}\right) = \tau_0 \sin\left(\frac{2\pi x}{a}\right) \cong \tau_0 2\pi\gamma \tag{1.68}$$

By definition (Section 6.2.3) of the shear modulus G,

$$\tau = G\gamma, \tag{1.69}$$

from which we have

$$\tau_0 \cong \frac{G}{2\pi} \tag{1.70}$$

This very simple evaluation of τ_0 gives a result which is too high. A more sophisticated calculation shows that $\tau_0 \approx G/30$. Despite this correction, the theoretical value remains as mentioned in the previous section, largely higher than the experimental values. In practice, values of τ_0 between $10^{-5}G$ and $10^{-4}G$ are currently found in metals.

1.5.6 Burgers Vector

The Burgers vector $\boldsymbol{b}$ characterizes the dislocations. Its orientation indicates the direction in which the atoms are displaced and its modulus measures the amount of this displacement.

The Burgers vector is determined by the procedure shown in Fig. 1.36 for the case of an edge dislocation.

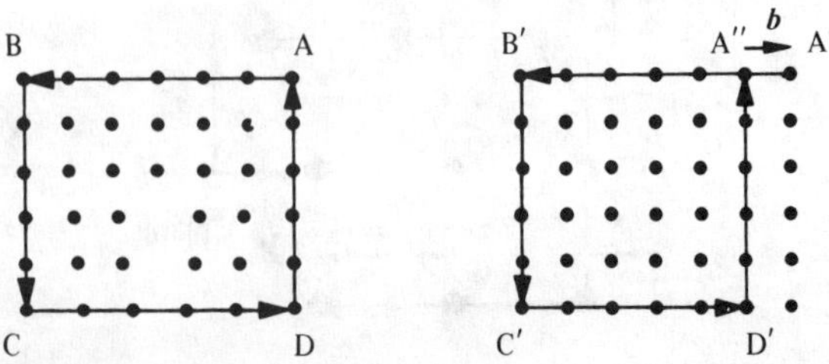

Fig. 1.36

A closed loop ABCDA is described in a plane perpendicular to ℓ by taking the direct trigonometric direction. This loop has to be sufficiently far from ℓ in order to be located in a crystal zone that is slightly deformed. Starting from A, the number of lattice points encountered over each rectilinear section AB, BC, CD, DA, is counted.

On the basis of this count, a homologous path A′B′C′D′A″ is followed in a perfect crystal.

The Burgers vector is defined as the closure vector $\boldsymbol{b}$ = A″A′.

The Burgers vector remains constant over the entire length of a dislocation.

1.5.7 Calculation of the Energy of a Dislocation

The calculation of the deformation energy associated with a dislocation is a useful step because it provides different kinds of information on the behavior of the dislocations and plastic deformation in general.

For this calculation, the crystal is considered as a continuous medium and the case is chosen of a screw dislocation. Under these conditions, the calculation of the dislocation energy reduces to that of the shear energy stored in a cylindrical ring surrounding the dislocation. This ring is shown in Fig. 1.37, in place on the left and unrolled on the right.

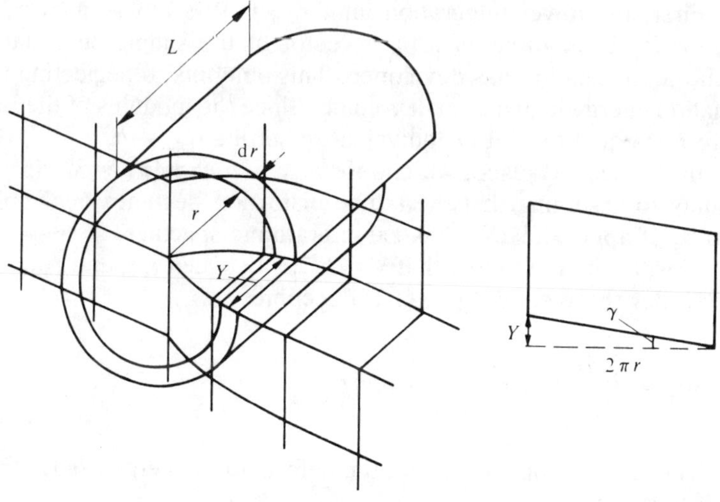

Fig. 1.37

The specific shear stress energy is given by the general relationship, when the shear angle x varies from zero to γ:

$$W = \int_0^\gamma \tau \, dx \tag{1.71}$$

By virtue of the definition of the shear modulus G (Section 6.2.3), we obtain

$$W = \int_0^\gamma Gx \, dx = \frac{G\gamma^2}{2} \tag{1.72}$$

The deformation energy in the cylindrical ring with thickness dr is therefore equal to

$$dW_c = \frac{G\gamma^2}{2} L 2\pi r \, dr \tag{1.73}$$

where L is the length of the dislocation.

Finally, the dislocation energy is given by

$$W_d = \int_{r_{min}=0}^{r_{max}=\infty} G\gamma^2 L\pi r \, dr \tag{1.74}$$

Three approximations make it possible to calculate this integral in a simple manner without changing the nature of the final result.

First, the lower integration limit $r_{min} = 0$ is replaced by $r_{min} = a$ where a is the modulus of a base vector of the simple cubic lattice in which the dislocation has developed. This amounts to neglecting the deformation energy in just a small volume. Since the modulus of the Burgers vector $\boldsymbol{b}$ is equal to a, it is equivalent to setting $r_{min} = b$.

In the second place, we choose $r_{max} = b \exp(4\pi) \approx 3 \cdot 10^5 b$. This amounts to assuming that the lattice distortion becomes negligible at a distance of approximately 300,000 interatomic spacings.

Thirdly, it is assumed that $\gamma = b/2\pi r$ between r_{min} and r_{max}.

Under these conditions, (1.74) becomes

$$W_d = \frac{Gb^2L}{4\pi} \ln r \Big|_b^{b \exp(4\pi)} = Gb^2L \tag{1.75}$$

This expression shows that the deformation energy associated with the dislocation is

- proportional to b^2. Now, the Burgers vector is always oriented in the direction of the shear stress. It is therefore easier to deform a crystal along directions in which the atoms are at a minimum distance from each other.

- proportional to the dislocation length. This will therefore tend to be rectilinear unless it is closed in on itself. In this case, its perimeter will have a tendency to diminish.
- proportional to the shear modulus. Since G increases with the modulus of elasticity E_Y (Section 6.2.2), this simply reflects the obvious fact that more energy is needed to deform plastically a material with high modulus of elasticity.

Procedures that can be used to reduce the mobility of dislocations and consequently increase the elastic limit of materials are studied in Section 6.3.6.

1.5.8 Grain Boundary

Between the grains (Section 1.41) forming a polycrystal, there is a fine junction zone in which the atoms have a bonding energy that is less than that of the atoms situated inside the grains in terms of absolute value. This zone called the *grain boundary* is a two-dimensional defect.

Various phenomena may be produced in grain boundaries. They affect the properties of polycrystalline materials. We can mention in particular,

- the *segregation of impurities,* i.e., the accumulation of them in the grain boundaries. The segregation is produced over a specific temperature range. If the temperature is too high, the impurities remain in solid solution (Section 5.2.2). If, on the contrary, the temperature is too low, the mobility of the impurities is insufficient to allow them to reach the grain boundary. Segregation produces fragility and intergranular corrosion effects.
- *nucleation,* i.e., the emergence of a new phase (Section 5.2), generally produced in grain boundaries.
- the *diffusion* of atoms from the surface of the specimen which is more rapid in the grain boundaries than elsewhere. This diffusion may also be responsible for fragility and grain-boundary corrosion.
- *fracture due to sliding* of the grain boundary.
- *blockage of dislocations* that cannot cross a grain boundary because of the lack of alignment of the base vectors of two neighboring grains

The size of the grains tends to increase with time, mainly in pure materials where the rearrangement of the atoms at the faces of the grain boundaries is not hampered by the presence of impurity atoms. When the size becomes considerable, this phenomenon impairs mechanical characteristics. The rapidity of growth strongly depends on temperature.

The grains of a polycrystal are described by their size, their shape, and their orientation. There are several standards specifying how to measure these characteristics.

1.6 OBSERVATION OF CRYSTALLINE STRUCTURES

1.6.1 Bragg's Law

A beam of x-rays is diffracted by a set of lattice planes provided that the beams emerging from each plane are in phase with respect to each other (Fig. 1.38).

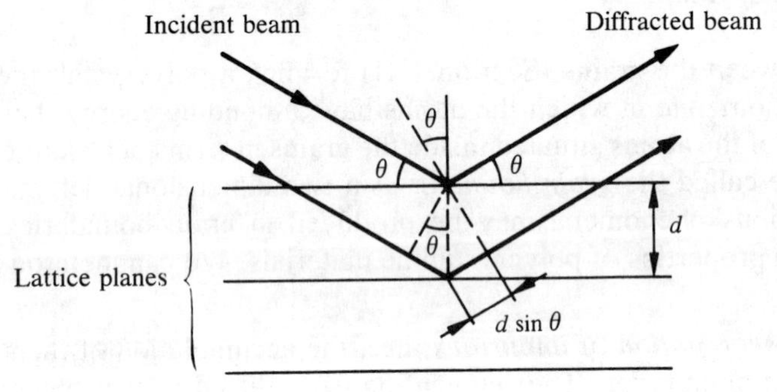

Fig. 1.38

This condition is expressed by Bragg's Law:

$$2d \sin \theta = n\lambda \tag{1.76}$$

where λ is the wavelength of the radiation and n is an integer.

1.6.2 Observation Methods

In order to determine a structure, it is necessary to observe a diffraction. Now, *a priori,* there is no reason for (1.76) to be satisfied. Three methods are currently used to obtain a diffraction.

In *Laue's method,* mainly used for determining the orientation of the crystalline axes in monocrystals, a beam with variable wavelength is used (Fig. 1.39).

As in the preceding method, the *rotating crystal method* can be applied to monocrystalline specimens. It consists of setting the specimen in rotation irradiated by a beam with fixed wavelength. For each revolution, a diffraction is produced each time a lattice plane is presented at a suitable angle.

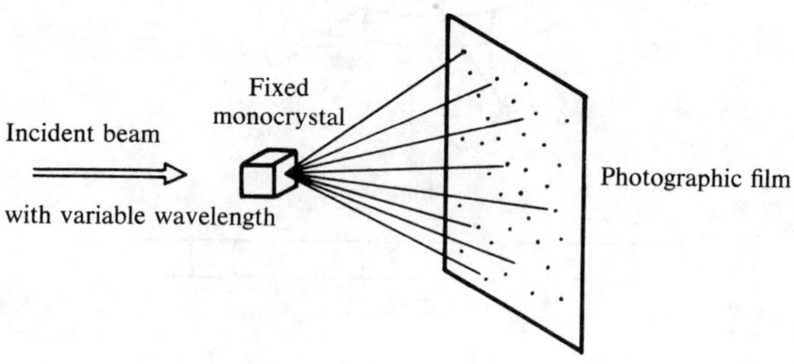

Fig. 1.39 Principle of Laue's Method

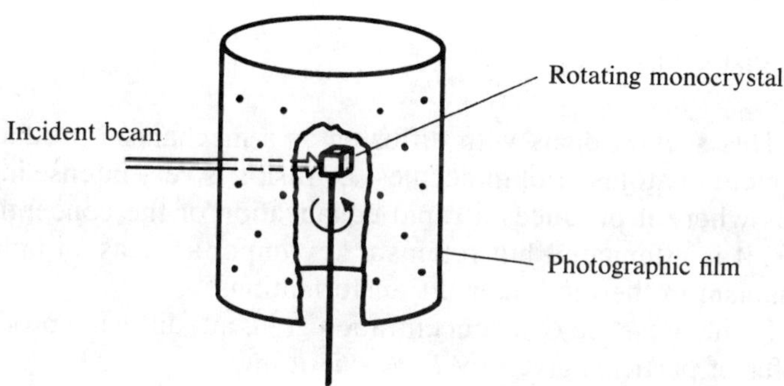

Fig. 1.40 Principle of the Rotating Crystal Method

The *powder method,* as its name suggests, uses a specimen in powder form. It is based on the principle of random orientation of the small monocrystals making up the specimen. There are always enough of them oriented to satisfy (1.76) so that a diffraction can be observed (Fig. 1.41).

Neutron diffraction and more rarely electron diffraction are also used (Section 3.5.2).

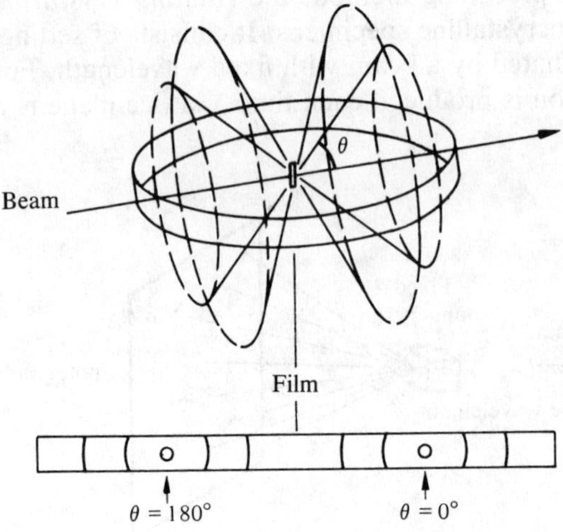

Fig. 1.41

1.7 DIFFUSION

1.7.1 Fick's First Law

This section deals with diffusion as a mechanism producing a displacement of atoms or of molecules. Diffusion is very intense in gases and liquids where it produces a rapid equalization of the concentrations. In solids, it is attenuated but retains a key importance as an indispensable mechanism in thermodynamic transformations.

In the presence of a concentration gradient, diffusion produces a net transfer of particles given by *Fick's first law:*

$$\boldsymbol{J} = -D\nabla N \tag{1.77}$$

The interpretation of this empirical expression similar to Fourier's heat equation is simpler if both sides are multiplied by an oriented surface element: $\mathrm{d}\boldsymbol{A}$. Thus,

$$\boldsymbol{J} \cdot \mathrm{d}\boldsymbol{A} = -D\nabla N \cdot \mathrm{d}\boldsymbol{A} \tag{1.78}$$

In (1.78), $\boldsymbol{J} \cdot \mathrm{d}\boldsymbol{A}$ represents the net number of particles crossing $\mathrm{d}\boldsymbol{A}$ per unit time $[\mathrm{J}] = \mathrm{m}^{-2}\mathrm{s}^{-1}$; D is the diffusion coefficient $[D] = \mathrm{m}^2\mathrm{s}^{-1}$ and N is the concentration of particles $[N] = \mathrm{m}^{-3}$.

Equation (1.77) shows that D is a measure of mobility. In order to move from one position to another, an atom (a molecule) must break the valence bonds which is comparable to surmounting a potential barrier. The probability of passing over such a barrier strongly depends on the kinetic energy of the atom and therefore on the temperature. For D, a dependence is often observed of the type:

$$D = D_0 \exp\left(-\frac{W}{k_B T}\right) \tag{1.79}$$

The constants D_0 and W are characteristics of the type of diffusion. W is called the *activation energy*.

1.7.2 Fick's Second Law

Fick's law expresses the balance of the number of particles in a volume Ω bounded by a surface $\boldsymbol{A}$:

$$\int_A \boldsymbol{J} \cdot \mathrm{d}\boldsymbol{A} = \int_\Omega -\frac{\partial N}{\partial t}\,\mathrm{d}\Omega \tag{1.80}$$

from which we have (1.77) and the divergence theorem:

$$\int_\Omega -\nabla \cdot (D\nabla N)\,\mathrm{d}\Omega = \int_\Omega -\frac{\partial N}{\partial t}\,\mathrm{d}\Omega \tag{1.81}$$

In the case when D does not depend on position, this equation reduces to *Fick's second law:*

$$D\nabla^2 N = \frac{\partial N}{\partial t} \tag{1.82}$$

1.7.3 One-dimensional Case

In the one-dimensional case, (1.82) reduces to

$$\frac{\partial^2 N}{\partial x^2} = \frac{1}{D}\frac{\partial N}{\partial t} \tag{1.83}$$

It can be shown [6] that the solution of (1.83) in the range $-\infty < x < +\infty$ with the initial condition

$$N(x, t = 0) = f(x) \tag{1.84}$$

can be written

$$N(x, t) = \frac{1}{2\sqrt{\pi D t}} \int_{-\infty}^{+\infty} f(x') \exp\left(-\frac{(x - x')}{4Dt}\right) \mathrm{d}x' \tag{1.85}$$

1.8 PROBLEMS

1.8.1 Write in the expanded form the wave function for the spherical orbit of the hydrogen atom defined by the quantum numbers $n = 2$, $l = 0$, $m = 0$. Calculate the two radii corresponding to the probability maxima of finding the electron.

1.8.2 The bond energy of KCl, an ionic crystal, is described by the expression $W = Ar^{-n} - Br^{-m}$ with $m = 1$, $n = 9$ and $B = 1.75\ e^2/4\pi\varepsilon_0$. The compressibility of KCl being $5.32 \cdot 10^{-11}\ m^2/N$, calculate the distance separating the ions K^+ and Cl^-.

1.8.3 Calculate the atomic packing factor for the following crystals: simple cubic, body-centered cubic, face-centered cubic, hexagonal close-packed structure.

1.8.4 Calculate the number of atoms in the planes (100) and (111) of copper. This metal crystallizes in the FCC structure, its density is $8.9 \cdot 10^3\ kg/m^3$ and its atomic weight is 63.5 kg/mole.

1.8.5 Silicon has the crystal structure of diamond. Calculate the density of silicon knowing that the length of a bond is 0.235 nm.

1.8.6 Zinc has diffused in a block of copper and at a given moment in time, the percentage of zinc atoms is equal to $10/(x + 0.12)$ where x is expressed in cm (Fig. 1.42). Calculate the number of zinc atoms crossing a parallel plane to S passing through $x = 1$ mm, and then $x = 1$ cm. Consider temperatures of 500°C and 1000°C. The activation energy for the diffusion of zinc in copper is 38,000 cal/mole and $D_0 = 0.033\ cm^2/s$.

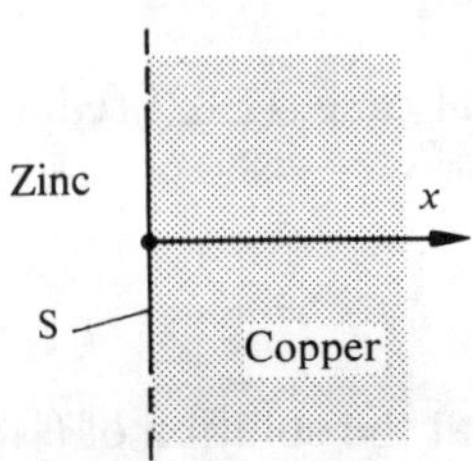

Fig. 1.42

1.8.7 Consider a semi-infinite medium bounded by a plane (Fig. 1.43) and characterized by a diffusion coefficient D. Find the concentration $N(x, t)$ satisfying the following conditions: $N(x, t = 0) = f(x)$ and $N(x = 0, t) = 0$. First of all consider any $f(x)$ and then $f(x)$ equal to a constant N_0. Method: consider an infinite medium and extend $f(x)$ to the left in order to obtain an odd function (Fig. 1.44).

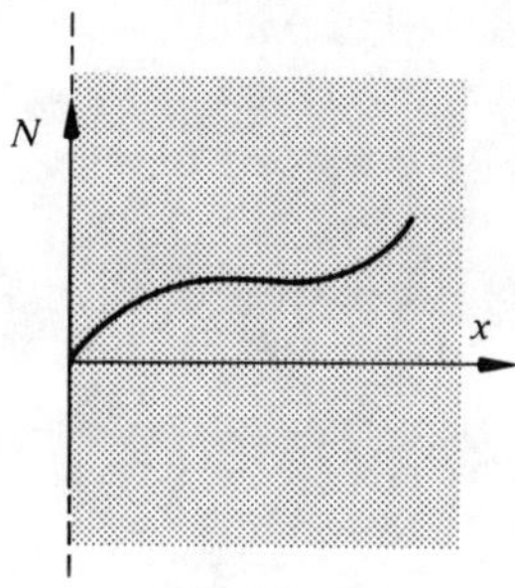

Fig. 1.43

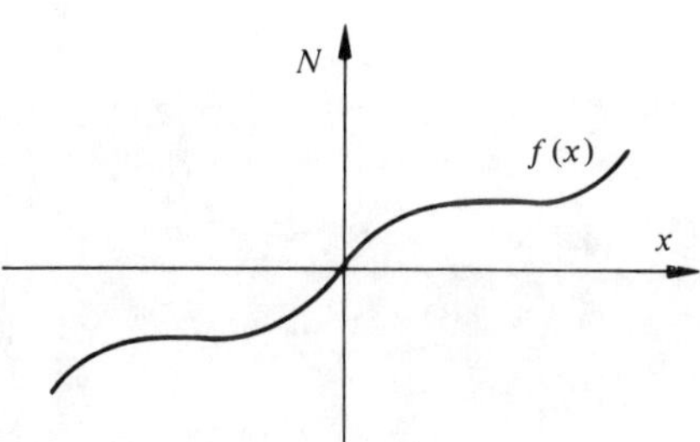

Fig. 1.44

Chapter 2
Electrical Conductivity

2.1 INTRODUCTION

2.1.1 Fundamental Questions

The electrical conductivity of matter is a measure of its capacity to carry an electric current when it is subjected to an electric field E. The current being defined as a flow of electric charges, the study of conductivity requires an answer to each of the following questions:

- What are the charge carriers?
- What charge does each of them transport?
- How do these carriers react when an external electric field is applied; in particular, what average velocity do they acquire?
- What is their total number?

2.1.2 Conduction Electrons: Definition

Certain electrons of the outer shells are capable of moving across matter under the action of an electric field. They are called *conduction electrons*.

2.1.3 A Choice of Four Models

In electrolytes and certain solids, the conductivity resulting from the movement of ions is added to the conductivity due to electrons. This chapter only deals with electron conduction but the formal treatment of the billiard balls model (Section 2.2) applies to the case of moving ions without any modification (Section 4.8.3). Within this framework, the answer to the first two questions is immediate. The same does not apply to the last two questions which have to be looked at from different points of view depending on the engineering problem to be solved. This is why four different models will be presented in turn. Each of them gives *its* answer to the following question, which by its very nature cannot have a totally satisfactory answer: *what is an electron?*

In turn, this particle will be considered as

- a small charged sphere obeying the laws of classical mechanics; this is the Drude model, also called the billiard balls model already mentioned;
- a free quantum object, i.e., without interaction with the medium in which it is moving except at the boundaries of this medium; this is the Sommerfeld model, also called the model of the free electron in a potential well (Section 2.4);
- a quantum object subjected to the action of the medium in which it is moving but having no effect itself on this medium, thus playing a passive role; this is the energy band model (Section 2.6);
- a quantum object subjected to the action of the medium and exerting an effect on it. This type of mutual action is found in the study of superconductivity. It is described in the model of Bardeen, Cooper, and Schrieffer (Section 2.8).

Of these models, which is the best, which is the one that gives the most exact results? *Formulated in this way, these questions have no sense,* because the quality of the results depends to a very large extent on the degree of match between the nature of the problem and the model used to solve it. It also frequently happens that it is necessary to use several models simultaneously.

These models will be presented in chronological order which is also the order of increasing complexity. Their particular areas of application, mainly depending on the starting assumptions, will be mentioned in each case.

2.2 THE BILLIARD BALLS MODEL

2.2.1 Introduction

This is the oldest and the most elementary model. The basis for this model was laid down by Drude in 1902, a little after the discovery of the electron by J.J. Thomson (1897).

Although it is inadequate to design and, *a fortiori,* to develop the solid state components which today make up most of the active elements used in electronics, the billiard balls model still has considerable interest:

- it is a useful tool for providing us with an image of phenomena of which we have no direct perception since they take place in the world of the infinitely small;
- the results for the engineer of more exact theories, such as the energy band theory, in particular, can be formulated using the same concepts as those belonging to the billiard balls model. Among these are the concentration and mobility of the electrons;
- However primitive it is, this model provides an interesting phenomenological interpretation of fundamental laws such as Ohm's law or Joule's law. It relates microscopic phenomena to certain observable quantities.

As its name suggests, this model compares electrons to minuscule billiard balls. These particles are therefore classical objects, simply governed by Newton's law and Maxwell's laws. This physical concept of the electron is also not totally contrary to the results of quantum mechanics in which a wave packet, a narrow zone where the function $|\Psi|^2 d\Omega$ has a sharp maximum, can always be interpreted as a particle with its mass and velocity.

In a cubic millimeter of copper, there are approximately 10^{20} electrons. There is, therefore, no question of treating them individually which would also be of no interest. It is the *average* behavior of the electrons we must study. Two types of interaction govern this behavior. These are

- the interaction of the electrons with the matter in which they are moving and to which they belong;
- the interaction of the electrons with electromagnetic fields applied from the outside.

2.2.2 Electron-Matter Interactions

Apart from a few exceptions, all conducting materials have a crystalline structure. Let us assume that, on the average, each atom releases one conduction electron. These electrons are immersed in the potential created by the cations forming the lattice.

If the crystal did not have any imperfection and if the cations were totally at rest, the electrons assumed not to interact with each other would move in a perfectly periodic potential and each of them would maintain a strictly constant energy. Their behavior would be comparable to that of heavy balls moving without friction on an undulating surface, on the average horizontal.

In reality, a crystal lattice is never perfect; the main types of imperfections that may occur were described in Section 1.5. The conduction electrons interact with these imperfections. It follows that the motion of these electrons is constantly perturbed by random processes of various types called *collisions* in the billiard balls model.

At room temperature and in most metals, the electrical resistance mainly depends on the interaction of the conduction electrons with phonons. At low temperatures, at a few kelvins, the interaction with other types of crystal imperfections becomes predominant.

The exchange of energy occurring during the collisions results in a thermodynamic equilibrium of the conduction electrons with the lattice. For example, the increase in kinetic energy of the conduction electrons due to an applied electric field causes an increase in the energy of the lattice corresponding to an increase in temperature.

2.2.3 Collision Time: Definitions

The path of an electron in matter can be shown in diagram form as follows:

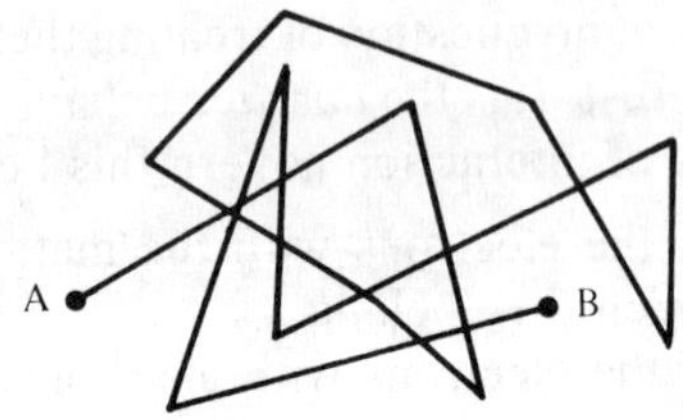

Fig. 2.1 Possible Path of an Electron Between Points A and B. Each Change in Direction Corresponds to a Collision

Let us follow the path of an electron, and let us take, as the origin of time, the moment the electron departs in a new direction when undergoing a collision. An important question is to know *when* the electron will undergo a new collision. The value of the electrical resistance directly depends on this moment in time.

The *collision time* is the interval that separates two consecutive collisions experienced by the same electron.

The collision time is a random variable. With all physical parameters remaining constant, this random variable is steady-state, and its mean value is called the *mean collision time* τ.

2.2.4 Calculation of the Collision Time

The random nature of electron movements suggests basing their study on the following *assumptions:*

- the number of collisions that an electron will experience over any time interval dt is equal to αdt, α being a constant;
- the collisions are isotropic;
- the energy that the electron has after the collision has no correlation with the energy before collision.

The first assumption can be found, for example, in the study of radioactive disintegration. The following two situations are comparable:

- observing an unstable atom awaiting decomposition
- observing an electron moving in matter awaiting a collision.

In both cases, the probability of observing the expected phenomenon does *not* depend on the time elapsed from the origin but only on the duration of the observation. A more sophisticated theory would show that the last two assumptions are acceptable. They reflect in particular the fact that the target atom is heavy with respect to the electron and that it has a kinetic energy that is distributed randomly at the moment of collision.

The mean collision time could be determined by the following imaginary experiment: a population of n_0 electrons is observed from a time $t = 0$ and for each electron i, the time t_i is noted after which it experiences its first collision. Then, for n_0 sufficiently large:

$$\tau = \frac{1}{n_0} \sum_{i=1}^{n_0} t_i \tag{2.1}$$

Following this experiment, the first assumption above makes it theoretically possible to calculate τ if the value of α is known. Let $n(t)$ be the number of electrons at time t not yet having experienced the first collision. The variation dn of $n(t)$ for dt is given by

$$dn = -n(t)\alpha dt \tag{2.2}$$

The positive quantity $-dn$ represents the number of electrons that have experienced their first collision between t and $t + dt$.

Taking into account the initial condition of the experiment, integration of (2.2) immediately gives

$$n(t) = n_0 \exp(-\alpha t) \tag{2.3}$$

From (2.3), we derive the probability for an electron of *not* experiencing collision before the moment t:

$$\frac{n(t)}{n_0} = \exp(-\alpha t) \tag{2.4}$$

The probability for an electron to experience its first collision between t and $t + \mathrm{d}t$ being given by (product of independent probabilities)

$$\exp(-\alpha t)\alpha \mathrm{d}t = -\frac{\mathrm{d}n}{n_0} \tag{2.5}$$

we obtain τ by the weighted mean of the collision time:

$$\tau = \int_0^\infty t \exp(-\alpha t)\alpha \mathrm{d}t = \frac{1}{\alpha} \tag{2.6}$$

2.2.5 Velocity of the Electrons: Definitions

The velocity of the electrons may be decomposed into two components:

$$\boldsymbol{v} = \boldsymbol{v}_{\mathrm{th}} + \boldsymbol{v}_{\mathrm{d}} \tag{2.7}$$

The *thermal velocity* of an electron $\boldsymbol{v}_{\mathrm{th}}$ is the component of its velocity due to the transfer of kinetic energy occurring during collisions. Only the last collision experienced determines $\boldsymbol{v}_{\mathrm{th}}$ in agreement with the third assumption of Section 2.2.4. The thermal velocity is a random variable retaining a constant value between collisions.

The mean $\bar{v}_{\mathrm{th}}$ of the modulus of $\boldsymbol{v}_{\mathrm{th}}$ is called the *mean thermal velocity* of the electrons.

The *drift velocity* of an electron $\boldsymbol{v}_{\mathrm{d}}$ is the component of its velocity due to the action of an applied electromagnetic field. This action is only manifested between collisions. The value taken by $\boldsymbol{v}_{\mathrm{d}}$ at the point of a collision is a random variable. An ensemble of electrons has a mean drift velocity denoted $\bar{\boldsymbol{v}}_{\mathrm{d}}$.

2.2.6 Estimation of Thermal Velocity

Within the framework of the billiard balls model, it is assumed that electrons behave as the atoms of a perfect gas. This is a crude assumption but sufficient for the time being. The energy distribution of electrons is studied more rigorously in Sections 2.4.5 and 2.6.2. Under these conditions, electrons follow a Maxwellian statistical distribution (Section 7.2). In particular, their mean velocity, which is equal to the mean thermal velocity defined in the preceding section, is given by

$$\bar{v}_{\text{th}} = \sqrt{\frac{8k_{\text{B}}T}{\pi m_{\text{n}}}} \tag{2.8}$$

where T represents the absolute temperature and m_{n} the mass of the electron. The proof of (2.8) is the object of Problem 2.9.1.

2.2.7 Calculation of the Mean Drift Velocity

We consider the case when only an electric field is applied. This is assumed uniform, steady-state, and directed along the x-axis. By taking the origin of time as the moment that immediately follows a collision, the equation for the trajectory of an electron can be written

$$\boldsymbol{r}(x, y, z) = \frac{1}{2}\boldsymbol{a}t^2 + \boldsymbol{v}_0 t + \boldsymbol{r}_0 \tag{2.9}$$

This equation is only valid until the next collision. The vector $\boldsymbol{r}_0$ gives the initial position of the electron. The reference origin is chosen so as to make $\boldsymbol{r}_0$ vanish. The initial velocity $\boldsymbol{v}_0$ of the electron is equal to the thermal velocity $\boldsymbol{v}_{\text{th}}$. The second and third assumptions of Section 2.2.4 therefore imply that $\boldsymbol{v}_0$ is independent of the field $\boldsymbol{E}$. The acceleration $\boldsymbol{a}$ of the electron whose value is equal to eE/m_{n} is directed along the x-axis. Consequently, an electron i characterized by a specific collision time t_i is displaced between two collisions in the direction of $\boldsymbol{E}$ *and only under the action of* $\boldsymbol{E}$ by an amount:

$$\boldsymbol{x}_i = \frac{1}{2}\boldsymbol{a}t_i^2 \tag{2.10}$$

For an electron belonging to an ensemble of electrons characterized by a collision time t (random variable), this displacement is on the average equal to

$$\bar{x} = \frac{1}{2}\,\boldsymbol{a}\overline{t^2} \tag{2.11}$$

A similar reasoning to that which made it possible to calculate τ using equation (2.6) allows us to write

$$\overline{t^2} = \int_0^\infty t^2 \exp(-\alpha t)\alpha \mathrm{d}t = \frac{2}{\alpha^2} = 2\tau^2 \tag{2.12}$$

Finally, the mean drift velocity of the electrons is obtained by the relationship:

$$\bar{\boldsymbol{v}}_{\mathrm{d}} = \frac{\bar{\boldsymbol{x}}}{\tau} = \boldsymbol{a}\tau = -\frac{e}{m_{\mathrm{n}}}\tau\boldsymbol{E} \tag{2.13}$$

2.2.8 Definition of the Mobility

The mobility μ of a charge carrier (μ_{n} for an electron) is the ratio of its mean drift velocity to the intensity of the electric field that produces it. The mobility is a variable that is always positive, by definition. Its value for an electron is given by

$$\mu_{\mathrm{n}} = \frac{|\bar{\boldsymbol{v}}_{\mathrm{d}}|}{|\boldsymbol{E}|} = \frac{e}{m_{\mathrm{n}}}\tau \tag{2.14}$$

The greater the mean collision time, the greater the velocity of the charge carrier in a given field $\boldsymbol{E}$. It is therefore logical that the mobility is proportional to τ.

2.2.9 Another Formulation

Instead of considering the electric current as a displacement of point charges, it can be compared to the flow of a charged fluid. The hydrodynamic model based on this idea leads to the same conclusions as the billiard balls model provided that in both cases the validity of the same fundamental laws of mechanics and electricity are accepted. The differences in formulation may, on the other hand, give to each model a preferred area of application.

In the hydrodynamic model, the dynamic equation relating to the fluid element corresponding to an electron is written

$$m_n \frac{d\boldsymbol{v}}{dt} = -e\boldsymbol{E} - m_n \xi \boldsymbol{v} \tag{2.15}$$

where $\boldsymbol{v}$ is the absolute velocity of the fluid and ξ a coefficient measuring its viscosity. Under permanent conditions (these conditions are established in a time on the order of τ, or approximately 10^{-14} seconds in a metal conductor), the force due to the field is balanced by the force of viscosity, from which we have

$$\boldsymbol{v} = -\frac{e}{m_n} \frac{1}{\xi} \boldsymbol{E} \tag{2.16}$$

This velocity plays the role of the drift velocity in the billiard balls model. The comparison with equation (2.13) shows that the equivalence of the two models can simply be written as

$$\xi = \frac{1}{\tau} \tag{2.17}$$

A low collision time corresponds to a high velocity.

2.3 INTERPRETATION OF SOME LAWS AND PHENOMENA USING THE BILLIARD BALLS MODEL

2.3.1 Ohm's Law and Conductivity

The current density $\boldsymbol{J}$ is defined as the net electric charge crossing a unit cross section placed perpendicular to the motion of the electrons during one second (Fig. 2.2).

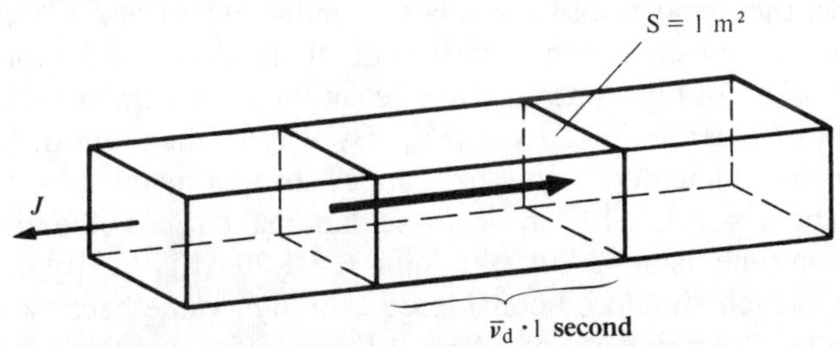

Fig. 2.2 Electrons Crossing S in the Second to Come are Contained in a Parallelepiped with Edge $\bar{v}_d \cdot 1$ s

We therefore have

$$J = -ne\bar{v}_d \tag{2.18}$$

where n is the concentration of the electrons. Taking into account (2.13), this expression can be written

$$J = \frac{ne^2\tau}{m_n}E \tag{2.19}$$

The conductivity σ can be immediately derived by comparing (2.19) with Ohm's law in the form $J = \sigma E$. We have

$$\sigma = \frac{ne^2\tau}{m_n} = ne\frac{e\tau}{m_n} = ne\mu_n \tag{2.20}$$

It can be seen that the conductivity contains the product of the concentration of electrons and their mobility. It is therefore necessary that at least one of these variables has a high value for a material to have a high conductivity.

Mobility is greater in semiconductors than in metals. However, this characteristic is completely masked by the ratio of the concentrations of the electrons—n is 10^6 to 10^8 times smaller in semiconductors than in metals, which explains the higher conductivity of the latter.

2.3.2 Linearity of Ohm's Law

According to (2.20), the conductivity should depend on the electric field via the collision time. In fact, the more E increases, the more the velocity of the electrons increases. Since the distance between possible impact points remains the same, the collision time, and consequently the conductivity, should diminish.

Now, the linearity of Ohm's law (σ independent of E) is an experimental fact established with precision in all standard conductors.

The origin of this contradiction lies in the considerable difference in the orders of magnitude of v_{th} and $\bar{v}_d$. By way of illustration, these two velocities are calculated below for copper. In this metal, $\sigma = 5.81 \cdot 10^7$ Ω^{-1} m^{-1} and $n = 1.16 \cdot 10^{29}$ m^{-3}. These data make it possible to calculate the collision time using (2.20), obtaining $\tau = 1.79 \cdot 10^{-14}$ s. By taking E = 0.32 V/m, which should be considered as a high value because this field produces a current density of $1.86 \cdot 10^7$ A/m^2, (2.13) gives

$$\bar{v}_d = 1 \cdot 10^{-3} \text{ m/s} \tag{2.21}$$

At 20°C, the thermal velocity of the electrons calculated by (2.8) is

$$\bar{v}_{th} = 1.1 \cdot 10^5 \text{ m/s} \tag{2.22}$$

Consequently, even in a strong electric field, the drift velocity only represents a tiny fraction of the thermal velocity:

$$\bar{v}_d \cong 10^{-8}\, \bar{v}_{th} \tag{2.23}$$

Since the thermal velocity does not depend on $\boldsymbol{E}$, it is found in practice that the velocity of the electrons is independent of the electric field. In other words, *the establishment of a current, even if it is intense, has an absolutely negligible effect on the velocity of the electrons!*

Therefore, the linearity of Ohm's law as an experimental reality is not in contradiction with the billiard balls model: the nonlinearity effect predicted by this model is smaller than the error margins of the best measurements.

2.3.3 Conductivity of Thin Films

In the overwhelming majority of cases, the dimensions of conductors are large compared to the average distance traversed by an electron between two consecutive collisions. The behavior of the electrons at the surface of the conductor is therefore of secondary importance. This is why the conductor medium is implicitly assumed as infinite throughout Section 2.2.

FET and MOS transistors (see Vol. VII) form an important exception in this matter. In these transistors, the current flows in a sufficiently thin film so that the mobility of the electrons is influenced by the diffusion of electrons to the boundary surfaces of this film.

2.3.4 Conductivity and Temperature

As was mentioned in Section 2.2.2, the electric resistance originates from perturbations in the movement of electrons produced by the presence of crystal imperfections. The study of the variation of conductivity as a function of temperature requires the separation of the effects of each type of imperfection. With this in mind, the imperfections considered in Section 1.5 are classified into the following three categories:

- phonons;
- chemical impurities (interstitial impurity atoms or substitutional impurity atoms);
- imperfections originating from mechanical deformations (mainly dislocations).

It is therefore also necessary to distinguish in the electron collision we have considered thus far, three types of collisions, each corresponding to one of the imperfection categories defined above. Let α_{ph}, $\alpha_{im(c)}$, and α_{imp} be the probabilities over an interval of unit time, for an electron to experience a collision with a phonon, a chemical impurity, and an imperfection caused by mechanical deformation, respectively. These three probabilities are independent of each other:

$$\alpha = \alpha_{ph} + \alpha_{im(c)} + \alpha_{imp} \tag{2.24}$$

The mean collision time for each type of collision is obtained by a calculation identical to that given in Section 2.2.4. With obvious notation, we find

$$\tau_{ph} = \frac{1}{\alpha_{ph}}; \qquad \tau_{im(c)} = \frac{1}{\alpha_{im(c)}}; \qquad \tau_{imp} = \frac{1}{\alpha_{imp}} \tag{2.25}$$

$$\frac{1}{\tau} = \frac{1}{\tau_{ph}} + \frac{1}{\tau_{im(c)}} + \frac{1}{\tau_{imp}} \tag{2.26}$$

By comparing (2.20) and (2.26) we discover that the partial resistivity resulting from each type of collision must be added to obtain the resistivity of the material in question. This is Mathiessen's rule. Thus,

$$\rho = \frac{m_n}{ne^2}\left(\frac{1}{\tau_{ph}} + \frac{1}{\tau_{im(c)}} + \frac{1}{\tau_{imp}}\right) \tag{2.27}$$

By definition of the partial resistivities ρ_{ph}, $\rho_{im(c)}$, ρ_{imp}, (2.27) can also be written

$$\rho = \rho_{ph} + \rho_{im(c)} + \rho_{imp} \tag{2.28}$$

Experiment shows that $\rho_{im(c)}$ and ρ_{imp} are independent of temperature provided that the concentration of imperfections remains small and that the temperature is not too high to act on the imperfections themselves, by chemical reaction involving impurities or by annealing of cold-working imperfections. Experiment and, to a certain extent, theory also show that

$$\lim_{T \to 0\text{K}} \rho_{ph} = 0 \tag{2.29}$$

Above a temperature where $\rho_{ph} \approx \rho_{im(c)} + \rho_{imp}$, we observe a linear variation of ρ_{ph} *versus* temperature. For temperatures close to room temperature and up to a few hundred degrees, we use an approximation of the type:

$$\rho(\theta) = \rho(\theta = 0)[1 + \alpha_\theta \theta] \tag{2.30}$$

where α_θ is the temperature coefficient of the resistivity and θ is the temperature in degrees centigrade. The coefficient α_θ is on the order of magnitude of $4 \cdot 10^{-3}$/°C in pure metals, often a little less in alloys.

The grid of straight lines in Fig. 2.3 confirms the validity of (2.30) and the additivity of the resistivities according to (2.28).

It can be seen from Fig. 2.3 that to add as little as 1% of arsenic to copper increases the resistivity of this metal by a factor of 5 at 20°C and by a factor of 17 at −173°C. In a more general manner, it should be kept in mind that the presence of certain elements as impurities may have an enormous effect on the resistivity.

The effect of a mechanical deformation is reflected in Fig. 2.3 by a displacement of the lines to the top without any appreciable change to their slope.

It is important to emphasize the empirical origin of expression (2.30) in this section devoted to applications of the billiard balls model. In fact, the billiard balls model does not provide an acceptable prediction of the variation in resistivity as a function of temperature. This derives from the assumption (Section 2.26) that states that the thermal velocity of the electrons follows a Maxwellian distribution. Equation (2.8) shows that $v_{th} \sim \sqrt{T}$, from which we have $\tau \sim 1/\sqrt{T}$ and $\rho \sim \sqrt{T}$ which contradicts experiment. A correction to this assumption will be made in Section 2.4.

2.3.5 Electron Collisions and Joule's Law

The flow of a current in a conductor is accompanied by the dissipation of energy. This phenomenon is described by Joule's law:

$$P_J = \sigma E^2 \tag{2.31}$$

where P_J is the dissipated power. This power is drawn from the current source and transferred to the conductor by electron collisions. The billiard balls model accounts for this process in a simple manner and makes it possible to establish (2.31).

Consider an electric field $\boldsymbol{E}_x$ applied along the x-axis. Also, let

$$\boldsymbol{v} = \boldsymbol{v}_x + \boldsymbol{v}_y + \boldsymbol{v}_z \tag{2.32}$$

be the velocity of an electron at the moment following a collision ($t = 0$). This electron experiences a new collision after a collision time t and at this time has a velocity:

$$\boldsymbol{v} = \boldsymbol{v}_x - \frac{e}{m_n}\boldsymbol{E}_x t + \boldsymbol{v}_y + \boldsymbol{v}_z \tag{2.33}$$

The increase in the electron's kinetic energy between the two collisions is

$$\Delta W(t) = \frac{1}{2} m_n \left[\left(v_x - \frac{e}{m_n} E_x t\right)^2 - v_x^2\right] = \frac{1}{2} m_n \left[\left(\frac{e}{m_n} E_x t\right)^2 - \frac{2e}{m_n} v_x E_x t\right] \tag{2.34}$$

Consider now an ensemble of N electrons each having the same collision time t. In this ensemble, the mean energy per electron increases when each electron has traversed the distance separating two consecutive collision points by the amount:

$$\overline{\Delta W(t)} = \frac{1}{N}\sum_{i=1}^{N} \frac{1}{2} m_n \left[\left(\frac{e}{m_n} E_x t\right)^2 - \frac{2e}{m_n} v_{xi} E_x t\right] \tag{2.35}$$

where v_{xi} represents the value of v_x for electron i. The assumption of isotropy of the collisions means that the term:

$$\sum_{i=1}^{N} v_{xi}$$

is zero, from which we have

$$\overline{\Delta W(t)} = \frac{1}{2} m_n \left(\frac{e}{m_n} E_x t\right)^2 \tag{2.36}$$

By finally considering the ensemble of all the electrons whose collision time distribution is given by (2.5), it is possible to calculate the mean increase in energy per electron between two collisions:

$$\overline{\overline{\Delta W}} = \int_{t=0}^{\infty} \overline{\Delta W(t)} \exp(-\alpha t)\alpha \mathrm{d}t = \frac{e^2E^2\tau^2}{m_\mathrm{n}} \tag{2.37}$$

Knowing that the n electrons contained in 1 m³ of matter experience an average of n/τ impacts per unit time, Joule's law can be derived directly from (2.37) and (2.20):

$$P_\mathrm{J} = \frac{n}{\tau}\overline{\overline{\Delta W}} = \frac{ne^2\tau}{m_\mathrm{n}} E^2 = \sigma E^2 \tag{2.38}$$

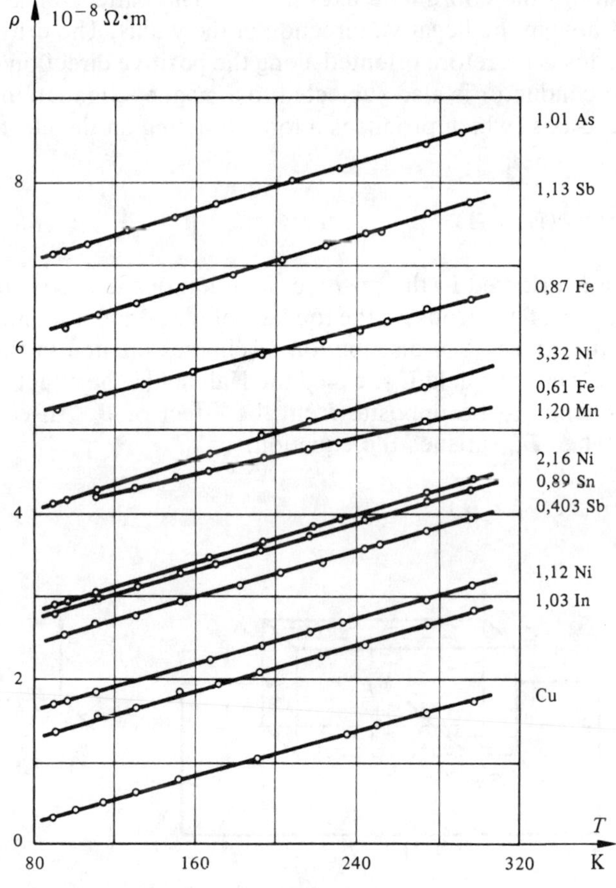

Fig. 2.3 **Resistivity of Copper and Some of its Alloys as a Function of Temperature, after [7]. The Percentage of Impurities is Given in Front of Their Chemical Symbols**

2.3.6 Definition of the Hall Effect

The *Hall effect* is the name given to the appearance of electric charges on the surface of a conductor traversed by a current, resulting from the fact that this conductor is subjected to a magnetic induction.

2.3.7 Study of the Hall Effect

Let us consider a parallelepiped conductor (Fig. 2.4) whose edges are parallel to the coordinate axes x, y, z. The battery makes a current density $\boldsymbol{J}$ flow in the negative direction of the y-axis. The drift velocity of the electrons is therefore oriented along the positive direction of the same axis. The conductor is also subjected to a magnetic induction $\boldsymbol{B}$ oriented along the x-axis, which produces a force $\boldsymbol{F}$ acting on the electrons, given by

$$\boldsymbol{F} = -e(\bar{\boldsymbol{v}}_{\mathrm{d}} \times \boldsymbol{B}) \tag{2.39}$$

The force $\boldsymbol{F}$ oriented in the positive direction of the z-axis, produces an accumulation of electrons on the top face of the conductor and a depletion on its bottom face. The distribution of charges created in this way produces a new electric field $\boldsymbol{E}_{\mathrm{H}}$, called the Hall field. The effect of $\boldsymbol{E}_{\mathrm{H}}$ on the electrons tends to be opposite from the effect of $\boldsymbol{B}$. An equilibrium is reached when $\boldsymbol{E}_{\mathrm{H}}$ satisfies the equation:

$$e\boldsymbol{E}_{\mathrm{H}} = e(\bar{\boldsymbol{v}}_{\mathrm{d}} \times \boldsymbol{B}) \tag{2.40}$$

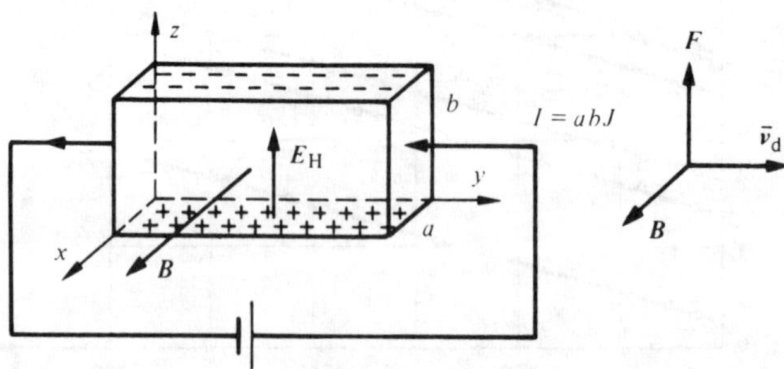

Fig. 2.4 The Hall Effect

By expressing $\bar{\boldsymbol{v}}_{\mathrm{d}}$ as a function of $\boldsymbol{J}$ by means of (2.18), equation (2.40) can be written

$$\boldsymbol{E}_{\mathrm{H}} = R_{\mathrm{H}}(\boldsymbol{J} \times \boldsymbol{B}) \tag{2.41}$$

with

$$R_{\mathrm{H}} = -\frac{1}{ne} \tag{2.42}$$

The quantity R_{H} is called the Hall constant. It is negative in metal conductors where the electrons are the only mobile charges. In intrinsic semiconductors (Section 2.7.1), the electrons and the holes are both involved in conduction at the same time. The expression for R_{H} is therefore more complex than (2.42): the value of this constant depends on the concentration of the two types of charge carriers as well as the ratio of their mobilities (Section 2.9.2). On the other hand, (2.42) remains valid for extrinsic semiconductors of the n type (Section 2.7.2). For semiconductors of the p type (Section 2.7.3), the minus sign of (2.42) must be replaced by a plus sign.

2.3.8 Applications of the Hall Effect

The Hall effect is currently used to measure various quantities such as the mobility, the concentration of charge carriers, and the magnetic field.

For example, the mobility of electrons is obtained by rewriting (2.41) in the form:

$$\boldsymbol{E}_{\mathrm{H}} = -\frac{1}{ne}\,\sigma\boldsymbol{E} \times \boldsymbol{B} \tag{2.43}$$

In this equation, $\boldsymbol{E}$ is the field created in the conductor by the battery. By replacing σ in (2.43) by its value according to (2.20), we obtain

$$\mu_{\mathrm{n}} = \frac{|\boldsymbol{E}_{\mathrm{H}}|}{|\boldsymbol{E} \times \boldsymbol{B}|} \tag{2.44}$$

The mobility of electrons in some metals is given in Table 2.5.

The concentration of electrons is directly derived from (2.41) and (2.42).

When the Hall constant is known, the Hall effect makes it possible to measure magnetic fields. In this application, it is preferable to use semiconductor materials instead of metal conductors. Due to their larger

Hall constant, they produce a higher value of $\boldsymbol{E}_H$ in a given magnetic field. The sensitivity of the measurement system is therefore improved.

Table 2.5
Hall Constants, after [5]. The Mobilities are Calculated by the Relationship $\mu_n = \sigma R_H$

Metal	R_H $\frac{Vm^3}{As}$	μ_n $\frac{m^2}{Vs}$
Cu	$-0.6 \cdot 10^{-10}$	$3.5 \cdot 10^{-3}$
Al	$-0.43 \cdot 10^{-10}$	$1.5 \cdot 10^{-3}$
Au	$-0.8 \cdot 10^{-10}$	$3.3 \cdot 10^{-3}$
Ag	$-1.0 \cdot 10^{-10}$	$6.3 \cdot 10^{-3}$
Li	$-1.89 \cdot 10^{-10}$	$2.0 \cdot 10^{-3}$
Na	$-2.3 \cdot 10^{-10}$	$4.9 \cdot 10^{-3}$

2.4 MODEL OF THE FREE ELECTRON IN A POTENTIAL WELL

2.4.1 Introduction

The main drawback of the billiard balls model is considering the electron as a classical particle. A system of such particles is not subject to Pauli's principle. One of the consequences of this is that all the conduction electrons are capable of increasing their energy, under the effect of an increase in temperature, for example. The theoretical variation of the resistivity as a function of temperature established on this basis is incorrect (Section 2.3.4). The estimates of specific heat are no better, whereas thermoelectronic emission, photoelectric emission, or field emission processes lie totally outside the scope of such a model.

These difficulties are largely removed by the model of the free electron in a potential well devised by Sommerfeld in 1928. In this model, the electrons subjected to Pauli's principle follow the Fermi-Dirac energy distribution (Section 7.4). Two important results follow from this:

- only a fraction of the electrons, close to 2%, is capable of having its energy vary under the effect of an external action, temperature, electromagnetic field, *et cetera;*
- even at absolute zero, the kinetic energy of the electrons is not zero. A considerable number of them have energies of a few eV which is

enormous. By way of comparison, the mean energy of a classical particle calculated from the Boltzmann distribution (Section 7.3) is only 0.038 eV at 20°C.

Despite the improvements that it brings, the Sommerfeld model does not give a satisfactory description of the electronic properties of solids in all cases. Its limitations derive from the fact that it does not take into account the structure of the materials. This model will therefore never make it possible to explain why one crystal is a conductor, another an insulator or a semiconductor. Its application must therefore *be limited to the case of the metals*. In fact, metallic bond puts the electrons in a situation that approximates that accepted by the assumptions of the model.

The Sommerfeld model provides a basis for constructing more specific theories. This is, therefore, not by itself a model dealing with an exact problem such as electric conduction or thermoelectronic emission. This basis is the electron energy distribution obtained by the product of two functions: the density of states and the Fermi-Dirac distribution. The formula giving the current density of thermoelectronic emission will be established as an application example.

2.4.2 Assumptions

There are two assumptions to the model:

- at the surface of metals, there is a potential barrier preventing electrons from leaving the material;
- inside the material, the electrons are subjected to a constant potential.

The first assumption rests on the following elementary observation: electrons moving in a metal do not go beyond the surfaces bounding the specimen, at least not at room temperature.

The second assumption seems rather crude. It is this assumption that removes the idea of the internal structure of matter from the model. It will be modified in the energy band model (Section 2.6). The assumption reflects that the electrons are considered to be free in the potential well.

The potential barrier has a finite width, i.e., in order to pass from the potential acting inside the matter to the potential acting outside takes a few interatomic distances. However, since the dimensions of the specimen are, in practice, always very large with respect to one interatomic distance, the potential barrier can be considered as infinitely steep. This simplifies the calculations.

2.4.3 Definition of the Density of States $Z(W)$

The *density of states* $Z(W)$ is a continuous function so that $Z(W)\mathrm{d}W$ represents the number of *possible* quantum states between the energies W and $W + \mathrm{d}W$ for free electrons contained in a unit volume of matter.

It should be noted that from a completely rigorous point of view, the function $Z(W)$ is discontinuous because it runs over discrete states. In practice, this function may always be considered as continuous (Section 2.4.4).

2.4.4 Establishment of the Density of States

The second assumption excludes electron-electron interaction. It is therefore sufficient to study the possible behavior of a single electron. In order to simplify the notation, this electron is placed in a one-dimensional sample of thickness L, theoretically represented by the potential well of Fig. 2.6.

Since a potential is always defined apart from an additive constant, we set the potential at the bottom of the well equal to zero. Inside the well, Schrödinger's equation takes the form:

$$\frac{\hbar^2}{2m_{\mathrm{n}}}\frac{\mathrm{d}^2\Psi}{\mathrm{d}x^2} + W\Psi = 0 \tag{2.45}$$

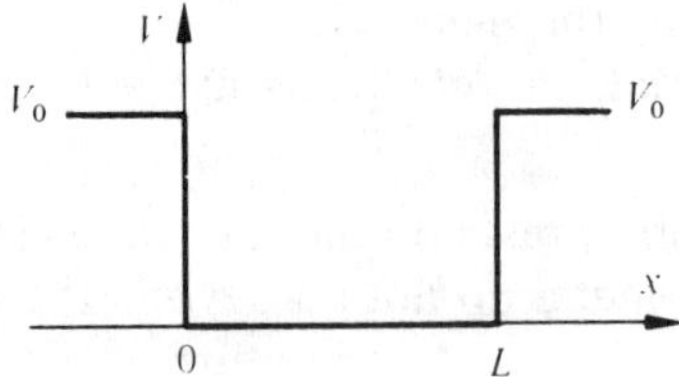

Fig. 2.6

The general solution of this equation can be written

$$\Psi = A \sin kx + B \cos kx \tag{2.46}$$

where

$$k = \sqrt{\frac{2m_{\mathrm{n}} W}{\hbar^2}} \tag{2.47}$$

A and B are integration constants.

The height V_0 of the barrier sufficiently large to confine the electrons in the well at normal temperatures may be assumed to be infinite. The conditions for boundary Ψ, which must be continuous across the surface delineating the limits of the well, can be expressed in a particularly simple manner. The probability of finding the electron outside the well being equal to zero, we have

$$\Psi(x = 0) = 0 \tag{2.48}$$

and

$$\Psi(x = L) = 0 \tag{2.49}$$

By imposing the conditions (2.48) and (2.49) on the function (2.46), we obtain, on the one hand,

$$B = 0 \tag{2.50}$$

and, on the other hand, after discarding the uninteresting solution $A = 0$:

$$\sin kL = 0 \tag{2.51}$$

This eigenvalue equation only has solutions if

$$kL = n_x \pi \tag{2.52}$$

where the quantum number n_x can take the values:

$$n_x = 1, 2, 3, \ldots \tag{2.53}$$

The value $n_x = 0$ should be excluded because it leads to the solution $\Psi \equiv 0$ signifying the absence of electrons in the well.

It follows from (2.52), (2.53), and (2.47) that the electron can only have discrete energies given by

$$W = \frac{k^2\hbar^2}{2m_n} = \frac{\hbar^2}{2m_n}\frac{n_x^2\pi^2}{L^2} = \frac{h^2}{8m_n L^2} n_x^2 \tag{2.54}$$

The expression corresponding to (2.54) for three-dimensional space can easily be obtained since the Schrödinger equation has separate variables. For a cube with side L, we have

$$W = \frac{h^2}{8m_n L^2}(n_x^2 + n_y^2 + n_z^2), \tag{2.55}$$

n_x, n_y, n_z being three quantum numbers. Each of these numbers defines fully the momentum of the electron along the corresponding axis. These three numbers may take any positive integer value.

The effect of the electron spin, neglected until now, may be introduced in a simple manner by establishing $Z(W)$. It is enough to remember that Pauli's principle allows any quantum state defined by n_x, n_y, n_z to be occupied by two electrons with opposite spins. The number of states obtained on the basis of (2.55) must therefore be multiplied by 2.

Equation (2.55) calls for two remarks.

- All the triplets (n_x, n_y, n_z), give the same value of n, where

$$n^2 = n_x^2 + n_y^2 + n_z^2 \tag{2.56}$$

correspond to the same energy.

- The energy of an electron is always a multiple of a unit energy W_u:

$$W_u = \frac{h^2}{8m_n L^2} \tag{2.57}$$

For a cube with edge 1 cm, $W_u = 3.76 \cdot 10^{-15}$ eV. The lowest energy levels therefore have a spacing of approximately 10^{-15} eV.

The maximum energy of the electrons in a metal is on the order of magnitude of a few eV. For a typical value of 3 eV, $n^2 = 7.98 \cdot 10^{14}$, and if $n_x = n_y = n_z$, then $n_x = 1.63 \cdot 10^7$. At 3 eV, two adjacent levels are separated by an energy interval given by:

$$\Delta W = \frac{h^2}{8m_n L^2}\{[n_x^2 + n_y^2 + n_z^2] - [(n_x - 1)^2 + n_y^2 + n_z^2]\}$$

$$\approx \frac{h^2}{8m_n L^2} 2n_x = 1.23 \cdot 10^{-7} \text{ eV} \tag{2.58}$$

Over the entire energy range of interest, the levels are therefore extremely close to each other, which justifies the replacement of (2.55) by the density of states $Z(W)$, a continuous function that is much easier to handle in practice.

Finally, the evaluation of $Z(W)$ is based on the following observations:

- in the space of the quantum numbers n_x, n_y, n_z, each point with integer coordinates corresponds to a quantum state. In this space,

there is therefore one quantum state per unit volume;
- in the space of quantum numbers, all states corresponding to the same energy are situated on a sphere of radius n.

The number of states corresponding to energies lying between W and $W + \mathrm{d}W$ is therefore equal to the volume $\mathrm{d}\Omega$ enclosed between two spheres with radius n and $n + \mathrm{d}n$, except for the following corrections:

- it is necessary to divide $\mathrm{d}\Omega$ by L^3 because, by definition, $Z(W)$ gives the number of states per unit volume, whereas (2.55) refers to a cube with edge L;
- it is necessary to multiply $\mathrm{d}\Omega$ by 2 in order to allow for the spin;
- it is necessary to divide $\mathrm{d}\Omega$ by 8 because in the space of quantum numbers only the octant corresponding to positive values of n_x, n_y, and n_z has any physical meaning.

Consequently,

$$Z(W)\,\mathrm{d}W = \frac{1}{L^3}\,2\,\frac{1}{8}\,4\pi n^2\,\mathrm{d}n \tag{2.59}$$

Equations (2.55) and (2.56) give

$$n^2 = \frac{8m_\mathrm{n}L^2}{h^2}\,W \tag{2.60}$$

By substituting (2.60) in (2.59), we obtain the expression:

$$Z(W) = \frac{4\pi(2m_\mathrm{n})^{3/2}}{h^3}\,\sqrt{W} \tag{2.61}$$

for the density of states.

2.4.5 Electron Energy Distribution

The density of states $Z(W)$ only provides the number of states *that can be occupied by the electrons* in a given energy interval. In order to determine the *total* number of electrons in this interval, we must also know what the *probability* is that these states are effectively occupied. The probability of occupation of a state characterized by an energy W is given by the Fermi-Dirac distribution $F(W)$, studied in Section 7.4. In fact, the electrons in a potential well form a system of quantum particles (subject to Pauli's principle), in thermodynamic equilibrium and without mutual interactions in the sense of Section 7.2.1.

Let $N(W)$ be the electron energy distribution defined so that $N(W)\mathrm{d}W$ represents the number of electrons per unit volume with an energy between W and $W + \mathrm{d}W$. This number of electrons is equal to the product of the number of available states in $\mathrm{d}W$ and the probability that these states are occupied:

$$N(W)\mathrm{d}W = Z(W)F(W)\mathrm{d}W \tag{2.62}$$

Taking into account (2.61) and (7.43), this equation can be expressed in the following manner:

$$N(W)\mathrm{d}W = C\left[1 + \exp\left(\frac{W - W_F}{k_B T}\right)\right]^{-1} \sqrt{W}\,\mathrm{d}W \tag{2.63}$$

where

$$C = \frac{4\pi(2m_n)^{3/2}}{h^3} \tag{2.64}$$

The three functions appearing in (2.62) are shown in Fig. 2.7.

2.4.6 Variation of the Fermi Energy W_F with Temperature

The variation of the Fermi energy as a function of temperature is very small. The calculation below will demonstrate this.

Consider a potential well containing a total number of N_n electrons. The Fermi energy corresponding to this number at temperature T is given by the expression:

$$N_n = C\int_{W=0}^{W=\infty} \frac{\sqrt{W}}{1 + \exp\left(\dfrac{W - W_F}{k_B T}\right)}\,\mathrm{d}W \tag{2.65}$$

This integral can be calculated by using the formula for integration by parts and by performing a Taylor expansion (truncated at the term $(k_B T)^2$) about W_F. We have

$$N_n = C\left(\frac{2}{3}W_F^{3/2} + \frac{\pi^2}{12}(k_B T)^2 W_F^{-1/2} + \ldots\right) \tag{2.66}$$

At 0 K, the Fermi-Dirac distribution takes the form of a step function (Section 7.4.3) so that (2.65) reduces to

$$N_n = C \int_0^{W_{F0}} \sqrt{W}\, dW = \frac{2}{3} C W_{F0}^{3/2} \tag{2.67}$$

where W_{F0} designates the Fermi energy at 0 K.

It can be easily verified by comparing (2.66) and (2.67) that

$$W_F = W_{F0} \left[1 - \frac{\pi^2}{12} \left(\frac{k_B T}{W_{F0}} \right)^2 \right] \tag{2.68}$$

Between 0 K and 293 K, the variation of W_F given by (2.68) is only $5.8 \cdot 10^{-5}$ eV. This value is very low compared to W_F, whose order of magnitude for a metal is about 3 eV. In what follows, W_F will therefore be considered to be independent of temperature.

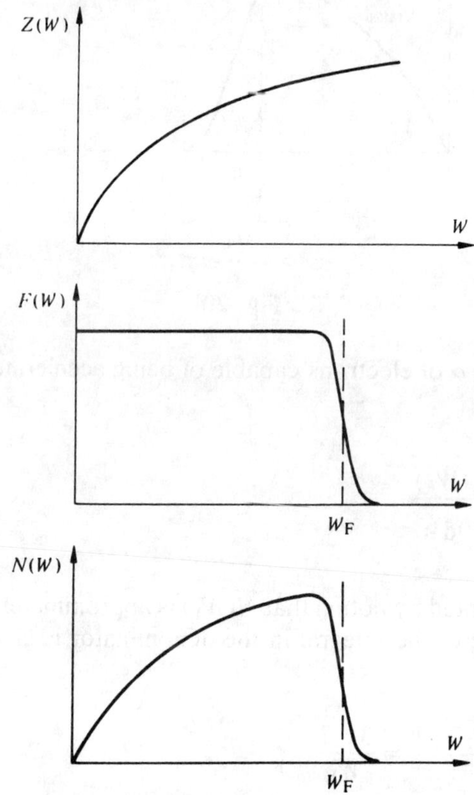

Fig. 2.7

2.4.7 Number of Electrons Capable of Being Accelerated

The energy that an electron may receive during a collision or under the action of an external field is on the order of $k_B T$. The only electrons capable of accepting such an increase in energy are those for which there is sufficient probability of finding an empty level with an energy of approximately $k_B T$ higher than their own. These electrons are situated in the neighborhood of W_F. It can be assumed as a first approximation that their total number corresponds to the hatched area in Fig. 2.8.

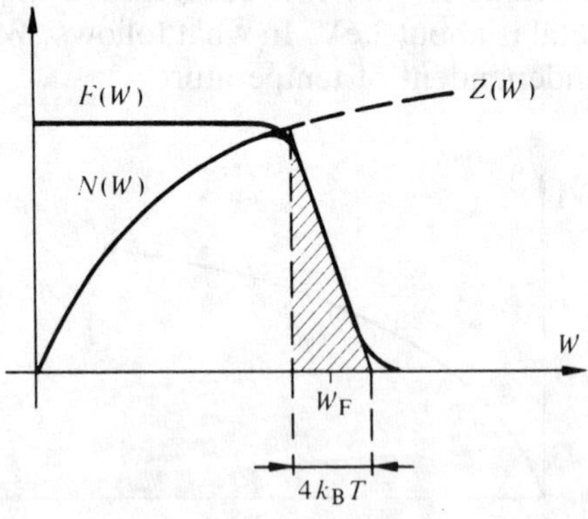

Fig. 2.8

The fraction α of electrons capable of being accelerated is therefore equal to

$$\alpha = \frac{4k_B T\, N(W_F)}{\int_0^\infty N(W)\mathrm{d}W} \tag{2.69}$$

This can be evaluated by noting that $N(W_F)$ is approximately equal to $0.5 \cdot Z(W_F)$. The value of the integral in the denominator is given by (2.67). Thus, we have

$$\alpha = \frac{2C\sqrt{W_F}\, k_B T}{\frac{2}{3}\, C W_F^{3/2}} = \frac{3k_B T}{W_F} \tag{2.70}$$

This fraction is very small; for $W_F = 3$ eV and $T = 293$ K, it is equal to 2.5%.

2.4.8 Mean Energy of the Electrons

At 0 K, the mean energy of the electrons is equal to

$$\overline{W} = \frac{\int_0^{W_F} WC\sqrt{W}\,dW}{\int_0^{W_F} C\sqrt{W}\,dW} = \frac{\frac{2}{5}CW_F{}^{5/2}}{\frac{2}{3}CW_F{}^{3/2}} = \frac{3}{5}W_F \tag{2.71}$$

This energy is 50 times higher for $W_F = 3$ eV than the mean energy derived from the Boltzmann distribution at 20°C! The discrepancy between the classical model (billiard balls) and the quantum model is therefore quite considerable at this point.

2.4.9 Comments on the Shape of the Potential Well

The potential well considered until now has an idealized form. It is infinitely deep and limited by vertical potential barriers. In reality, it is obvious that the potential well has a finite depth and that the transition of the potential from the inside to the outside takes place over a nonzero distance.

The study by quantum mechanics of an electron in a potential well with finite depth and vertical barriers does not present any difficulty. The wave function (2.46) remains valid inside the well but it no longer vanishes at $x = 0$ and $x = L$. At these two values of x, it is coupled to the wave functions valid at the outside of the well. The wave functions have an exponential character. The energy levels resulting from these new boundary conditions cannot be expressed in such a simple manner as in the case of the infinitely deep well. However, the density of states $Z(W)$ given by (2.61) remains a good approximation of the actual density of states. It will therefore be retained for what follows.

The distance over which the potential varies across the surface of the metal is usually much smaller than the thickness of the sample. Consequently the form of the potential barrier does not affect the density of states. This form affects electron emission, mainly when an electric field is applied to the surface of the sample (Schottky effect, Section 2.5.6).

The form of the potential barrier is found by studying the force acting on an electron leaving an infinite metal plate. By emitting the electron, the initially neutral plate acquires a charge e. As long as the electron is at a distance z from the plate, which is large with respect to an interatomic distance, it may be considered as subjected to the action of a plane carrying a surface charge. The action of this plane is identical to that of a charge e situated at the image of the electron in the plane (Fig. 2.9).

The electron therefore experiences a force:

$$F(z) = -\frac{e^2}{4\pi\varepsilon_0(2z)^2} \tag{2.72}$$

This force gives the electron a potential energy $W(z)$ equal to

$$W(z) = -\int_z^\infty F(z)\mathrm{d}z = -\frac{e^2}{16\pi\varepsilon_0 z} \tag{2.73}$$

In problems involving the finite height of the barrier as well as its shape, it is worth setting to zero the potential energy of an electron situated at infinity. This convention assumed in (2.73) will be retained in Section 2.5.

In reality, $W(z)$ does tend toward $-\infty$ at $z = 0$, as predicted by (2.73). This equation ceases to be valid in the immediate vicinity of the plate, a region in which the electron no longer "sees" a charged plane but a lattice of ions. By only considering the four closest ions to the electron (Fig. 2.10), we obtain the following approximation for the potential energy:

$$W(z) = -\frac{e^2}{4\pi\varepsilon_0\sqrt{z^2 + \dfrac{a^2}{2}}} \tag{2.74}$$

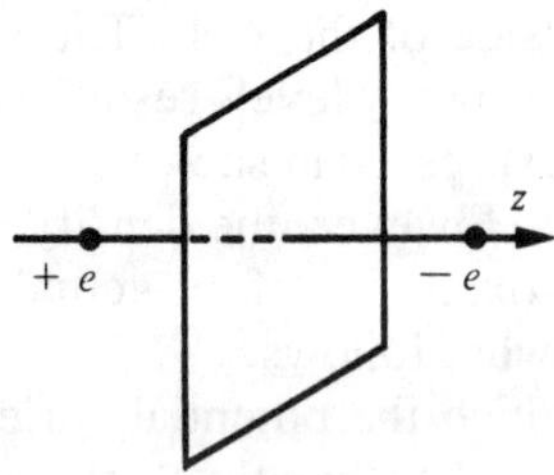

Fig. 2.9

When $z^2 \gg a^2/2$, (2.74) tends toward (2.73), except for a factor of 4, which follows from the fact that we have only considered the four nearest ions to the electron. This simplification results in overestimating the mean charge attributed to each of these ions. The form of the exact solution is compared with (2.73) in Fig. 2.11.

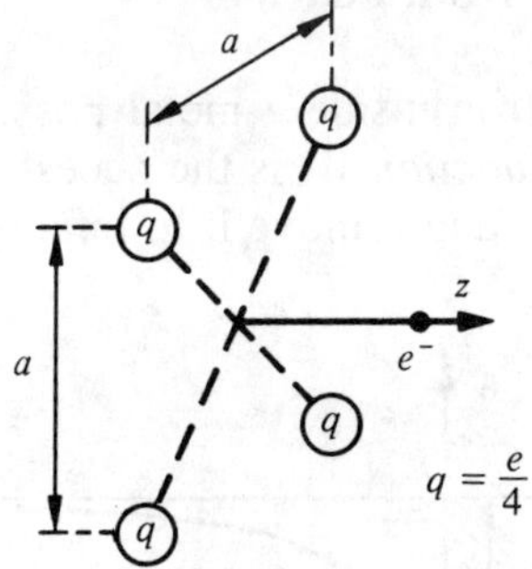

Fig. 2.10

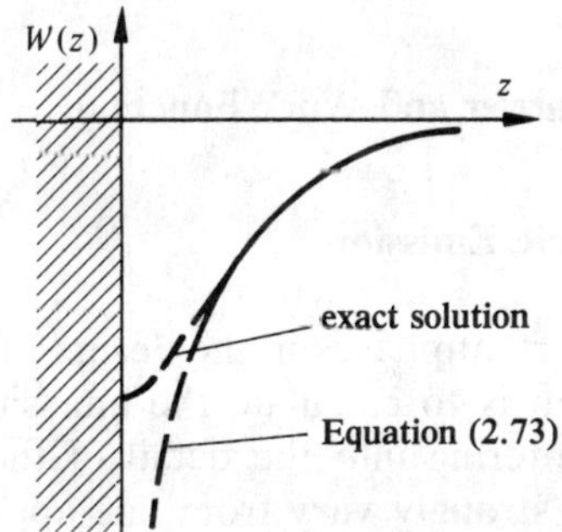

Fig. 2.11

2.5 ELECTRON EMISSION

2.5.1 Introduction

The operation of several devices is based on the emission of electrons at the surface of metals. We can mention TV tubes, photomultiplier tubes, electron microscopes, *et cetera* as application examples for the model of the free electron in a potential well; thermoelectronic emission, field emission, and photoelectric emission are studied in the following sections.

2.5.2 Definition of the Work Function

Consider an electron inside a metal possessing exactly the Fermi energy W_F. The *work function* W_s is the necessary energy to extract this electron from the metal and remove it to infinity (Fig. 2.12).

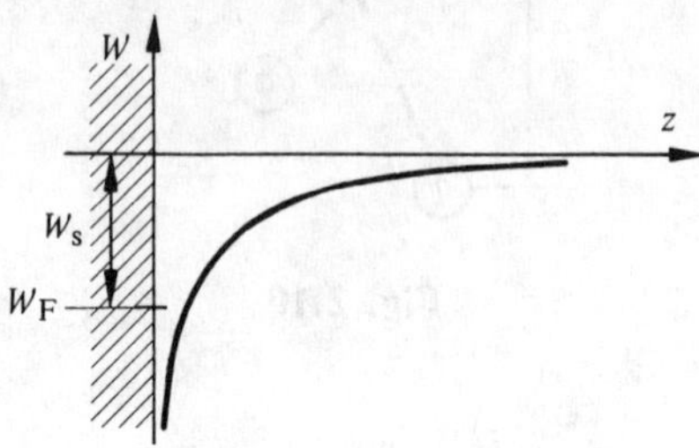

Fig. 2.12 Potential Barrier and Work Function

2.5.3 Thermoelectronic Emission

One of the important phases in the design of any device using thermoelectronic emission is to calculate the emission current that can be achieved. Without entering into the detail of the design of thermionic cathodes, which can strongly vary from one application to another, we shall establish the relationship between the emission current density and the temperature, known as Dushman's equation.

Consider a plate with infinite extension along x and y.

For an electron to be able to escape from this plate, it is necessary that its momentum along z, denoted p_z is larger than a minimum value p_{z0} defined by

$$\frac{p_z^2}{2m_n} > W_F + W_s = \frac{p_{z0}^2}{2m_n} \tag{2.75}$$

In writing (2.75), we are implicity making the assumptions that the tunneling effect (Vol. VII, Ch. 3), and the backscattering of electrons having $p_z > p_{z0}$ due to the presence of the potential barrier, are negligible, which is reasonable here.

The density J of the emission current, by generalization of (2.18) can be written

$$J = -e \int_{p_{z0}}^{\infty} \frac{p_z}{m_n} N(p_z) \mathrm{d}p_z \tag{2.76}$$

with

$$N(p_z)\mathrm{d}p_z = \int_{p_x}\int_{p_y} N(p_x, p_y, p_z)\mathrm{d}p_x\, \mathrm{d}p_y\, \mathrm{d}p_z \tag{2.77}$$

$N(p_x, p_y, p_z)$ being the distribution of the momentum of the electrons. As in the case where the energy is the independent variable, $N(p_x, p_y, p_z)$ is the product of a density of states $Z(p_x, p_y, p_z)$ and the Fermi-Dirac distribution.

By analogy with (2.59):

$$Z(p_x, p_y, p_z)\mathrm{d}p_x\, \mathrm{d}p_y\, \mathrm{d}p_z = \frac{1}{L^3}\, 2\, \frac{1}{8}\, \mathrm{d}n_x\, \mathrm{d}n_y\, \mathrm{d}n_z \tag{2.78}$$

The momentum along an axis depends only on the value of the quantum number related to this axis. In fact, the solutions of Schrödinger's equation along the three axes are independent of one another. Consequently (2.55) makes it possible to write

$$W = \frac{h^2}{8m_\mathrm{n}L^2}(n_x^2 + n_y^2 + n_z^2) = \frac{1}{2m_\mathrm{n}}(p_x^2 + p_y^2 + p_z^2) \tag{2.79}$$

with

$$n_x = \frac{2L}{h}p_x; \qquad n_y = \frac{2L}{h}p_y; \qquad n_z = \frac{2L}{h}p_z \tag{2.80}$$

by taking the derivatives of (2.80) and substituting the result in (2.78), we obtain

$$Z(p_x, p_y, p_z) = \frac{2}{h^3} \tag{2.81}$$

from which we have

$$N(p_x, p_y, p_z) = \frac{2}{h^3}\left\{\exp\left[\left(\frac{1}{2m_\mathrm{n}}(p_x^2 + p_y^2 + p_z^2) - W_\mathrm{F}\right)\Big/ k_\mathrm{B}T\right] + 1\right\}^{-1} \tag{2.82}$$

and

$$N(p_z)\mathrm{d}p_z = \frac{2}{h^3}\mathrm{d}p_z \int_{-\infty}^{\infty}\int_{-\infty}^{\infty} \frac{\mathrm{d}p_x\,\mathrm{d}p_y}{\exp\left[\left(\frac{1}{2m_\mathrm{n}}(p_x^2+p_y^2+p_z^2)-W_\mathrm{F}\right)\Big/ k_\mathrm{B}T\right]+1} \tag{2.83}$$

The calculation of this integral may be considerably simplified because of the order of magnitude of W_s: a few eV (Table 2.13). In order to be emitted, an electron must have an energy $p_z^2/2m_\mathrm{n}$ so that $p_z^2/2m_\mathrm{n} - W_\mathrm{F} > W_\mathrm{s} \gg k_\mathrm{B}T$ which makes it possible to neglect the unit in front of the exponential in the denominator:

$$N(p_z)\mathrm{d}p_z = \frac{2}{h^3}\exp\left(\frac{W_\mathrm{F}}{k_\mathrm{B}T}\right)\exp\left(-\frac{p_z^2}{2m_\mathrm{n}k_\mathrm{B}T}\right)\mathrm{d}p_z$$

$$\int_{-\infty}^{\infty}\exp\left(\frac{-p_x^2}{2m_\mathrm{n}k_\mathrm{B}T}\right)\mathrm{d}p_x \int_{-\infty}^{\infty}\exp\left(\frac{-p_y^2}{2m_\mathrm{n}k_\mathrm{B}T}\right)\mathrm{d}p_y \tag{2.84}$$

By means of (7.8.1):

$$N(p_z) = \frac{4\pi m_\mathrm{n}k_\mathrm{B}T}{h^3}\exp\frac{W_\mathrm{F}}{k_\mathrm{B}T}\exp\left(-\frac{p_z^2}{2m_\mathrm{n}k_\mathrm{B}T}\right) \tag{2.85}$$

By substituting this expression in (2.76) we obtain

$$J = \frac{-e4\pi k_\mathrm{B}T}{h^3}\exp\left(\frac{W_\mathrm{F}}{k_\mathrm{B}T}\right)\int_{p_{z0}}^{\infty} p_z \exp\left(\frac{-p_z^2}{2m_\mathrm{n}k_\mathrm{B}T}\right)\mathrm{d}p_z \tag{2.86}$$

whose integration is immediate. Taking into account (2.75):

$$J = A_0T^2\exp(-W_\mathrm{s}/k_\mathrm{B}T) \tag{2.87}$$

This is *Dushman's equation*. The constant:

$$A_0 = \frac{4\pi e m_\mathrm{n}k_\mathrm{B}^2}{h^3} = 1.20\cdot 10^6\ \mathrm{Am^{-2}\,K^{-2}} \tag{2.88}$$

is called the *thermoelectronic emission coefficient*.

2.5.4 Value of the Thermoelectronic Emission Coefficient

It is found experimentally that the thermoelectronic emission coefficient is not only smaller than that predicted by (2.88) but that it changes from one metal to another (Table 2.13).

Table 2.13
Mean Values of A_0 and W_s for Polycrystalline Samples, after [8–10].

Element	A_0 $10^4\ Am^{-2}\ K^{-2}$	W_s *eV*	*Melting Point* °*C*
Ni	30	5.03	1 455
Fe	26	4.48	1 535
Th	60	3.35	1 845
Pt	32	5.32	1 773
Cr	48	4.60	1 890
Mo	55	4.20	2 620
Ta	55	4.19	3 000
W	60	4.52	3 370

However, the variation of the values of A_0 and W_s does not mean that (2.87) is of no use because the *form* of this equation is well verified by experiment.

The parameters A_0 and W_s are very sensitive to the nature and condition of the emitting surface. In the case of a monocrystal, the work function increases when the emitting face coincides with a crystal plane with a higher surface density of ions (Table 2.14).

Table 2.14
Work Function of Monocrystalline Tungsten (CC) as a Function of the Orientation of the Emitting Plane, after [10].

Emitting Face	*Surface Density of Ions (Relative Value)*	W_s *eV*
⟨111⟩	0.577	4.39
⟨100⟩	1.000	4.52
⟨110⟩	1.414	4.68

The effect of surface impurities is even greater and finds an important industrial application in the manufacture of thorium cathodes. Tungsten, which is most often used as the base metal because of its high melting point and its low vapor pressure, is covered by a monoatomic layer of thorium. Because thorium has a lower work function than that of tungsten, the thorium atoms transfer an electron to the base metal and form a layer of ions strongly attached by electrostatic attraction. The positive charge formed in this way favors the emission of electrons. The work function of such a cathode is on the order of 2.6 eV. Even lower values of W_s, approximately 1 eV, are obtained by covering the base metal with a layer of strontium and barium oxides. Cathodes constructed in this way already supply a large current around 1000 K and are largely used in TV tubes, for example. Their operating mechanism is still not completely understood today.

2.5.5 Comments

The application of an electric field to the cathode is, in practice, always necessary to remove the emitted electrons that otherwise would form a space-charge impeding emission.

The presence of the temperature and work function in the exponential of (2.87) reveals the high sensitivity of the emission current to these two variables. At 1900 K, for example, the lowering of W_s from 4.52 eV to 2.6 eV, produced by a layer of thorium on tungsten, has the effect of multiplying J by a factor of 10^5. In a similar manner, the lowering of the temperature of a thorium cathode from 1900 K to 1710 K (−10%) is sufficient to reduce J by the ratio of 3.7 to 1.

By writing (2.87) in the form:

$$\ln \frac{J}{T^2} = \ln A_0 - \frac{1}{k_B T} W_s \tag{2.89}$$

we see that W_s and A_0 can easily be determined by plotting the experimental points (J and T) in a diagram of $\ln J/T^2$ as a function of $1/T$. The slope of the straight line gives W_s and its intersection with the y-axis, $\ln A_0$.

2.5.6 The Schottky Effect

The application of an electric field $\boldsymbol{E}$ perpendicular to an emitting surface produces a lowering of the potential barrier called the *Schottky effect*. Let us consider the planar case shown in Fig. 2.15.

As soon an an electron leaves the plate it experiences the action of an electric field which is manifested by a force $-e\boldsymbol{E}$ tending to remove it from the plate. Its potential energy W is therefore the sum of two terms, one resulting from $\boldsymbol{E}$ (curve b), the other from the potential barrier (curve a). In the region where (2.73) applies,

$$W = -\frac{e^2}{16\pi\varepsilon_0 z} - e\boldsymbol{E}z \tag{2.90}$$

This function, represented by the curve c of Fig. 2.15, has a maximum at

$$z_m = \sqrt{\frac{e}{16\pi\varepsilon_0 \boldsymbol{E}}} \tag{2.91}$$

At this point, the potential energy of the electron is

$$W(z_m) = -e\sqrt{\frac{e\boldsymbol{E}}{4\pi\varepsilon_0}} \tag{2.92}$$

The work function is therefore reduced by a quantity $\Delta W_s = e\sqrt{e\boldsymbol{E}/4\pi\varepsilon_0}$ and Dushman's equation (2.87) takes the form:

$$J = A_0 T^2 \exp\left[-\left(W_s - e\sqrt{\frac{e\boldsymbol{E}}{4\pi\varepsilon_0}}\right)\Big/ k_B T\right] \tag{2.93}$$

By taking the logarithm of both sides of (2.93), we have

$$\ln J = \ln A_0 T^2 - \frac{W_s}{k_B T} + \frac{e}{k_B T}\sqrt{\frac{e\boldsymbol{E}}{4\pi\varepsilon_0}} \tag{2.94}$$

At constant temperature, $\ln J$ is theoretically a linear function of $\sqrt{\boldsymbol{E}}$. In practice, we observe a dependence of the type shown in Figure 2.16.

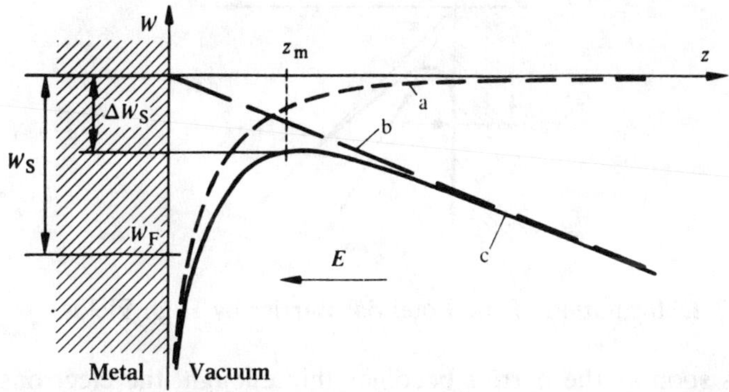

Fig. 2.15 (a) Barrier in the Absence of *E*; (b) Potential Energy Resulting from *E*; (c) Barrier in the Presence of *E*

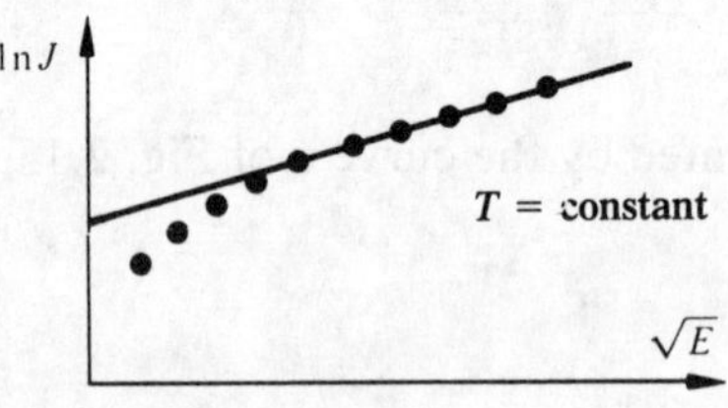

Fig. 2.16

The straight line passing through the points plotted for the large values of $\boldsymbol{E}$ is called the Schottky line. The parameters of this line can be used to calculate A_0 and W_s. For small values of $\boldsymbol{E}$, the measured current density departs from the Schottky line. The appearance of a space charge in the vicinity of the emitting surface is responsible for this phenomenon.

2.5.7 Field Emission

By increasing the electric field applied to the surface of a conductor up to values on the order of 10^9 to 10^{10} V/m the deformation of the barrier sketched in Fig. 2.15 is accentuated. Curve (a) varies much more rapidly than curve (b) in the vicinity of the surface. It follows that for large values of E, it is mainly the thickness of the barrier that varies rather than its height (Fig. 2.17).

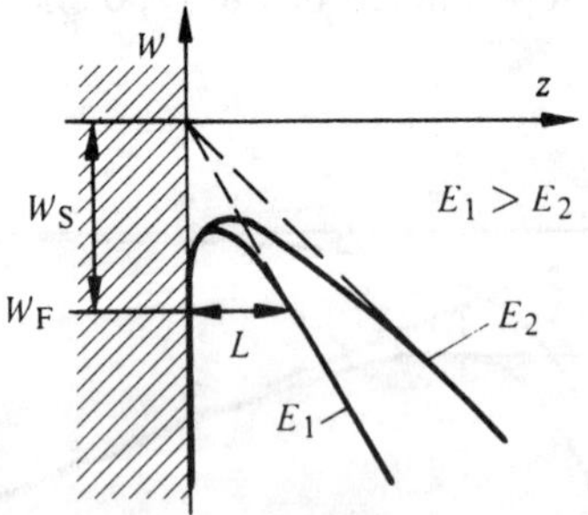

Fig. 2.17 Deformation of the Potential Barrier by High Fields

As soon as the barrier becomes thin enough, the electrons have a new route for leaving the metal: instead of passing over the barrier, they can cross it directly by the tunnel effect. This process is called *field emission* which can lead to relatively high current densities. In fact, all the electrons are capable of leaving the metal by the tunnel effect (with a variable probability it is true) whereas in thermoelectronic emission, only electrons occupying states situated in the vicinity of and above the Fermi energy have this possibility.

An approximation of the law governing field emission may be obtained in the following manner. Let P be the probability for an electron with kinetic energy W to cross the barrier defined by Fig. 2.18. In the case $W < W_0$ corresponding to the tunnel effect, we have [1]:

$$P = \frac{1}{1 + \dfrac{W_0^2}{4W(W_0 - W)} \sin h^2 \left(\dfrac{\sqrt{2m_n}}{\hbar} L\sqrt{W_0 - W} \right)} \tag{2.95}$$

At the moment when E becomes high enough so that (2.95) in practice differs from zero, the unity present in the denominator may still be neglected and the sin h^2 represented by an exponential:

$$P \approx \frac{16W(W_0 - W)}{W_0^2} \exp \left(- 2 \frac{\sqrt{2m_n}}{\hbar} L\sqrt{W_0 - W} \right) \tag{2.96}$$

The most important contribution to the field emission current is made by electrons possessing energies close to the Fermi energy because $N(W)$ has a maximum at this point. At higher energies, P would be higher but $N(W)$ decreases very rapidly. At lower energies, P and $N(W)$ decrease. It is therefore reasonable to take the width of the barrier at W_F for L. By referring to Fig. 2.17, we easily find

$$L = W_s/eE \tag{2.97}$$

From which we deduce, from (2.96), that the required distribution is of the type:

$$J \approx \exp \left(- 2 \frac{\sqrt{2m_n}}{\hbar} \frac{W_s^{3/2}}{eE} \right) \tag{2.98}$$

A more detailed analysis, comparable in certain aspects to that of Section 2.5.3, confirms this result.

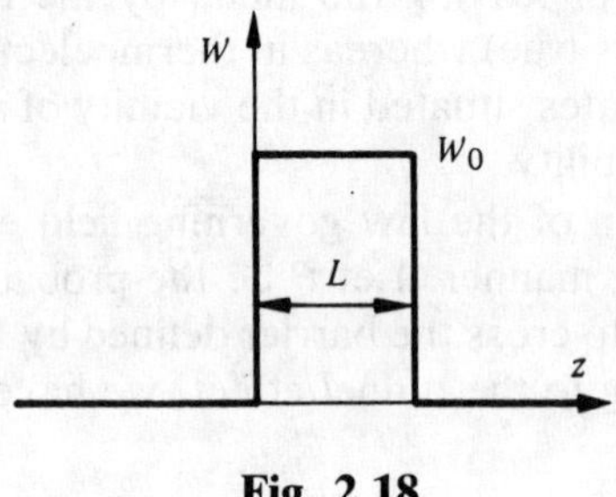

Fig. 2.18

2.5.8 Photoelectric Emission

A photon penetrating into a conductor or semiconductor may transfer energy to a conduction electron. If the energy acquired in this way by the electron and the direction in which it is moving allows the electron to cross the potential barrier, it will be emitted. This process is called the *photoelectric effect* and the emitted electron is called a *photoelectron*.

The maximum energy W_{max} of a photoelectron is given by Einstein's equation:

$$W_{max} = h\nu - W_s \tag{2.99}$$

where ν is the frequency of the photon and $h\nu$ represents its energy.

The energy W_{max} is obtained when an electron at the Fermi level is excited, but all electrons with an energy $W \geqslant W_F + W_s - h\nu$ may be emitted, at the limit with a zero energy (Fig. 2.19).

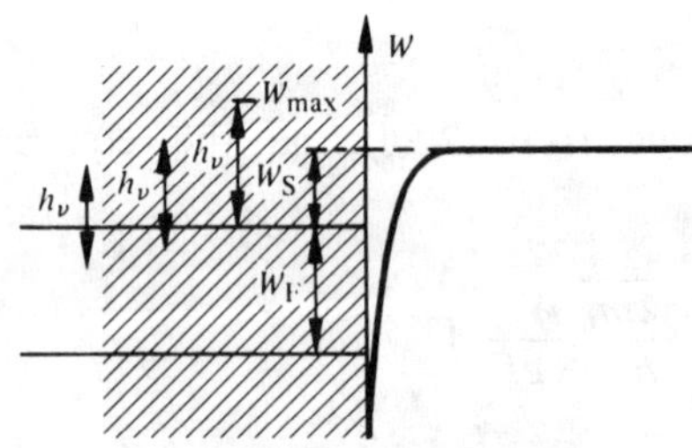

Fig. 2.19

Einstein's equation also shows that W_{max}, a linear function of the frequency of the light, does not depend on the intensity of the light. On the contrary, only light with frequency $\nu \geqslant \nu_{lim}$ with

$$\nu_{\lim} = \frac{1}{h} W_s \tag{2.100}$$

is capable of producing the photoelectric effect. The measurement of $\nu_{\lim}$ is one way of determining W_s.

The photoelectric yield η_{ph}, measured by the number of electrons emitted per incident photon, varies considerably from one material to another. In metals, it is limited for two reasons. First, metals are shiny. They therefore reflect a considerable part of the light which is thus lost for photoelectric emission. Second, the thickness of the useful layer is limited to a few tens of nm, corresponding to the order of magnitude of the free path of the electrons in a metal. By comparison, the penetration depth of photons is 100 to 1000 times greater.

The behavior of alkali metals is typical in this respect (Fig. 2.20). Their η_{ph} has a maximum for a certain frequency ν_{max}. Below ν_{max}, the yield is lowered by an increase in the reflection coefficient and above ν_{max}, it is lowered by an increase in the penetration depth of the photons.

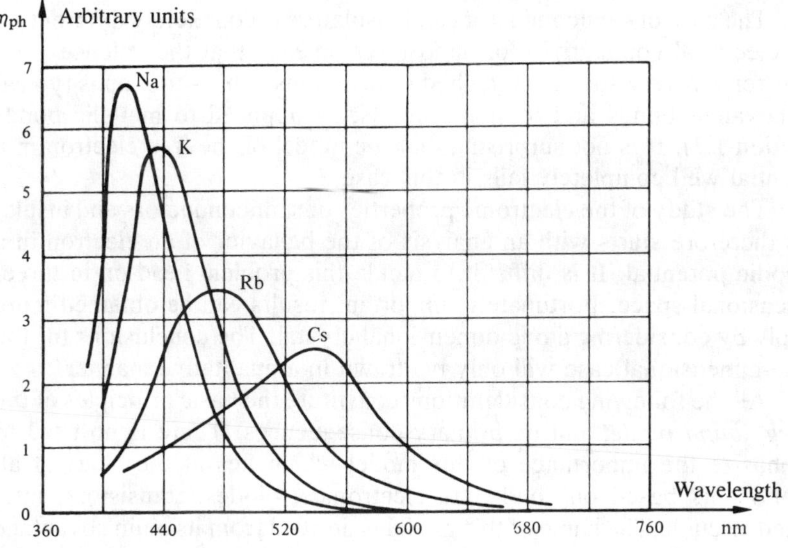

Fig. 2.20 (After [8].)

Larger values of η_{ph} are obtained by means of cathodes formed from a substrate covered by several layers (oxides, alkali metals, alloys) reducing the reflection coefficient and the work function.

The exploitation of the photoelectric effect in semiconductor junctions is the basis for a range of very important components: photodiodes, phototransistors, *et cetera* (Vol. VII, Ch. 9).

2.6 ENERGY BAND MODEL

2.6.1 Introduction

The Sommerfeld model gives a satisfactory representation of most of the electronic properties of metals. We have just seen a few examples of this in the preceding section. This justifies *a posteriori* the assumption of this model that the electrons move freely; i.e., they are subjected to a constant potential in matter. However, it is well known, by x-ray diffraction experiments in particular, that metals have a crystal structure. The electrons should therefore be subjected to a periodic potential with a period that should be related to the size of the unit cell. We must therefore conclude that this periodic variation in potential is small in metals and can be neglected.

The case of semiconductors and insulators is completely different. A low electrical conductivity of almost zero means that the valence electrons remain very strongly attached to the atoms. This situation is typical for covalent bonds and for ionic bonds, as opposed to metallic bonds (Section 1.3). It is not surprising that the model of the free electron in a potential well completely fails in this case.

The study of the electronic properties of semiconductors and insulators therefore starts with an analysis of the behavior of an electron in a periodic potential. It is difficult to tackle this problem head-on in three-dimensional space. Fortunately, important results can be obtained more simply by considering a one-dimensional crystal. The conclusions for the three-dimensional case will only be drawn in a qualitative manner.

All the following considerations constitute the basic principles of the *energy band model* and its primary consequences. There is no need to emphasize the importance of this model which lies at the heart of all components based on solid-state electronics: diodes, transistors, integrated circuits. The name of this model is derived from its main advantage which is to show that the possible energies of electrons in a crystal are grouped in a certain number of allowed energy bands separated from one another by forbidden energy bands. Starting here, a unified treatment of metals, semiconductors, and insulators, is possible. The position of the Fermi energy in the sequence of allowed and forbidden bands is the key parameter here.

2.6.2 Electrons in a Periodic Potential: One-Dimensional Case

The potential energy W_{pot} of an electron in a one-dimensional crystal with base vector $\boldsymbol{a}$ has the form shown in Fig. 2.21. The behavior of this electron is governed by Schrödinger's equation which here takes the form:

$$\frac{\hbar^2}{2m_n}\frac{d^2\Psi}{dx^2} + (W - W_{pot}(x))\Psi = 0 \tag{2.101}$$

This equation has functions of the type:

$$\Psi(x) = U_k(x)\exp jkx \tag{2.102}$$

as solutions, called Bloch functions. The function $U_k(x)$ is periodic with the same period a as the crystal:

$$U_k(x + a) = U_k(x) \tag{2.103}$$

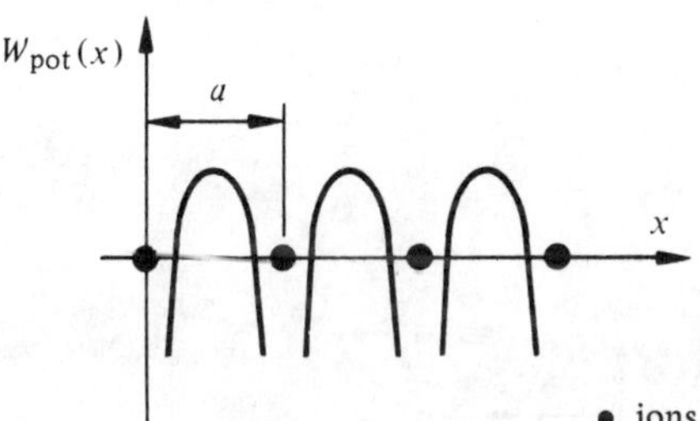

Fig. 2.21

$U_k(x)$ depends on the form chosen for $W_{pot}(x)$. All reasonable forms of $W_{pot}(x)$, however, lead to comparable results. We shall take the approximation of the model of Kronig and Penney for $W_{pot}(x)$. This approximation is very rough (Fig. 2.22) but it retains the main characteristics of the actual potential: the same periodicity as the lattice, a potential maximum between the ions, and a minimum in the vicinity of the ions.

The parameter b, the width of the potential barrier separating two neighboring ions, makes it possible to study the effect of a more or less strong bonding of the electrons with the atoms.

The integration of (2.101) with the function shown in Fig. 2.22 for $W_{pot}(x)$ is simple in principle although the calculations are rather laborious. We will just describe the procedure here. Equation (2.101) is first solved separately for the regions where $W_{pot}(x) = +W_0/2$ and $W_{pot}(x) = -W_0/2$. The obtained wave functions as well as their derivatives with respect to x are then coupled (continuity condition) at the boundaries of the regions in which the potential energy is constant. In this way, an eigenvalue equation is obtained whose form is considerably simplified in the calculation when the potential barrier separating two ions takes the form of a Dirac function, W_0 tending to infinity whereas $W_0 b$ remains constant. Under these conditions, we have

$$\cos ka = P \frac{\sin \alpha a}{\alpha a} + \cos \alpha a \tag{2.104}$$

where

$$P = \frac{m_n a}{\hbar^2} W_0 b \tag{2.105}$$

and

$$\alpha = \frac{1}{\hbar} \sqrt{2 m_n W} \tag{2.106}$$

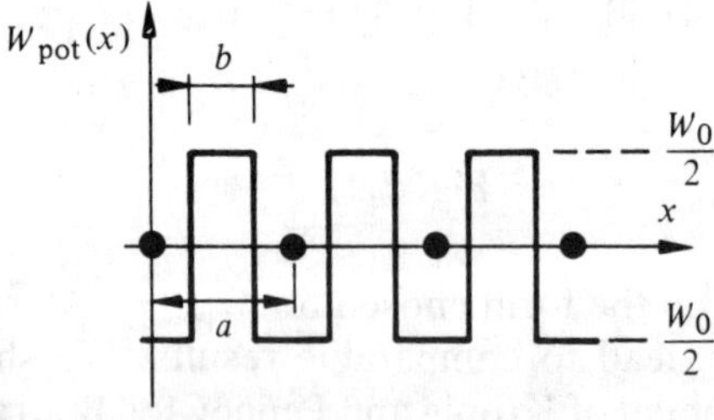

Fig. 2.22 Periodic Potential Assumed in the Model of Kronig and Penney

The meaning of (2.104) appears clearly when the right-hand side of this equation is expressed graphically as a function of αa (Fig. 2.23).

Only the sections of this function lying between the two horizontals at ±1 have to be retained as solutions to (2.104) because of the cosine on

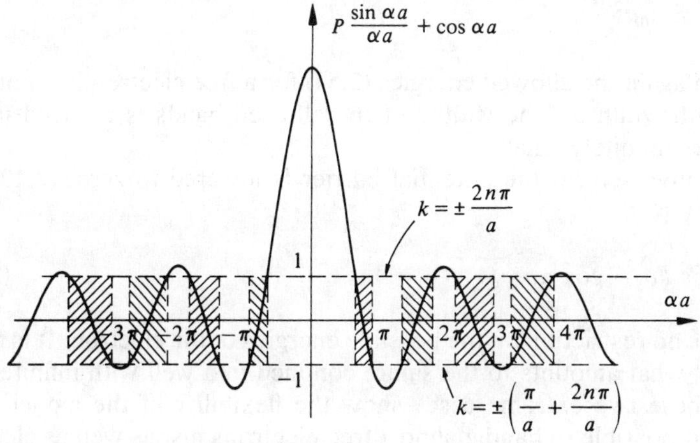

Fig. 2.23

the left-hand side. The αa axis therefore exhibits an alternation of segments inside which (2.104) in turn does or does not have solutions with a physical meaning.

The possible energies of an electron subjected to a periodic potential are therefore defined in what are called *allowed energy bands* separated from each other by *forbidden energy bands*.

2.6.3 Limiting Cases

The width of the allowed bands depends on the height of the potential barrier separating two ions. The higher this barrier (for large W_0b), the larger the amplitude of the oscillating function in Fig. 2.23, which causes a reduction in the width of the allowed bands. At the limit, for W_0b tending to infinity, (2.104) reduces to

$$\sin \alpha a = 0 \tag{2.107}$$

from which

$$\alpha a = n\pi \qquad \text{with } n = 1, 2, 3, \ldots \tag{2.108}$$

and, by means of (2.106):

$$W \frac{h^2}{8m_n a^2} n^2 \tag{2.109}$$

We find again the allowed energies (2.54) for a free electron in a potential well with width a. The widths of the allowed bands is reduced until it becomes infinitely small.

Conversely, if the potential barrier is lowered to zero, (2.104) becomes

$$\cos ka = \cos \alpha a \tag{2.110}$$

There is no restriction on the possible energies of the electron. It is totally free, or what amounts to the same, confined in a well with infinite size.

These two extreme cases show the flexibility of the model which makes it possible to handle almost free electrons just as well as electrons belonging to deeper shells strongly attached to the nucleus.

2.6.4 The $W(k)$ Functions

For what follows, it is worth examining equation (2.104) from another angle and, in particular, studying the functions $W(k)$ inside the allowed bands. In Fig. 2.23, the upper horizontal corresponds to the following values of k:

$$k = \pm \frac{2n\pi}{a} \qquad \text{where } n = \mathbf{0, 1, 2,} \ldots \tag{2.111}$$

The lower horizontal corresponds to

$$k = \pm \left(\frac{\pi}{a} + \frac{2n\pi}{a}\right) \qquad \text{where } n = \mathbf{0, 1, 2,} \ldots \tag{2.112}$$

By remaining within the same allowed band, when k increases or decreases continuously, the corresponding part of the function $P\sin(\alpha a)/\alpha a + \cos(\alpha a)$ is expressed. The concomitant variation of αa shows that in each allowed band, the electron energy is an even function of k. The period is $2\pi/a$. The $W(k)$ functions therefore have the form shown in Fig. 2.24.

For the sake of comparison, we plotted the parabola $W(k)$ of the Sommerfeld model (2.54) in Fig. 2.24 and marked the parts of the $W(k)$ functions of the band model close to this parabola in bold. It can be seen

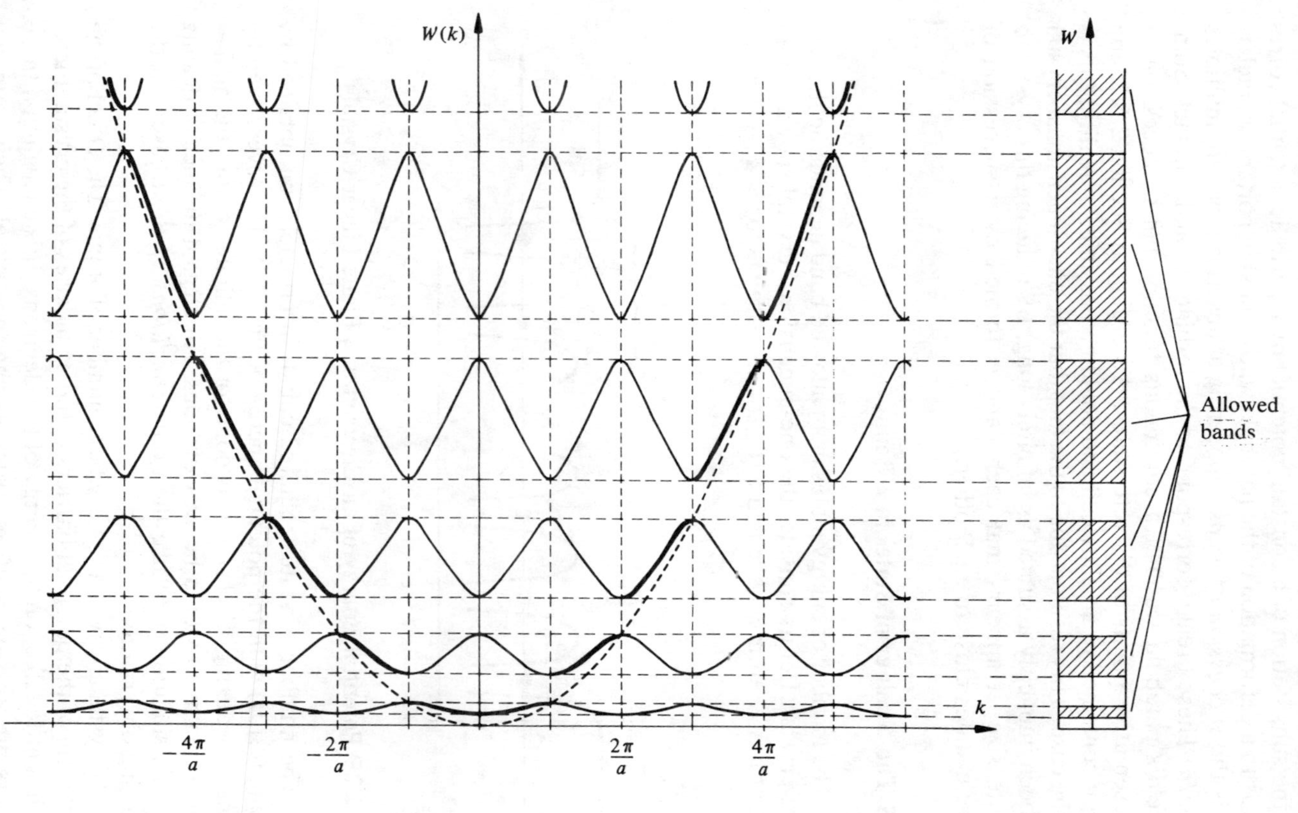

Fig. 2.24

that the substitution of a constant potential by a periodic potential corresponds to a deformation of the parabola whose most important characteristic is the *succession of breaks* appearing at abscissae that are multiples of $\pm\pi/a$. These breaks correspond to the forbidden bands. In fact, Sommerfeld's parabola was not a continuous function but a succession of representative points of discrete states. According to Section 2.6.2, any energy contained in an allowed band could be possible. In reality, this is not the case. This result is due to the fact that the one-dimensional crystal has been implicitly assumed as infinitely long. By reducing this crystal to a finite size, it appears that each band is formed by a succession of discrete states close to each other.

2.6.5 The Number of States in a Band

The number of energy states in an allowed band depends on the size of the crystal. Let us consider the one-dimensional crystal shown in Fig. 2.25. It has N atoms, its length is L, and its base vector is $\boldsymbol{a}$.

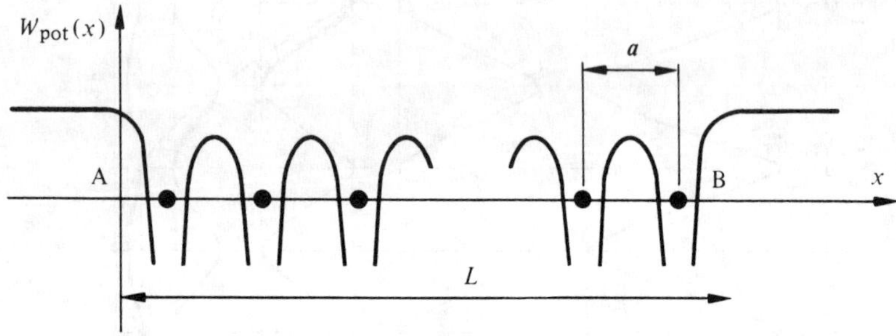

Fig. 2.25 Potential Energy of an Electron in a Finite Linear Crystal

The number of states required could, in principle, be obtained by solving (2.101) with the potential energy of Fig. 2.25. However, even by strongly idealizing $W_{pot}(x)$ as was done for the infinite crystal, the mathematical treatment would be very arduous. An elegant manner of dealing with this situation is to use the *cyclic conditions of Born* based on the following observation.

A crystal always has a very large number of atoms. The effect of the exact form of the potential barriers at the boundaries of the crystal therefore has little effect on the energy of the electrons. It is enough that these barriers prevent the electrons from leaving the crystal. Their form can therefore be chosen so as to simplify calculations as much as possible.

With this aim, we imagine that we bend the crystal until its end A is in contact with its end B (on the atomic scale, this operation is only accompanied by an infinitesimal deformation because of the enormous ratio between the radius of curvature and the base vector of the crystal). In this way, the wave function in a finite crystal recovers the periodic character that it has in an infinite crystal:

$$\Psi(x + L) = \Psi(x) \tag{2.113}$$

This equation expresses the *Born cyclic condition*. By applying (2.113) to the general solution (2.102), we have

$$U_k(x + L) \exp [jk(x + L)] = U_k(x) \exp (jkx) \tag{2.114}$$

Now, U_k is periodic with period a, and

$$L = Na \tag{2.115}$$

Equation (2.114) therefore entails

$$kL = 2\pi l, \tag{2.116}$$

l being any integer.

At the two ends of a band (Fig. 2.24), k takes the limiting values:

$$k_{\min} = \frac{n\pi}{a}, \tag{2.117}$$

$$n = 0, \pm 1, \pm 2, \ldots$$

$$k_{\max} = \frac{(n + 1)\pi}{a}, \tag{2.118}$$

The number N_e of possible values of k between these two limits corresponds to the number of possible energy states in the band. Because k varies in steps of $2\pi/L$, as (2.116) shows, we have

$$N_e = \frac{k_{\max} - k_{\min}}{2\pi/L} = \frac{N}{2} \tag{2.119}$$

This result must be multiplied by 2 to allow for the spin and then again by 2 because we have neglected negative values of k. Finally,

$$N_e = 2N \tag{2.120}$$

The number of states in the band is therefore proportional to the number of atoms in the crystal. This result is important because it shows that there are always enough places to accommodate the electrons of a crystal of a given nature, regardless of its size, without violating Pauli's principle or resorting to higher energy bands.

2.6.6 Effective Mass of the Electron: Definition

The calculation of the response of an electron to an external field is not immediate, at the instant that this electron is no longer free, but is subjected to the potential created by the ions of a crystal. The nature of a key problem, such as the determination of a current density, changes fundamentally, even if, ultimately, a formal approach identical to that adopted in the billiard balls model may be established.

If we consider the electron as a classical particle, Newton's equation can be applied, thus,

$$\boldsymbol{F} = m_{\mathrm{n}}\boldsymbol{a} \tag{2.121}$$

The force $\boldsymbol{F}$ acting on the electron has two components:

$$\boldsymbol{F} = \boldsymbol{F}_{\mathrm{int}} + \boldsymbol{F}_{\mathrm{ext}} \tag{2.122}$$

$\boldsymbol{F}_{\mathrm{ext}}$ is the force due to the external field, $\boldsymbol{F}_{\mathrm{int}}$ is the force deriving from the potential created by the ions. The form of (2.121) is very simple but its application to calculate $\boldsymbol{a}$ remains difficult because $\boldsymbol{F}_{\mathrm{int}}$ is not known.

This difficulty is overcome by introducing the *effective mass of the electron* m_{n}^*, defined by the following relationship:

$$\boldsymbol{F}_{\mathrm{ext}} = m_{\mathrm{n}}^*\boldsymbol{a} \tag{2.123}$$

The value of the procedure obviously lies in the fact that m_{n}^* may be easily calculated.

2.6.7 Calculation of the Effective Mass of the Electron

In order to determine m_{n}^* we must use both the results of classical physics and of quantum physics. The link between these two theories is made when it is assumed that an electron considered as a classical particle is present where $|\Psi(x)|^2$ has a maximum, i.e., at the center of the wave packet. The velocity of the electron is therefore equal to the group velocity $\boldsymbol{v}_{\mathrm{g}}$ of the wave that represents it.

It is well known from wave theory that

$$\boldsymbol{v}_g = \frac{\partial\omega}{\partial\boldsymbol{k}} \tag{2.124}$$

where ω is the wave angular frequency. The wave vector $\boldsymbol{k}$ is reduced in a one-dimensional model to the wave number k already encountered. On the other hand, here, ω only depends on k. We will therefore use (2.124) in the form:

$$v_g = \frac{d\omega}{dk} \tag{2.125}$$

The electron energy is related to the wave angular frequency by Einstein's relationship:

$$W = \hbar\omega \tag{2.126}$$

which makes it possible to write (2.125) in the form:

$$v_g = \frac{1}{\hbar}\frac{dW}{dk} \tag{2.127}$$

By taking the derivative of (2.127) with respect to time, we obtain the group acceleration a_g which is nothing other than the modulus of the acceleration appearing on the right-hand side of (2.123). We have

$$a_g = \frac{1}{\hbar}\frac{d}{dt}\left(\frac{dW}{dk}\right) = \frac{1}{\hbar}\frac{d^2W}{dk^2}\frac{dk}{dt} \tag{2.128}$$

It remains to make F_{ext} appear in (2.128) in order to be in the position to compare this equation with (2.123) and to derive m_n^* from it. Let us write the work of F_{ext} on the electron as

$$F_{ext}v_g dt = dW \tag{2.129}$$

from which we have

$$F_{ext} = \frac{1}{v_g}\frac{dW}{dt} = \frac{1}{v_g}\frac{dW}{dk}\frac{dk}{dt} \tag{2.130}$$

Due to (2.127):

$$\frac{dk}{dt} = \frac{1}{\hbar} F_{ext} \tag{2.131}$$

Substituting (2.131) in (2.128), after identifying the expression obtained with relationship (2.123), we have

$$m_n^* = \hbar^2 \left(\frac{d^2W}{dk^2} \right)^{-1} \tag{2.132}$$

The effective mass of the electron therefore depends on its energy through a derivative of the function $W(k)$ shown in Fig. 2.24.

2.6.8 Comments on the Effective Mass of the Electron

In the Sommerfeld model, the energy of the electron is related to the wave number by (2.54). From this equation, we obtain

$$\frac{d^2W}{dk^2} = \frac{\hbar^2}{m_n} \tag{2.133}$$

from which we have, using (2.132):

$$m_n = m_n^* \tag{2.134}$$

This result is consistent with the fact that in this model the electron is not considered as being subjected to the periodic potential of the ions.

In the energy band model, the conclusions are more interesting. Let us consider the lowest energy band and the lowest values of k (Table 2.26). Similar results would be obtained for any allowed band and any interval of k.

At the bottom of an energy band, the electrons have a positive effective mass. This mass increases as the energy increases and becomes infinite in the middle of the band. In this situation, an external force can no longer impart acceleration to the electron. Further up in the band, the effective mass becomes negative! This should remind us that the effective mass is only a parameter having the dimension of a mass and calculated so that (2.123) applies. It is therefore not a quantity of matter, which obviously could not be negative. When $m_n^* < 0$, the acceleration is opposed to the external force. The consequences of this are examined in the following sections.

Table 2.26
Qualitative Deduction of the Variation in the Effective Mass of an Electron as a Function of Its Energy

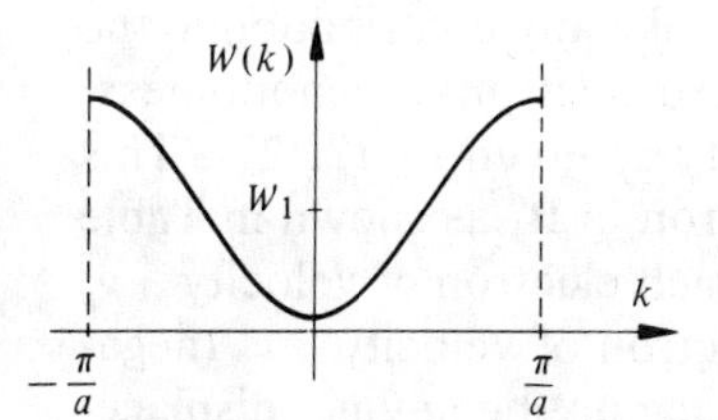	This function extracted from Fig. 2.24 is of the type: $W = W_1 - 2A \cos ka$ (2.135) A = constant
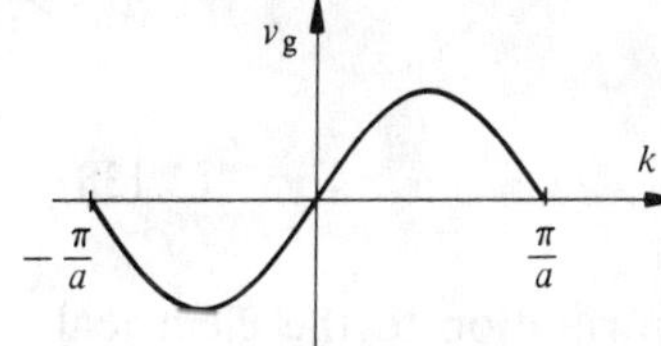	From (2.127) and (2.135) $v_g = \frac{2Aa}{\hbar} \sin ka$ (2.136)
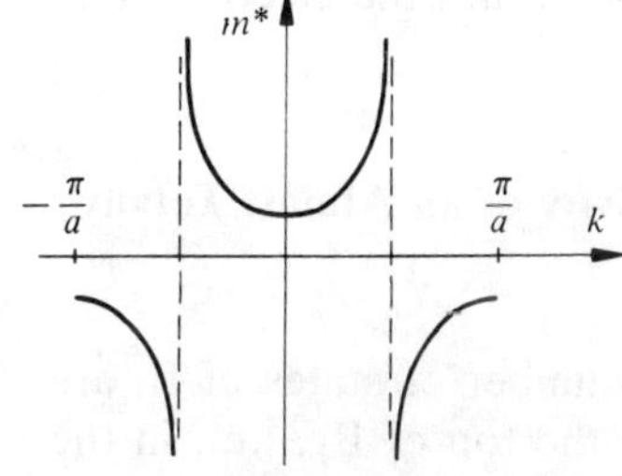	From (2.132) and (2.136) $m_n^* = \frac{\hbar^2}{2Aa^2} \frac{1}{\cos ka}$ (2.137)

2.6.9 Contribution of a Totally Occupied Band to Electrical Conductivity

Let B_l be the allowed energy band with all states effectively occupied by electrons. Let B_2 be the allowed energy band close to B_l in the direction of increasing energy. It is assumed that B_2 does not contain any electrons. Let B_I be the forbidden band separating B_l and B_2.

Band B_l does not make any contribution to the electrical conductivity. In other words, the application of an electric field $\boldsymbol{E}$ has no effect on the electrons of B_l for two reasons:

- by assumption, B_l does not have any vacant levels to accept electrons that could be accelerated by $\boldsymbol{E}$. Therefore, the electrons of B_l can not be subject to any acceleration;

- the width of the forbidden band is too large (apart from exceptions) for an electron transfer from B_1 to B_2 via B_I to take place *under the action of the electric field only*.

The fact that the electrons of B_1 do not make any contribution to the electrical conductivity does not mean that these electrons are motionless. On the contrary, each of them has a velocity v_g given by (2.127). This velocity depends on the position of the electron in B_1 as shown in Table 2.26. Since the band B_1 is totally occupied, each electron of velocity $+v_g$ (positive value of k) has a corresponding electron of velocity $-v_g$ (negative value of k) so that in this situation there cannot be any net displacement of charges due to the electrons of B_1.

Consequently, we can write

$$\boldsymbol{J}_{\mathrm{B}(N,N)} = -e \sum_{i=1}^{N} \boldsymbol{v}_{\mathrm{g}i} = 0 \qquad (2.138)$$

In this expression, $\boldsymbol{J}_{\mathrm{B}(N_1,N_2)}$ represents the contribution to the electrical conductivity of a band B with N_1 allowed states of which N_2 are effectively occupied by the electrons. The index i identifies the electrons and v_{gi} represents the velocity of each of them.

2.6.10 Contribution to the Electrical Conductivity of an Almost Totally Occupied Band

Let us consider the case when a small number of states of B_1 are vacant. These states are naturally situated at the top of B_1, i.e., in the region of the highest energies.

The presence of vacant states allows the electrons of B_1 to be accelerated and thus, in a certain manner, to take part in the conductivity.

Let us suppose that only electron j is missing. The contribution of B_1 to the conductivity can be written

$$\boldsymbol{J}_{\mathrm{B}(N,N-1)} = -e \sum_{\substack{i=1 \\ i \neq j}}^{N} \boldsymbol{v}_{\mathrm{g}i} \qquad (2.139)$$

or

$$\boldsymbol{J}_{\mathrm{B}(N,N-1)} = -e \sum_{i=1}^{N} \boldsymbol{v}_{\mathrm{g}i} + e\boldsymbol{v}_{\mathrm{g}j} \qquad (2.140)$$

Allowing for (2.138), we have

$$\boldsymbol{J}_{\mathrm{B}(N,N-1)} = e\boldsymbol{v}_{\mathrm{g}j} \qquad (2.141)$$

Equation (2.141) shows that the conductivity resulting from the movements of the $N-1$ electrons of B_1, is equivalent to that of a single particle with a charge $+e$.

If not just one but l electrons are missing at the top of B_1, l remaining very small with respect to the total number of states in B_1, we have

$$\boldsymbol{J}_{B(N,N-1)} = le\boldsymbol{v}_{gj} \tag{2.142}$$

2.6.11 Concept of a Hole: Definition

In summary, the preceding section shows that the absence of l electrons at the top of an energy band is equivalent, from the point of view of conductivity to the presence of l mobile particles with a charge $+e$.

These particles are called *holes*. A hole may be considered as an ordinary particle playing a comparable role to that of the electron in the billiard ball model, for example. The effective mass of the hole is equal to the absolute value of the effective mass relative to the corresponding vacant state.

The concept of a hole is extremely useful, in particular in the study of semiconductor devices (Vol. VII). We shall look at a few applications in this section.

It should not be totally forgotten (which is very easy to do!) that *the hole is an imaginary particle*. It is a model that, in a very simple but valid manner, illustrates the contribution to conductivity made by electrons in an almost totally occupied band.

2.6.12 Structure of Bands in Conductors, Semiconductors, and Insulators: Definitions

A fundamental question has not yet been answered, which is where the Fermi energy is situated in the sequence of allowed and forbidden bands. Two cases are possible:

- the Fermi energy lies in an allowed band (Fig. 2.27) and we are dealing with a metallic conductor. The band containing W_F is called the *conduction band;* the band next to it, toward the decreasing energies, is called the *valence band;*
- the Fermi energy is situated in a forbidden band and we are dealing with an insulator or a semiconductor (Fig. 2.28). Here, the valence band is the first band situated below W_F whereas the conduction band (empty or scarcely occupied) is situated immediately above W_F.

The two band structures just mentioned are the only possible structures in a one-dimensional crystal. Another type of structure may appear in real crystals (Section 2.6.15).

2.6.13 Band Structure and Resistivity of Metals versus Temperature

It is well known that the resistivity of metals increases with temperature. The band structures shown in Fig. 2.27 qualitatively explain this experimental fact.

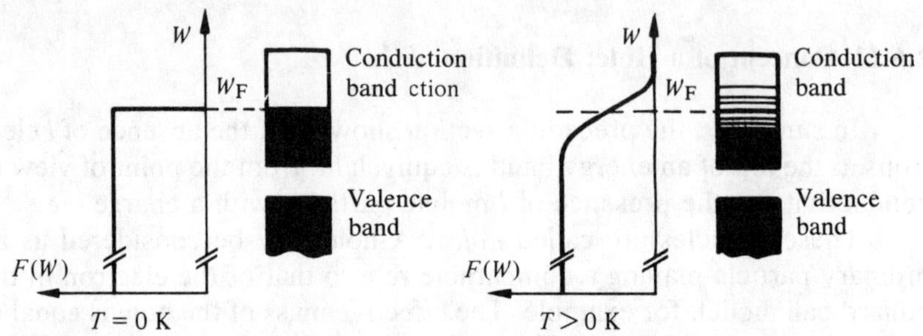

Fig. 2.27 Position of the Fermi Energy in a Metal. The Occupied States Are Shown in Black

At 0 K, all the levels situated below W_F are occupied and all the levels above W_F are empty. When the temperature increases, the thermal energy allows a certain number of electrons to acquire an energy higher than W_F.

Let us suppose that W_F is in the middle of the conduction band. The electrons transferred to $W > W_F$, not only no longer provide the contribution to the conductivity that they made when they had an energy less than W_F, but they lower the current created by the electrons with $W < W_F$. In fact, since their effective mass has become negative (Table 2.26), they move in the opposite direction to the electrons situated below the Fermi energy. Therefore, for a given field, the net current has decreased and consequently the resistivity has increased.

The variation in effective mass shown in Table 2.26 allows us, by similar arguments to the above, to arrive at the same conclusions when W_F does not occupy the exact middle of the conduction band.

2.6.14 Band Structure and Resistivity of Semiconductors and Insulators versus Temperature

A perfect insulator can only exist at 0 K. In fact, at any finite temperature, the probability of transition of an electron from the valence band to the conduction band is not zero (Fig. 2.28). Electrical conductivity occurs first due to the presence of electrons in the conduction band, and second due to the holes that the departure of these electrons has created in the valence band.

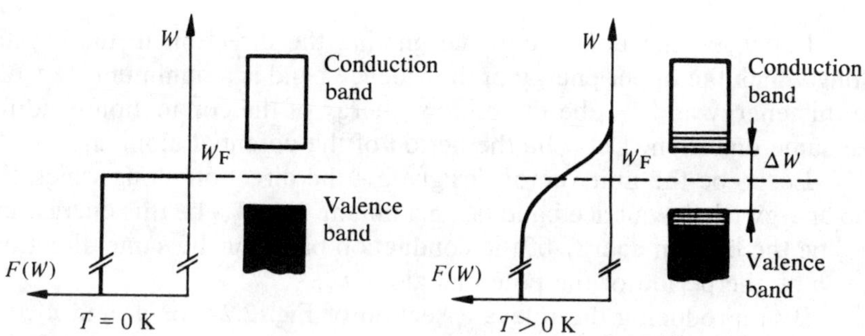

Fig. 2.28 Position of the Fermi Energy in an Insulator or Semiconductor

The magnitude of this conductivity very strongly depends on the width ΔW of the forbidden band. Only the value of ΔW distinguishes a semiconductor from an insulator. A semiconductor is an insulator with a narrow forbidden band. Table 2.29 summarizes the electrical behavior of materials according to their value of ΔW at room temperature.

Table 2.29

ΔW	*Behavior*
>5 eV	Insulator
~1 eV	Semiconductor
<0.1 eV	Metal

2.6.15 Critical Remarks on the One-Dimensional Model

It is obvious that real crystals have three dimensions and not just one as considered above. The purpose of this section is to show to what extent the results obtained for the one-dimensional case apply to the real case.

Regardless of the direction in which an electron is moving inside a crystal, it is subject to a periodic potential. A solution of the one-dimensional type can therefore be applied to it. The difficulty of the three-dimensional problem lies in the fact that the position of the energy bands varies with the orientation of the movement with respect to the crystal axes.

Let $\boldsymbol{u}_1$ be the unit vector designating the direction in the crystal along which the upper energy of the valence band is a minimum. Let W_{v1} be this energy and W_{c1} be the bottom energy of the conduction band for the same direction. Let a be the period of the potential along $\boldsymbol{u}_1$.

Let $\boldsymbol{u}_2$ be the unit vector designating the direction along which the top energy of the valence band is a maximum. Let W_{v2} be this energy and W_{c2} be the bottom energy of the conduction band for the same direction. Let b be the period of the potential along $\boldsymbol{u}_2$.

By reproducing the relevant section of Fig. 2.24 for $\boldsymbol{u}_1$ and $\boldsymbol{u}_2$, we find that there are two possible cases (Fig. 2.30).

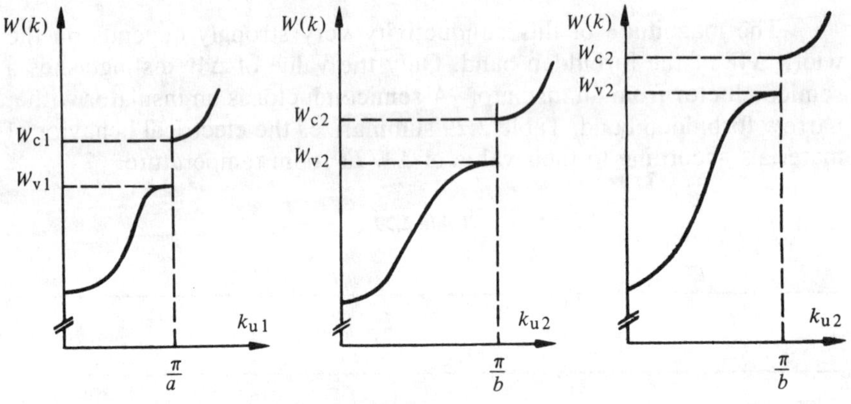

Fig. 2.30

If $W_{v1} < W_{v2} < W_{c1}$, there is a forbidden band in the crystal with width $\Delta W = W_{c1} - W_{v2}$. This crystal is therefore a semiconductor or an insulator depending on the value of ΔW (Fig. 2.31).

On the other hand, if $W_{v2} > W_{c1}$, the forbidden band disappears. The conduction band will therefore be populated before the valence band is filled. There is overlapping of the valence and conduction bands and the crystal is a conductor (Fig. 2.31).

Since the width of the forbidden band varies with the orientation, it is the minimum value of ΔW which is given in the tables unless otherwise specified.

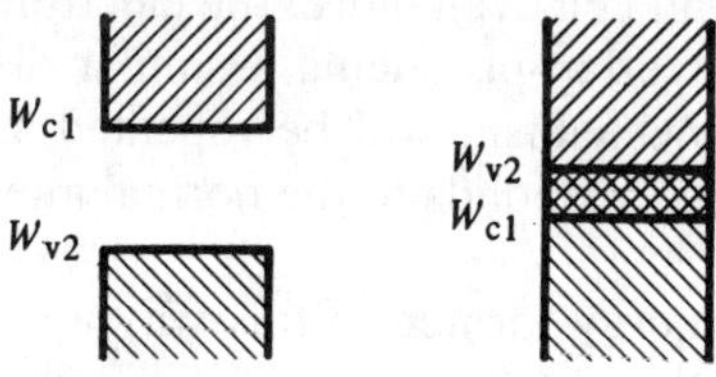

Fig. 2.31

Summarizing, the results of the one-dimensional model remain valid in three-dimensional space but a new structure appears in this space characterized by the overlapping of allowed bands.

2.6.16 Brillouin Zones and Fermi Surface

Figure 2.24 shows that the limits of the energy bands are situated at the surface of the Brillouin zones (Section 7.5) of the one-dimensional crystal. The same applies to a three-dimensional crystal.

In the three-dimensional reciprocal space, the functions $W(k_x, k_y, k_z) = C$, where C is a constant, are surfaces. If $C = W_F$, the corresponding surface is called the *Fermi surface*.

The detailed analysis of the electronic properties of a crystal therefore involves the study of the position of the Fermi surface with respect to the Brillouin zones in three-dimensional geometry.

2.6.17 Relationship Between the Allowed Bands and the States of the Isolated Atom

Let us imagine a pseudocrystal formed from N identical atoms with a unit cell so large that the atoms do not have any mutual interaction. The electronic states of this pseudocrystal are those of an atom multiplied N times which is not contrary to Pauli's principle because the atoms are isolated.

Let us imagine that the unit cell of this crystal is reduced until the orbits of the outer electrons overlap. Then, these electrons form a *system* and their possible energy levels are distributed over N neighboring levels each of which can accept two electrons with opposite spin. The energy range occupied by these levels does not depend on N but only on the degree of overlap of the orbits, i.e., the intensity of interaction between these electrons. The greater the interaction, the more extensive the energy range.

If we reduce the unit cell even more, the electron orbits belonging to the deeper shells will overlap producing a similar distribution of corresponding levels. This mechanism will be repeated a certain number of times until the unit cell corresponds to the normal interatomic distance of the crystal (Fig. 2.32).

The intervals of energy formed in this way are actually the energy bands described in Section 2.6.2.

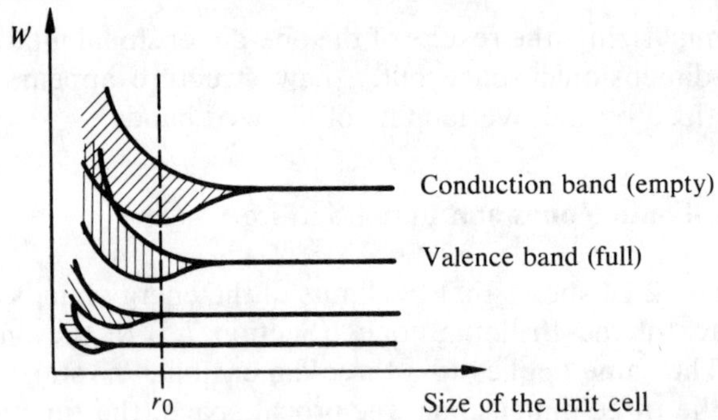

Fig. 2.32 Formation of Energy Bands in a Crystal (Case of an Insulator)

2.7 PROPERTIES OF SEMICONDUCTORS

2.7.1 Intrinsic Semiconductors

The band structure shown in Fig. 2.28 is that of a pure semiconductor which is also called an *intrinsic semiconductor* or an *undoped semiconductor*. The resistivity of such a semiconductor is high and it is mainly used as a starting material for the manufacture of extrinsic semiconductors (Sections 2.7.2 and 2.7.3).

The conductivity mechanism in an intrinsic semiconductor is shown in the diagram of Fig. 2.33.

At stage (a), a valence bond is broken by the thermal energy. An electron e_1 is thus liberated. In other words, it has just made a transition from the valence band to the conduction band, leaving a vacant level in the valence band. The positive charge situated in the region of the broken valence bond corresponds to a hole, denoted h_1. This region provides a low energy level to any electron moving in its vicinity so that the probability that e_1 returns to occupy this level is high. In the presence of an applied field, e_1 will more easily move away from h_1, leaving an electron coming from the left on the figure to occupy the vacant level.

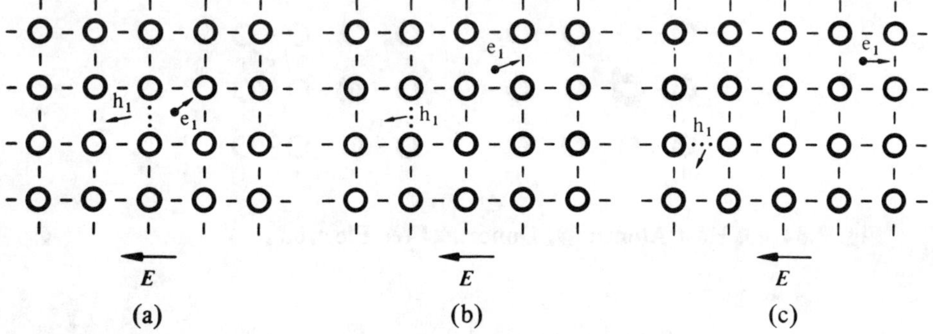

Fig. 2.33

Stages (b) and (c) of Fig. 2.33 illustrate the successive displacements of e_1 and h_1 under the action of the electric field.

It can be seen that by its nature, an *intrinsic semiconductor has the same number of electrons as holes*.

2.7.2 Extrinsic Type n Semiconductors

An extrinsic semiconductor is obtained by *doping,* i.e., by adding a very small amount of impurities to an intrinsic semiconductor.

If the impurities have a higher number of valence electrons than that of the host (>4 in the case of silicon and germanium), we talk of *extrinsic semiconductors of type* n because the negative mobile charges (electrons) are in the majority. Atoms of phosphorus, arsenic or antimony with five valence electrons are currently used as dopants in type n semiconductors.

These impurities are called *donors* because they provide electrons to the conduction band. The donors are situated in substitution positions in the lattice (Fig. 2.34).

Their fifth valence electron which cannot form a bond with the host atoms remains at a higher energy level than the four others. This electron will therefore easily leave the donor, making the latter a positive ion.

Doping is reflected by the appearance of a very narrow allowed energy band situated at a level called the *donor level,* just below the conduction band (Fig. 2.35). The new band represents the energy of the fifth electrons bound by electrostatic attraction to the donor atoms. The small energy gap separating the level of donor from the bottom of the conduction band, typically 0.01–0.1 eV, shows the fragility of these bonds and the fact that the fifth electrons are almost free.

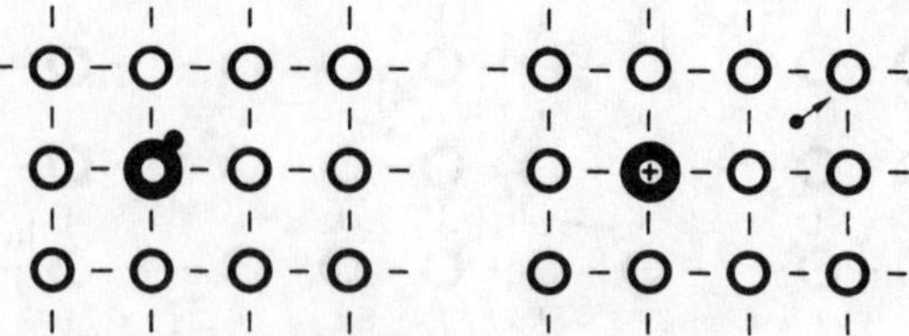

Fig. 2.34 ○, Host Atoms; ◎, Donor; ·, Free Electron

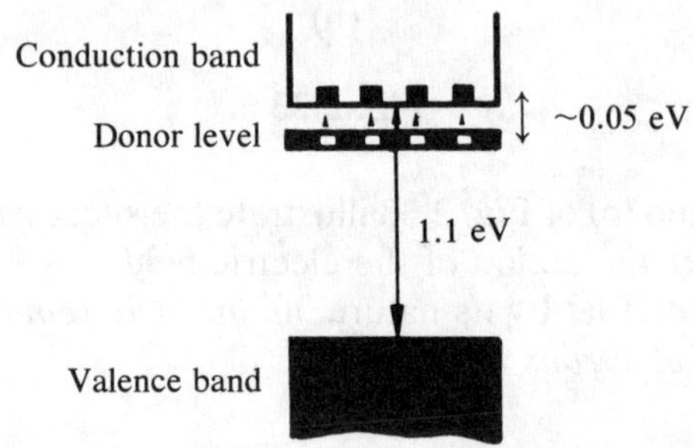

Fig. 2.35 Band Structure in an Extrinsic Type N Semiconductor (Silicon)

2.7.3 Extrinsic Type p Semiconductors

Another type of extrinsic semiconductor is obtained by doping an intrinsic semiconductor using impurities with a number of valence electrons less than that of the host. Boron, aluminum, gallium, and indium with three valence electrons are currently used for this purpose.

These impurities are called *acceptors* because they accept electrons from the valence band, creating vacant levels there. The holes thus become the majority of mobile charge carriers and the semiconductor resulting from this doping is called a *type p semiconductor*.

The acceptors also occupy a substitution position but this time there is a lack of one valence electron to completely bind the impurity to its neighbors (Fig. 2.36). In other words, a low-energy electron state remains unoccupied in the vicinity of the impurity. The probability is high that this state will subsequently be occupied by an electron from a neighboring valence bond moving under the effect of thermal motion and possibly an applied field. This transfer corresponds to the displacement of a hole.

The electron state left vacant by the impurity corresponds to a very narrow energy band situated at an energy level called the *acceptor level*

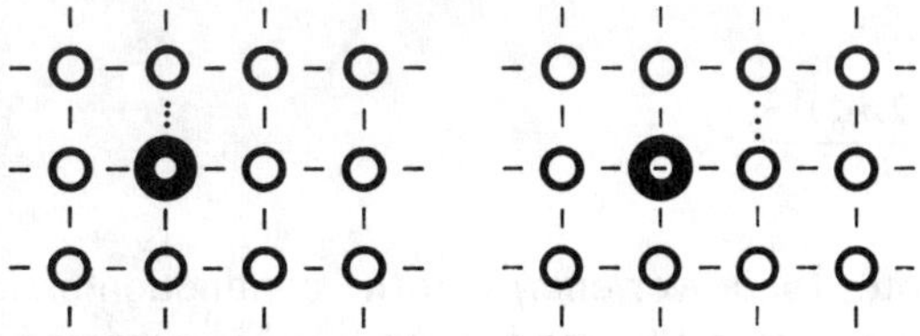

Fig. 2.36 ○, Host Atoms; ◎, Acceptor; ⋮, Position of the Hole

situated just above the top of the valence band (Fig. 2.37). The small energy gap separating the acceptor band from the top of the valence band, typically 0.01–0.1 eV, explains the large transfer of electrons from the valence band to the acceptor band.

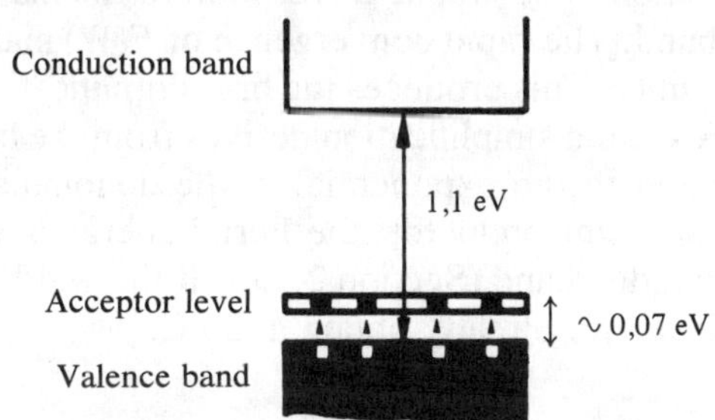

Fig. 2.37 Band Structure in an Extrinsic Type P Semiconductor (Silicon)

2.7.4 Density of Electrons in the Conduction Band of an Intrinsic Semiconductor

Only the bottom part of the conduction band can actually contain electrons because of the rapid convergence of the Fermi-Dirac distribution at higher energies.

In this part of the band, the function $W(k)$ (Fig. 2.24) is close to a parabola, while the effective mass m_n^* of the electron is virtually constant (Table 2.26). Locally, we find the same conditions for which the density of states was calculated for the case of the free electron. Here, (2.61) takes the form:

$$Z(W) = C_n \sqrt{W - W_c} \tag{2.143}$$

with

$$C_{\mathrm{n}} = \frac{4\pi(2m_{\mathrm{n}}^{*})^{3/2}}{h^3} \tag{2.144}$$

where W_{c} denotes the lower energy of the conduction band. The density of electrons n present in the conduction band is expressed by

$$n = \int_{W_{\mathrm{c}}}^{\infty} Z(W)F(W)\,\mathrm{d}W = C_{\mathrm{n}} \int_{W_{\mathrm{c}}}^{\infty} \sqrt{W - W_{\mathrm{c}}}\,\frac{1}{1 + \exp\left(\dfrac{W - W_{\mathrm{F}}}{k_{\mathrm{B}}T}\right)}\,\mathrm{d}W \tag{2.145}$$

The upper integration limit should correspond to the maximum energy of the conduction band. The rapid convergence of $F(W)$ makes it possible to fix this limit at infinity. This produces the first simplification in the calculation of (2.145). A second simplification derives from the fact that 1 may be neglected compared to the exponential in the denominator of $F(W)$. In fact, in an intrinsic semiconductor, the Fermi energy is very close to the center of the forbidden band (Section 2.7.6). If the width of the forbidden band is only 0.7 eV, we already obtain at 20°C:

$$\exp\left(\frac{W - W_{\mathrm{F}}}{k_{\mathrm{B}}T}\right) \cong 10^6 \tag{2.146}$$

Equation (2.145) therefore becomes

$$n = C_{\mathrm{n}} \int_{W_{\mathrm{c}}}^{\infty} \sqrt{W - W_{\mathrm{c}}}\,\exp\left(-\frac{W - W_{\mathrm{F}}}{k_{\mathrm{B}}T}\right)\mathrm{d}W \tag{2.147}$$

By making the change of variable:

$$u = \frac{W - W_{\mathrm{c}}}{k_{\mathrm{B}}T} \tag{2.148}$$

we have

$$n = C_{\mathrm{n}}(k_{\mathrm{B}}T)^{3/2}\exp\left(-\frac{W_{\mathrm{c}} - W_{\mathrm{F}}}{k_{\mathrm{B}}T}\right)\int_0^{\infty}\sqrt{u}\,\exp(-u)\mathrm{d}u \tag{2.149}$$

from which, due to the mathematical relation (7.8.3), we obtain

$$n = N_c \exp\left(-\frac{W_c - W_F}{k_B T}\right) \tag{2.150}$$

with

$$N_c = 2\left(\frac{2\pi m_n^* k_B T}{h^2}\right)^{3/2} \tag{2.151}$$

The exponential in (2.150) represents a very good approximation to the probability of the occupation of a state with energy W_c. We can therefore interpret equation (2.150) in the following manner: the conduction band provides a density of electrons equal to that provided by a single level situated at W_c and possessing N_c possible states per unit volume. At 20°C and for $m_n^* = m_n$, we have

$$N_c = 2.42 \cdot 10^{25}\ \mathrm{m}^{-3} \tag{2.152}$$

Silicon has $5.0 \cdot 10^{28}$ atoms per cubic meter. For this material where the width of the forbidden band is 1.12 eV, expression (2.150) gives, at 20°C, $n = 5.7 \cdot 10^{15}\ \mathrm{m}^{-3}$. At this temperature, therefore, pure silicon only has one free electron per $9 \cdot 10^{12}$ atoms.

2.7.5 Density of Holes in the Valence Band of an Intrinsic Semiconductor

The calculation of the density of holes is similar to the procedure of the previous section in all points. Only the top of the valence band may have vacant states. In this region, the function $W(k)$ plotted in Fig. 2.24, still had the form of a parabola. By analogy with (2.143), we can therefore write

$$Z(W) = C_p \sqrt{W_v - W} \tag{2.153}$$

where

$$C_p = \frac{4\pi(2m_p^*)^{3/2}}{h^3} \tag{2.154}$$

W_v denotes the maximum energy of the valence band, and m_p^* is the effective mass of the hole, which is virtually constant throughout the considered energy interval (Table 2.26).

The total number of holes p in the valence band is given by the expression:

$$p = \int_0^{W_v} Z(W)[1 - F(W)]\, dW \tag{2.155}$$

In $F(W)$, the term $\exp[(W - W_F)/k_B T]$ is sufficiently small to set

$$1 - F(W) \cong \exp\left(\frac{W - W_F}{k_B T}\right) \tag{2.156}$$

so that (2.155) takes the form:

$$p = \int_0^{W_v} C_p \sqrt{W_v - W} \exp\left(-\frac{W_F - W}{k_B T}\right) dW \tag{2.157}$$

This equation can be integrated by changing the variable:

$$u = \frac{W_v - W}{k_B T} \tag{2.158}$$

which allows the mathematical relation (7.8.3) to be used, and leads to

$$p = N_v \exp\left(-\frac{W_F - W_v}{k_B T}\right) \tag{2.159}$$

with

$$N_v = 2\left(\frac{2\pi m_p^* k_B T}{h^2}\right)^{3/2} \tag{2.160}$$

2.7.6 Fermi Energy in an Intrinsic Semiconductor

In an intrinsic semiconductor, the density of electrons in the conduction band is equal to the density of holes in the valence band. Due to (2.150) and (2.159), this equality can be written

$$N_c \exp\left(-\frac{W_c - W_F}{k_B T}\right) = N_v \exp\left(-\frac{W_F - W_v}{k_B T}\right) \tag{2.161}$$

Because the energies W_c and W_v are known, this expression makes it possible to find W_F. By taking the natural logarithm of both sides of (2.151), we have

$$\ln N_c - \frac{W_c - W_F}{k_B T} = \ln N_v - \frac{W_F - W_v}{k_B T} \tag{2.162}$$

from which we have, allowing for (2.151) and (2.160):

$$W_F = \frac{W_c + W_v}{2} + \frac{3}{2}\frac{k_B T}{2} \ln \frac{m_p^*}{m_n^*} \tag{2.163}$$

Even for a relatively large difference between m_p^* and m_n^*, the natural logarithm in (2.163) remains small compared to $(W_c + W_v)/2$. It follows from this that in an intrinsic semiconductor, the Fermi energy is very close to the middle of the forbidden band.

2.7.7 Remark

It is interesting to note that we need not know W_F in order to calculate n or p in an intrinsic semiconductor. For such a semiconductor, let us set

$$n_i = n = p \tag{2.164}$$

Due to (2.150) and (2.159), we have

$$n_i^2 = N_c N_v \exp\left(-\frac{W_c - W_v}{k_B T}\right) \tag{2.165}$$

The Fermi energy has disappeared in this expression. However, the width of the forbidden band plays an important part here, as is to be expected.

2.7.8 *pn* Product Rule

An extrinsic semiconductor is distinguished from an intrinsic semiconductor by the position of the Fermi energy in the forbidden band, as we will see in the following section. Relationship (2.165), independent of W_F, is also valid for an extrinsic semiconductor. In such a semiconductor, n is different from p (Section 2.7.11) but as (2.165) shows, the product pn

is a constant independent of the concentration of the impurities (doping) *at a given temperature:*

$$pn = n_i^2 \tag{2.166}$$

This result constitutes the *pn* product rule. It is extremely useful in the study of extrinsic semiconductors.

2.7.9 Fermi Energy in an Extrinsic Semiconductor

Let us consider the most general case of a semiconductor receiving donors *and* acceptors. Before doping, the semiconductor is electrically neutral. The atoms of the added dopant are also neutral so that the doped semiconductor must itself be neutral as well.

This statement makes it possible to determine the Fermi level in a doped semiconductor. Let

- N_d be the concentration of donors and W_d be the energy at which the donor level is situated;
- N_a and W_a be the corresponding values for the acceptors.

We find negative charges

- in the conduction band (electrons). Their concentration is given by (2.150);
- in the acceptor band (ions), their concentration is

$$N_a^- = N_a F(W_a) \tag{2.167}$$

We find positive charges

- in the valence band (holes). Their density is given by (2.159);
- in the donor band (ions), their density is

$$N_d^+ = N_d[1 - F(W_d)] \tag{2.168}$$

Expressions (2.167) and (2.168) may be evaluated by using the same approximations for the Fermi-Dirac distribution as those assumed for the calculation of n and p.

Having done this, the neutrality condition for the semiconductor takes the following form:

$$N_c \exp\left(-\frac{W_c - W_F}{k_B T}\right) + N_a \left[1 - \exp\left(-\frac{W_F - W_a}{k_B T}\right)\right]$$
$$= N_v \exp\left(-\frac{W_F - W_v}{k_B T}\right) + N_d \left[1 - \exp\left(-\frac{W_d - W_F}{k_B T}\right)\right] \quad (2.169)$$

The numerical solution of (2.169) does not cause any particular difficulty although it is worth having available an approximate analytical solution. This is obtained at the cost of two approximations.

We first assume that $(W_F - W_a)$ and $(W_d - W_F)$ are sufficiently large with respect to $k_B T$ so that (2.169) can be put in the following form:

$$N_c \exp\left(-\frac{W_c - W_F}{k_B T}\right) + N_a = N_v \exp\left(-\frac{W_F - W_v}{k_B T}\right) + N_d \quad (2.170)$$

It can be seen from Fig. 2.38 that this simplification is acceptable over a large temperature range. In a more compact form, (2.170) can also be written

$$n + N_a = p + N_d \quad (2.171)$$

Second, we limit ourselves to the only interesting case in practice corresponding to $N_d \neq N_a$. There only needs to be a small difference between N_d and N_a for n and p to differ from each other by several orders of magnitude (Section 2.9.6). If $N_d > N_a$, the density of holes can be neglected compared to the density of electrons. Then, the solution of (2.170) is very simply obtained by taking the natural logarithm of both sides. We have

$$W_F = W_c - k_B T \ln \frac{N_c}{N_d - N_a} \quad (2.172)$$

If $N_a > N_d$, it is the density of electrons that is negligible compared to the total number of holes. In this case, (2.170) leads to the solution:

$$W_F = W_v + k_B T \ln \frac{N_v}{N_a - N_d} \quad (2.173)$$

It can be seen from Fig. 2.38, that over a wide range of temperatures, W_F varies linearly with the temperature according to (2.172) and

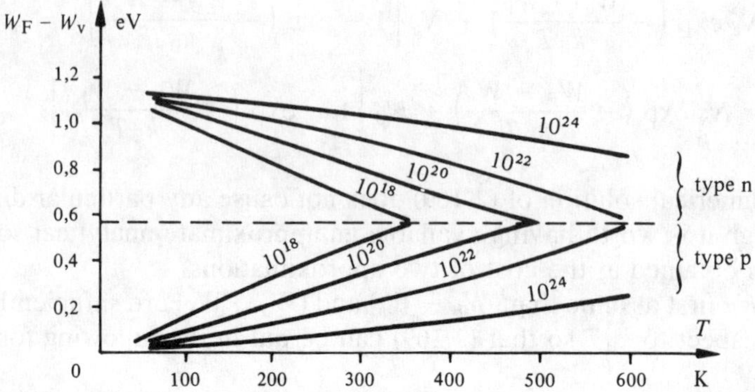

Fig. 2.38 Fermi Energy in Extrinsic Silicon as a Function of Temperature. The Parameters are the Concentration of Donors and the Concentration of Acceptors. (After [11].)

(2.173). However, the slope $\mathrm{d}W_F/\mathrm{d}T$ given by these equations is less than that of the figure.

The Fermi-Dirac function appearing in (2.167) and (2.168) should be modified with respect to (7.43) because of the degeneracy of the donor and acceptor levels [12]. The simplifications introduced in the solution of (2.169) have made this irrelevant.

2.7.10 Behavior of Extrinsic Semiconductors at Extreme Temperatures

The variation of W_F shown in Fig. 2.38 explains the behavior of extrinsic semiconductors at extreme temperatures.

At low temperatures, the Fermi energy in a type n semiconductor lies between the bottom of the conduction band and the donor level. The semiconductor therefore behaves as an intrinsic semiconductor with a donor level that acts as the valence band. In a type p semiconductor, W_F is situated between the top of the valence band and the acceptor level. The semiconductor therefore behaves as an intrinsic semiconductor with an acceptor level that acts as the conduction band.

At high temperatures, the Fermi energy goes back to the middle of the forbidden band both in type n and type p semiconductors. These semiconductors therefore behave as intrinsic semiconductors. In the conduction band (type n), the number of electrons from the valence band appreciably exceeds the number of electrons coming from the donors. Conversely, in the valence band (type p), the number of holes corresponding to the departure of electrons to the conduction band appreciably exceeds the number of holes formed because of the acceptors.

2.7.11 Occupation of Bands in a Semiconductor: General Case

The occupation of bands in a semiconductor determined by the density of states, the Fermi-Dirac distribution, and the value of W_F is shown in the diagrams of Fig. 2.39. In particular, it is worth noting the appearance of a large number of electrons in the conduction band of type n semiconductors and of holes in the valence band of type p semiconductors.

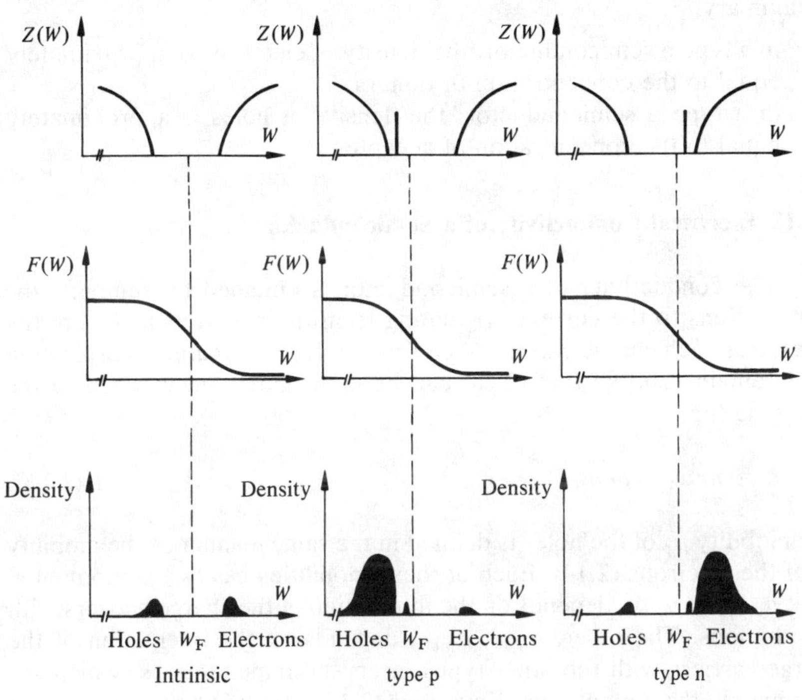

Fig. 2.39

The density of electrons and holes in extrinsic semiconductors can be easily calculated. Let n_n, p_n and n_p, p_p be these quantities in type n and type p semiconductors, respectively.

We first consider the case of type n semiconductors. Equations (2.166) and (2.171) form a system with two unknowns n, n_n, and p, p_n. By taking n_n from this system, we obtain

$$n_n = \frac{1}{2}\left(N_d - N_a + \sqrt{(N_d - N_a)^2 + 4n_i^2}\right) \tag{2.174}$$

We know from Section 2.9.6 that $\sqrt{n_i^2} \ll n$. On the other hand, N_d is very large compared to N_a. Consequently,

$$n_n \cong N_d \quad \text{and} \quad p_n = n_i^2/N_d \tag{2.175}$$

It can be shown in the same manner that

$$p_p \cong N_a \quad \text{and} \quad n_p = n_i^2/N_a \tag{2.176}$$

In summary,

- in a type n semiconductor, the density of electrons is approximately equal to the concentration of donors;
- in a type p semiconductor, the density of holes is approximately equal to the concentration of acceptors.

2.7.12 Electrical Conductivity of a Semiconductor

The conductivity of a semiconductor is obtained by summing the contributions to the current originating from the electrons and from the holes. For each charge carrier taken separately, the arguments of Section 2.3.1 remain valid so that (2.20) can be immediately generalized to the following form:

$$\sigma = ne\mu_n + pe\mu_p \tag{2.177}$$

The mobility μ_p of the holes is defined in the same manner as the mobility μ_n of the electrons (2.14). Each of these mobilities has two components. One, $\mu_{n,ph}$, or $\mu_{p,ph}$, depends on the interaction of the charge carriers with the phonons. The other, $\mu_{n,im}$, $\mu_{p,im}$, depends on the interaction of the charge carriers with the other types of crystal imperfections, which are here mainly the dopant ions. Equations (2.14) and (2.26) give the rule for adding these quantities:

$$\frac{1}{\mu_s} = \frac{1}{\mu_{s,ph}} + \frac{1}{\mu_{s,im}} \tag{2.178}$$

The index s represents the indices n or p. The calculation of $\mu_{s,ph}$, and $\mu_{s,im}$, is a complicated matter. As a function of temperature, these quantities approximately obey the following laws:

$$\mu_{s,ph} \sim T^{-3/2} \tag{2.179}$$

$$\mu_{s,im} \sim N_{im} T^{3/2} \tag{2.180}$$

In (2.180), N_{im} represents the total number of foreign atoms. Expressions (2.179) and (2.180) agree with the considerations discussed in Section 2.2.2 with regard to metals, i.e., that the interaction of the charge carriers with impurities predominates at low temperatures whereas the interaction with the phonons predominates when the temperature increases (Fig. 2.40).

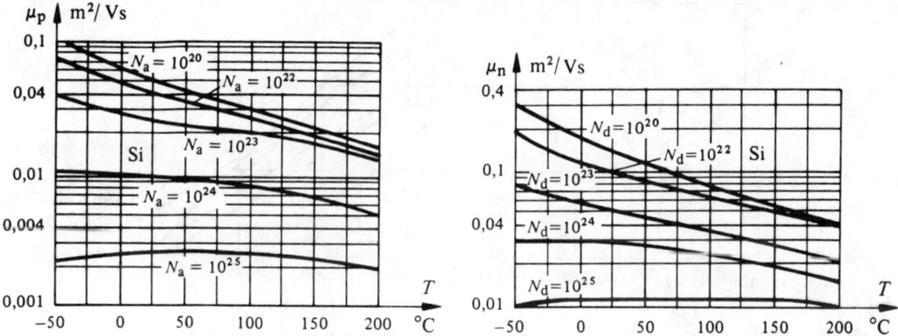

Fig. 2.40 Mobility of the Electrons and Holes in Silicon as a Function of Temperature and the Concentration of Dopants in m^{-3}. (After [13].)

At ordinary temperatures, the logarithm of the resistivity varies quite linearly as a function of the logarithm of the concentration of impurities, as is shown in Fig. 2.41.

2.7.13 Semiconductor Materials

The band structure, and in particular the width of the forbidden band, depends as much on the crystal structure as on the nature of the atoms. The effect of the crystal structure is shown by the classical example of carbon, which in the form of diamond is an insulator and in the form of graphite is a conductor.

Semiconductor materials can be divided basically into the following classes:

- elements of groups IVB of the periodic table. They possess the diamond structure (Fig. 1.22);
- compounds formed from the elements of groups IIIB and VB or

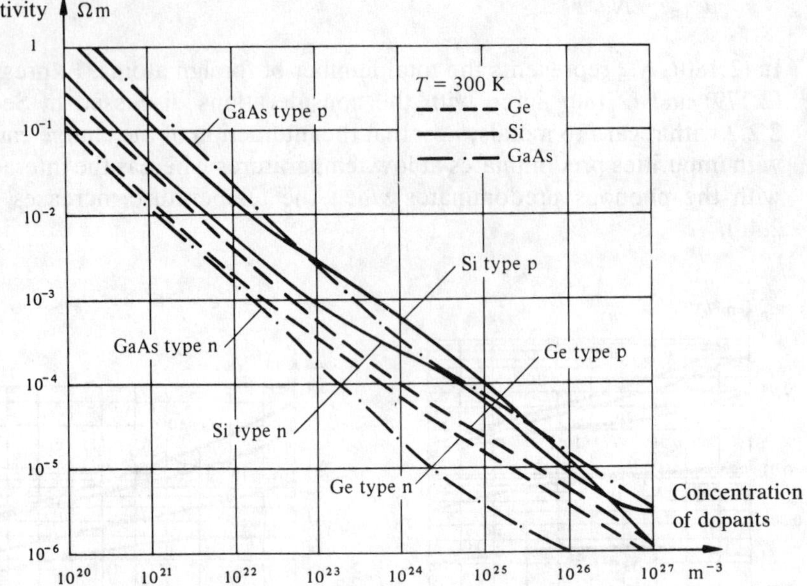

Fig. 2.41 Resistivity of Silicon, Germanium, and Gallium Arsenide at 300 K as a Function of the Concentration of Dopants. (After [13].)

from groups IIB and VIB. These compounds have the blende structure (Fig. 2.44) or that of hexagonal zinc sulphide, both close to the diamond structure.

The first class contains germanium and silicon the economic importance of which is well known. In these materials, the bonds are covalent. The bond energy is lower in germanium than in silicon and this is why the width ΔW of the forbidden band is smaller in germanium than in silicon (Table 2.42).

Table 2.42

Element	*Ge*	*Si*	*C* *(diamond)*
ΔW eV (at 20°C)	0.66	1.12	5.3

Germanium and silicon are doped with elements of group IIIB for type p semiconductors and group VB for type n semiconductors. Electronic components using doped silicon can operate at higher temperatures (~200°C) than those using germanium (~100°C). In fact, the wider forbidden band of silicon increases the temperature at which this material in the doped state returns to intrinsic semiconductor behavior, making it unsuitable for the operation of a junction (Vol. VII).

The use of compounds formed from elements of groups IIIB and VB increases the choice of the ΔW values and the available mobilities as shown by Table 2.43. All the compounds of Table 2.43 have the blende structure derived from the diamond structure by replacing the carbon atoms by atoms of groups IIIB and VB, so that each atom of a group is at the center of a tetrahedron, the vertices of which are formed from atoms of the other group (Fig. 2.44).

Table 2.43

Material	ΔW *eV*	μ_n $m^2/V\ sec$	μ_p $m^2/V\ sec$
Ga	2.25	0.045	0.002
AlSb	1.60	0.040	0.020
GaAs	1.42	0.85	0.045
InP	1.27	0.60	0.016
GaSb	0.70	0.50	0.085
InAs	0.33	2.3	0.010
InSb	0.18	8.0	0.070

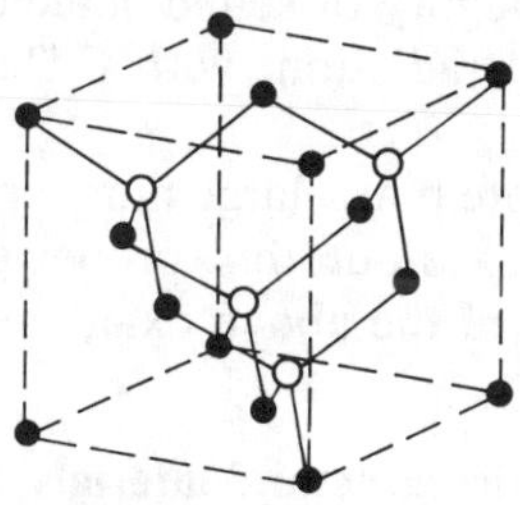

Fig. 2.44 Blende Structure

In the blende structure, the valence bonds still have a predominant covalent character although with a slight ionic component (5–10% of the bonding energy). This component is responsible for the increase in ΔW in a compound with respect to the value of ΔW in the corresponding semiconductor of group IVB. In gallium arsenide (GaAs), for example, $\Delta W = 1.42$ eV, whereas in germanium (Ge), an element situated between gallium and arsenic in the periodic table, $\Delta W = 0.66$ eV.

The production of semiconductor compounds is more critical than the production of semiconductors of group IVB because added to the problem of the chemical purity is that of the relative atomic concentrations of the two constituents. The difference between these concentrations must not exceed a value approximately equal to the total number of impurities that can be tolerated. In fact, if the concentration of atoms of group IIIB is slightly less than that of group VB, vacancies appear at the sites normally occupied by the atoms of group IIIB. These vacancies behave as acceptors. Conversely, the presence of vacancies at sites normally occupied by atoms of group VB is equivalent to doping by donors.

Gallium arsenide is one of the most interesting semiconductor compounds. It is widely used in optoelectronic components (Vol. VII, Ch. 9). Its advantages lie in a relatively high value of ΔW, associated with donor and acceptor levels very close to the valence and conduction bands. This means that high operating temperatures can be used, but the efficiency of the dopants at room temperature is very high. Gallium arsenide may be doped with germanium which can be substituted by a suitable thermal treatment, either at a site normally occupied by an IIIB atom (GaAs type n) or at a site normally occupied by a VB atom (GaAs type p).

In compounds formed from elements belonging to groups IIB and VIB, the contribution of the bonding energy corresponding to ionic valence increases with respect to that of covalent bonding as ΔW increases, as has just been considered. The compounds CdS, CdSe, and CdTe are representatives of this category of semiconductors, which are frequently used as detectors of light radiation, with CdTe being used for infrared radiation.

In conclusion, we note that a large number of other substances have semiconductor behavior without this property always possessing the technological importance of the above examples.

2.7.14 Production of Semiconductor Materials

Semiconductor materials used in the manufacture of electronic components must have two properties:

- chemical purity, without which the effects of doping performed at very low concentrations cannot be realized;
- the best possible crystalline perfection, because structural imperfections cause additional energy levels to appear locally. For example, the presence of a grain boundary in a pn junction (Vol. VII) is enough to reduce the rectification effect or even to destroy it completely. A high number of dislocations is likely to have the same effect.

Chemical purity is obtained by pyrolysis or by the melting zone method. In this method (Fig. 2.45), a cylindrical specimen is placed in a graphite crucible. A slowly moving coil induces eddy currents which make the semiconductor melt locally. In its movement, the liquid zone carries with it the impurities of the specimen, leaving the material that solidifies behind it in a state of higher purity.

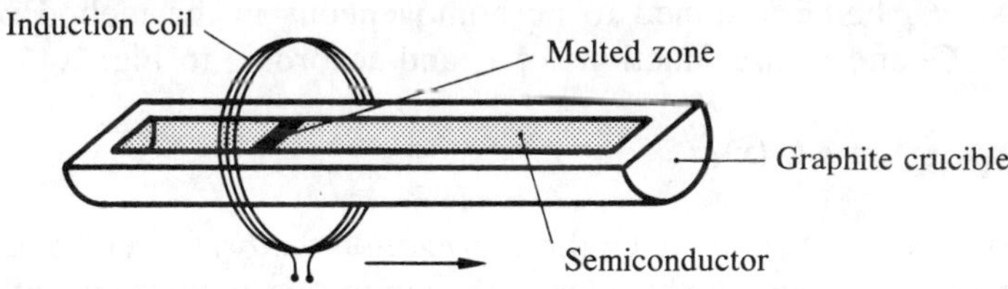

Fig. 2.45 Principle of the Melting Zone Method

The operation takes place in a controlled atmosphere to prevent any external pollution of the semiconductor. It has to be repeated a certain number of times so that sufficient purity is obtained. Industrially, these repetitions are avoided by placing several induction coils one after the other. The purification process will be described for the case of a semiconductor that only has one type of impurity, lowering, for example, the melting point. It is assumed that the degree of initial purity is relatively high so that it is sufficient to consider a small fraction of the phase diagram (Chapter 5) corresponding to the low concentrations of the impurities. In this region, the liquidus and the solidus can be approximated by two converging straight lines (Fig. 2.46).

Let C_0 be the initial concentration of the impurity. In the initial liquid zone, still motionless, the concentration of impurities is also C_0.

The displacement of the inductor by a value $\mathrm{d}x$ to the right causes solidification on the left of an area of thickness $\mathrm{d}x$ in which the impurity concentration is lowered to C_{s0}, whereas a new zone melts on the right.

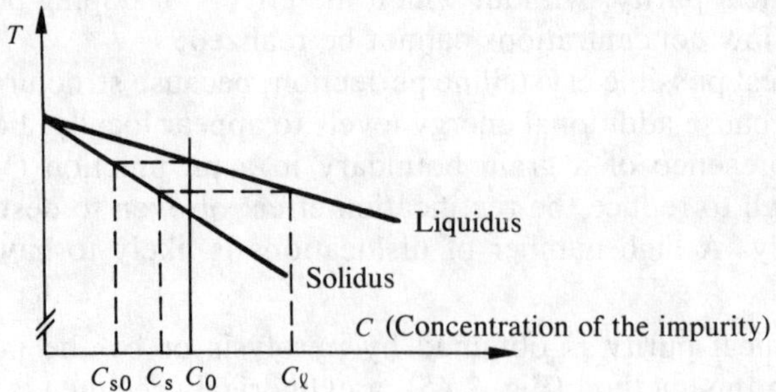

Fig. 2.46

Let C_s be the impurity concentration of the solid phase in contact with the liquid phase and C_ℓ be the impurity concentration of the liquid phase, C_ℓ being assumed to be homogeneous in the melted zone (Fig. 2.47). C_ℓ and C_s are functions of x, and according to Fig. 2.46, we have

$$C_s(x) = KC_\ell(x), \tag{2.181}$$

where K is a constant called the *segregation factor*. On a displacement dx of the melted zone, the balance of the impurities in this zone with length L can be written

$$L\mathrm{d}C_\ell = C_0\,\mathrm{d}x - C_s(x)\,\mathrm{d}x \tag{2.182}$$

or by expressing C_ℓ in terms of C_s, which is the variable of interest:

$$\mathrm{d}C_s = \frac{K}{L}(C_0 - C_s)\,\mathrm{d}x \tag{2.183}$$

from which we have

$$\int_{C_{s0}}^{C_s} \frac{\mathrm{d}C_s}{C_0 - C_s} = \frac{K}{L}\int_0^x \mathrm{d}x \tag{2.184}$$

and by integration:

$$\frac{C_s(x)}{C_0} = 1 - (1 - K)\exp\left(-\frac{K}{L}x\right) \tag{2.185}$$

This expression makes it possible to evaluate the distance x for which the process has a significant yield.

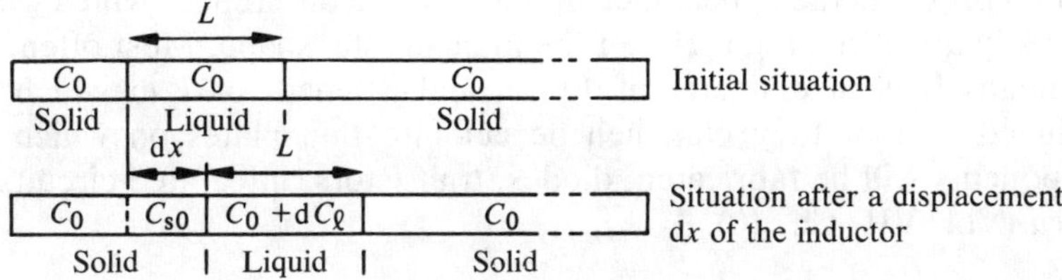

Fig. 2.47

Once the required chemical purity has been achieved, it remains to ensure adequate crystalline perfection, i.e., to turn the specimen into the form of a monocrystal. The method of Czochralski (Fig. 2.48) is usually used for this operation.

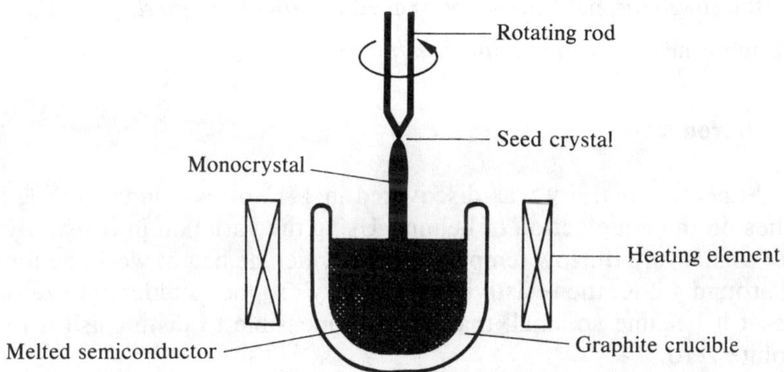

Fig. 2.48 Drawing of a Monocrystal by the Czochralski Method

The semiconductor is placed in a crucible and brought to a temperature very slightly above the melting point. A seed crystal, i.e., a small monocrystal of the same semiconductor, correctly oriented and fixed at the end of a rod, is placed in contact with the surface of the liquid. To begin with, the seed crystal is left to melt very slightly so that it is perfectly wetted by the liquid semiconductor. Then, the rod extracts the quantity of heat just necessary for the liquid in contact with the seed crystal to progressively solidify by prolonging the crystal axes of the seed crystal. During this phase, the rod is made to rotate and is slowly pulled upwards so as to keep the liquid-solid interface close to the surface. The rotation prevents the appearance of temperature differences which would adversely affect the regularity of the monocrystal shape. Most often, the specimens have a diameter of 10 cm and a length of 1 m which are produced so that they can then be cut into thin plates on which the components will be fabricated: diodes, transistors, integrated circuits, *et cetera* (Vol. VII, Ch. 10).

2.8 SUPERCONDUCTIVITY

2.8.1 Definitions

Superconductivity is the property belonging to certain materials having a zero electrical resistance to direct current when

- the temperature is less than a *critical temperature* T_c;
- the magnetic field does not exceed a *critical magnetic field* H_c.

Such materials are called *superconductors*.

2.8.2 Introduction

Superconductivity was discovered in 1911 by K. Onnes during his studies on the liquefaction of helium. Using the variation in resistivity in order to measure the low temperatures at which he had to work, he found that around 4.2 K, the resistivity of mercury dropped suddenly to zero or at least it became so small that it was impossible to distinguish it from absolute zero.

Soon afterwards other elements and alloys were found with the same property but in spite of considerable efforts the maximum values of T_c never exceeded about 20 K until 1986. That year was to be a milestone in the history of superconductivity. Indeed, the work of A. Müller and G. Bednorz of the IBM Laboratories, Rüschlikon (Switzerland), who were awarded the Nobel prize in 1987, opened up an entirely new field of research [71]. They demonstrated that certain ceramics could become superconductive at temperatures definitely higher than those obtained in metals so far. The limit corresponding to the temperature of boiling nitrogen (77.4 K) was passed at the beginning of 1987 [72] with Y-Ba-Cu-O compounds. The possibility of using liquid nitrogen as cooling fluid instead of liquid helium and the relative simplicity of making and using superconductive ceramics considerably increase the field of potential applications of superconductivity. Considering the fact that it does not appear totally impossible one day to obtain superconductors working at room temperature, it becomes evident that the revolution of "hot superconductors" could equal or even exceed the revolution of semiconductors. (The work of Japanese researchers, reported in January 1988, will indeed hasten the advent of superconductivity.)

A large number of phenomenological theories were developed to account for superconductivity but it was 46 years before Bardeen, Cooper, and Schrieffer developed the first microscopic theory of superconductivity. Known today as *BCS theory,* it won the Nobel prize for its authors in 1972.

From a practical point of view, the BCS theory, which continues to undergo significant developments, has not made all the phenomenological theories which preceded it irrelevant. It has made it possible to establish what were up until now just postulates or to place them in context.

Although the BCS theory is undeniably the most detailed model, it still remains too general to predict, for example, which of the elements or alloys are superconductors and which are not. Experimental techniques are of great importance if not of key importance in research involving superconductivity. In this section, we shall describe the most remarkable thermodynamic and electrodynamic behavior of superconductors through experiments and macroscopic explanations before giving a qualitative microscopic interpretation based on the BCS theory.

2.8.3 Resistivity and Critical Temperature

The transition from the normal state to the superconducting state (Fig. 2.49) is sudden in pure metals where ΔT may drop to 10^{-3} K. In alloys, ΔT may reach a few kelvins.

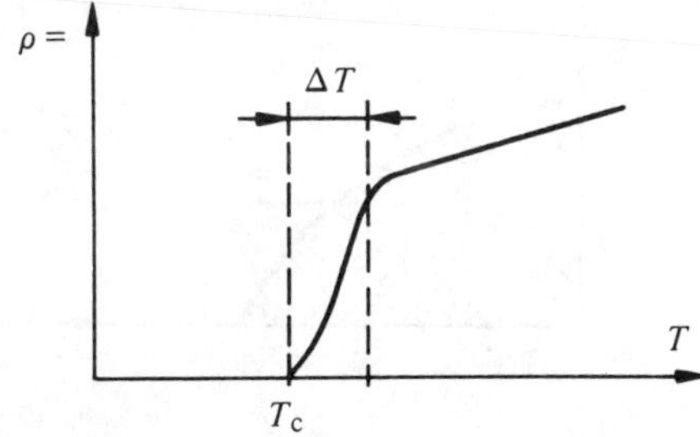

Fig. 2.49 ρ **= Direct Current Resistivity**

In many respects, this transition is comparable to an ordinary phase transition. The expressions *normal phase* and *superconducting phase* are also currently used instead of normal state and superconducting state.

Is the resistivity below T_c strictly zero? It is not possible to confirm this because this would suppose that the experimental error could also be zero. It has been shown today that $\rho = <10^{-21}\ \Omega m$, which is extremely low compared, for example, with the resistivity of copper at temperatures of the same order: $\rho \approx 10^{-10}\ \Omega m$. Under these conditions, a direct current can flow in a superconducting ring for several years without any detectable attenuation!

2.8.4 Critical Temperature and Critical Field

The critical temperature T_c and the critical field H_c are related by an empirical relationship of the type:

$$H_c = H_{c0}\left[1 - \left(\frac{T_c}{T_{c0}}\right)^2\right], \tag{2.186}$$

shown in Fig. 2.50. In this relationship, H_{c0} denotes the critical field at zero temperature and T_{c0} is the critical temperature at zero field. The area S (Fig. 2.50) contained between the function and the coordinate axes corresponds to the superconducting state. As soon as we leave S, by increasing H, or T, or both at once, the material immediately returns to the normal state. The critical temperatures and the critical fields are given for a few elements and alloys in Table 2.51. The critical induction B_{c0} is obtained from H_{c0} by the relationship $B_{c0} = \mu_0 H_{c0}$ where μ_0 is the permeability of vacuum.

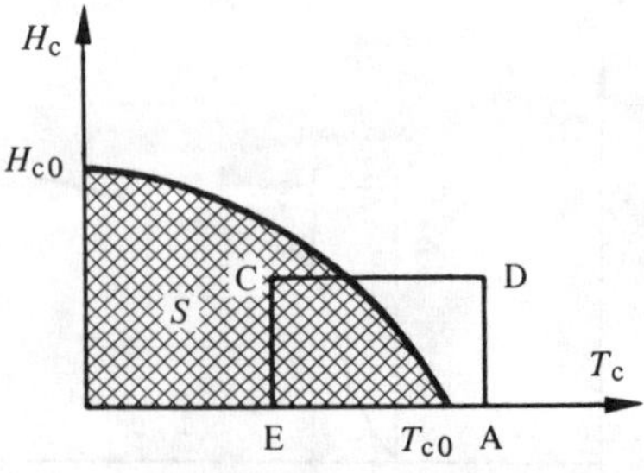

Fig. 2.50

Table 2.51
Characteristics of Some Superconductors. (After [14].)

Element	*Type**	T_{c0} *K*	H_{c0} *A/m*	B_{c0} *T*	*Alloy*	*Type**	T_{c0} *K*	H_{c0} *A/m*	B_{c0} *T*
Al	1	1.19	$8 \cdot 10^3$	0.010	$Nb_{0.44}Ti_{0.56}$	2	10	$9.5 \cdot 10^6$	12
In	1	3.41	$23 \cdot 10^3$	0.029	Nb_3Sn	2	18	$18 \cdot 10^6$	22
Sn	1	3.72	$25 \cdot 10^3$	0.032	$Nb_{0.75}Zr_{0.25}$	2	11	$6.4 \cdot 10^6$	8
Pb	1	7.18	$65 \cdot 10^3$	0.082	V_3Ga	2	15	$20 \cdot 10^6$	25
V	1	5.30	$105 \cdot 10^3$	0.132	$Nb_3(Al_{0.8}Ge_{0.2})$	2	20.7	$32 \cdot 10^6$	40
Nb	2	9.46	$156 \cdot 10^3$	0.196					

* *See* Section 2.8.6.

2.8.5 Meissner Effect

Consider a cylindrical specimen (Fig. 2.52), initially in the normal state corresponding to point A in Fig. 2.50. By lowering its temperature to the point E, it is made superconducting. When a field H is applied to it (point C), eddy currents are generated on its surface having the effect of preventing any variation in magnetic flux inside this specimen (Lenz's law). In a normal conductor, these currents are very rapidly attenuated and the magnetic field penetrates into the specimen. Here, the eddy currents do not encounter any resistance so that the magnetic induction remains permanently zero in the specimen.

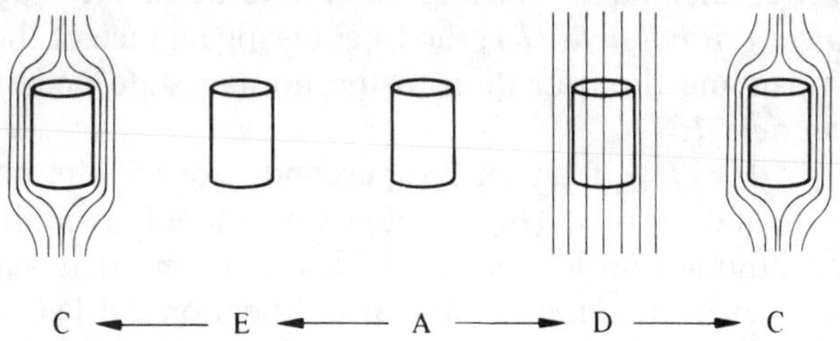

Fig. 2.52 Illustration of the Meissner Effect (the Letters Correspond to the Points in Fig. 2.50)

What happens if the point C is reached by the route A → D → C? At the point D, an induction B is present in the specimen. It would be expected that this induction would be preserved during the transition up to C. In reality, this will not be the case. At the time of the transition, eddy currents will be spontaneously developed that will exactly cancel the effect of H in the specimen and thus bring the magnetic induction to zero. It is often said that the eddy currents have excluded the external magnetic field.

This property of superconductors of excluding an external magnetic field is called the *Meissner effect.*

The existence of the Meissner effect makes the transition between normal and superconducting states a reversible phenomenon in the thermodynamic sense: the state at point C does not depend on the path taken to reach this point. This statement must be amended in the case of superconductors of the second kind (Section 2.8.6).

It is convenient to represent the Meissner effect by considering the superconductor as ideally diamagnetic. The energy necessary to expel H, in particular, is calculated much more easily. It is obvious that the origin of this diamagnetism has nothing to do with the mechanism discussed in Section 3.3.1.

2.8.6 Superconductors of the First and Second Kind: Definitions

Superconductors of the first kind or *type 1 superconductors* is the name given to superconductors in which the Meissner effect stops suddenly when H reaches H_c.

Superconductors of the second kind or *type 2 superconductors* is the name given to superconductors in which there is progressive disappearance of the Meissner effect starting from a value of the magnetic field called the *lower critical field* H_{c1}, the total disappearance of the Meissner effect accompanying the transition to the normal state occurring at the *upper critical field* H_{c2}.

For $H_{c1} < H < H_{c2}$, the type 2 superconductor has alternating zones where $B = 0$ and $B \neq 0$ (Fig. 2.68). Considered overall, it is still diamagnetic, although no longer in an ideal manner. It is said that the superconductor is in the *intermediate state* (Section 2.8.18).

The fields H_{c1} and H_{c2} are often very different from each other; for the alloy Nb_3Sn, for example, $\mu_0 H_{c1} = 0.01T$ and $\mu_0 H_{c2} = 22T$. When we simply speak of the critical field of a type 2 superconductor, it is H_{c2} that is meant. The form of (2.186) is valid for H_{c1} and H_{c2}.

2.8.7 Thermodynamic Study of the Transition

Thermodynamics gives a useful model of the transition from the normal state to the superconducting state. It will first be shown that the free enthalpy is constant during the transition. This result will then be used to calculate the latent heat of transition and finally the discontinuity in the specific heat accompanying the transition when $H_c = 0$. All these considerations apply to type 1 and type 2 superconductors provided that $H < H_{c1}$.

The free enthalpy G, also called the Gibbs free energy, is defined by the following relationship:

$$G = U + W - TS, \tag{2.187}$$

in which U denotes the internal energy of the system, T its absolute temperature, S its entropy and W the work which it performs. In this case, the system is of course a specimen of superconducting material.

The variation of internal energy is given by the expression:

$$\mathrm{d}U = \mathrm{d}q - \mathrm{d}W, \tag{2.188}$$

in which $\mathrm{d}q$ is the amount of heat supplied to the system. Since the transition is reversible, we have

$$\mathrm{d}q = T\,\mathrm{d}S \tag{2.189}$$

There may be two types of work in the system: mechanical and magnetic. If the specimen is subjected to a constant pressure P, the mechanical work is $P\mathrm{d}V$ where $\mathrm{d}V$ is the variation in volume. Experiment shows that $\mathrm{d}V$ is negligible during the transition. The magnetic work apart from the sign is equal to the magnetic energy of the specimen. The density of magnetic energy w_{mag} is equal to (Section 3.2.6):

$$w_{\mathbf{mag}} = \int \boldsymbol{H} \cdot \mathrm{d}\boldsymbol{B} = \mu_0 \int \boldsymbol{H} \cdot \mathrm{d}\boldsymbol{H} + \int \boldsymbol{H} \cdot \mathrm{d}\boldsymbol{I} \tag{2.190}$$

On the right-hand side, the first term represents the contribution of w_{mag} in vacuum, and the second term is the contribution of w_{mag} in matter. Only the last term interests us here. By assuming $\boldsymbol{H}$ and $\boldsymbol{I}$ are constant over the entire volume V of the specimen,

$$\mathrm{d}W = -\mathrm{d}w_{mag}\,V = -VH\,\mathrm{d}I, \tag{2.191}$$

because $\boldsymbol{H}$ and $\boldsymbol{I}$ are assumed parallel. Relationships (2.187) and (2.188) therefore take the form:

$$G = U - VHI - TS \tag{2.192}$$

and

$$\mathrm{d}U = T\,\mathrm{d}S + VH\,\mathrm{d}I \tag{2.193}$$

By calculating the total differential of G, we obtain, allowing for (2.193):

$$\mathrm{d}G = -VI\,\mathrm{d}H - S\,\mathrm{d}T \tag{2.194}$$

This equation shows that G remains constant throughout the transition because we know that the transition occurs at constant temperature and magnetic field.

The variation of G along an isotherm is deduced simply from (2.194).

Let n and s be the indices denoting the variables relating to the normal and superconducting states. In the superconducting phase, the magnetic induction is zero, consequently (Section 3.2.6):

$$I_{\mathrm{s}} = -\mu_0 H \tag{2.195}$$

from which we have

$$\mathrm{d}G_{\mathrm{s}} = \mu_0 VH\,\mathrm{d}H \tag{2.196}$$

and

$$G_{\mathrm{s}} = \int_H \mathrm{d}G_{\mathrm{s}} = G_{\mathrm{s}}(H = 0) + \frac{1}{2}\,\mu_0 H^2 V \tag{2.197}$$

In the normal phase, the free enthalpy is a constant because

$$I_{\mathrm{n}} = 0 \tag{2.198}$$

from which we have

$$G_{\mathrm{n}} = G_{\mathrm{s}}(H = 0) + \frac{1}{2}\,\mu_0 H_{\mathrm{c}}^2 V \tag{2.199}$$

The actual phase always corresponds to the smallest value of G. It can be seen from (2.197) and (2.199) that $H < H_{\mathrm{c}}$ results in $G_{\mathrm{s}} < G_{\mathrm{n}}$, and the material is consequently in the superconducting state. Conversely, $H > H_c$ results in $G_s > G_n$, and the material is then in the normal state. $G_s = G_n$ corresponds to the limit $H = H_c$, when the transition from one state to another is in progress.

Once the Meissner effect is known, thermodynamics provides a very simple global model of the transition. It is used below to determine the latent heat L of the transition.

Let T_c be the transition temperature and H_c be the corresponding value of the magnetic field. In order to calculate L it is necessary to consider an infinitesimal displacement on the characteristic $H_c(T_c)$ during the transition (Fig. 2.53).

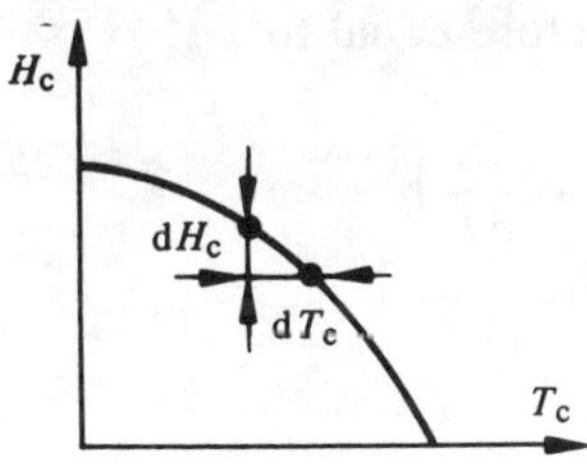

Fig. 2.53

For this displacement:

$$dG_s = dG_n \tag{2.200}$$

From (2.194), we have

$$-VI_s\, dH_c - S_s\, dT_c = -VI_n\, dH_c - S_n\, dT_c \tag{2.201}$$

Due to (2.195) and (2.198), expression (2.201) reduces to

$$S_n - S_s = -V\mu_0 H_c \frac{dH_c}{dT_c} \tag{2.202}$$

Now, by definition of L:

$$L = T\Delta S = T(S_n - S_s) \tag{2.203}$$

from which

$$L = -\mu_0 T_c V H_c \frac{dH_c}{dT_c} \tag{2.204}$$

This heat is certainly positive. In fact, it can be seen (Fig. 2.53) that $dH_c/dT_c < 0$. Experiment confirms the validity of (2.204). This equation also shows that $L = 0$ when H_c or T_c vanish. A phase transformation with latent heat equal to zero (second-order transformation) is characterized by a zero variation of the entropy and a discontinuity in the specific heat C. By definition:

$$C = \frac{dq}{dT} = T\frac{dS}{dT} \tag{2.205}$$

The discontinuity is therefore equal to

$$C_n - C_s = T_c\left(\frac{dS_n}{dT_c} - \frac{dS_s}{dT_c}\right) \tag{2.206}$$

or, because of (2.202):

$$C_n - C_s = -V\mu_0 T_c\left[H_c\frac{d^2H_c}{dT_c^2} + \left(\frac{dH_c}{dT_c}\right)^2\right] \tag{2.207}$$

$H_c = 0$ corresponds to $T_c = T_{c0}$, therefore,

$$C_n - C_s = -V\mu_0 T_{c0}\left(\frac{dH_c}{dT_c}\right)^2 \tag{2.208}$$

By way of illustration, Fig. 2.54 shows the discontinuity observed in tin.

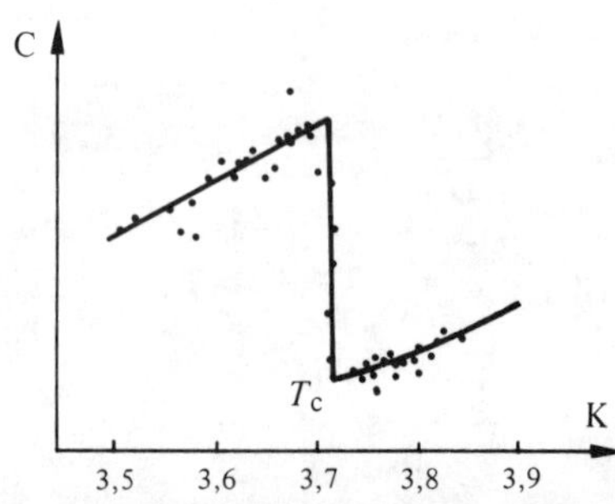

Fig. 2.54 Specific Heat of Tin in the Neighborhood of T_c. (After [15].)

2.8.8 Electrodynamic Behavior of a Superconductor

The BCS theory has shown that in a superconductor, only a fraction of the conduction electrons move without dissipation of energy. The index sc will denote the quantities relating to these electrons called *superconducting electrons*. The remaining fraction of conduction electrons behaves as in a normal metal. The quantities relating to these electrons called *normal electrons* are denoted by an index number.

Some phenomenological theories have already accounted for this fact, in particular the two-fluid theory of Gorter and Casimir [16] as well as London's theory which gives the classical description of the electrodynamic behavior of superconductors.

Maxwell's equations and the expressions that can be derived from them do not contradict superconductivity but they are inadequate to describe it. Let us consider, for example, Ohm's law:

$$\boldsymbol{J} = \sigma \boldsymbol{E} \tag{2.209}$$

The current density is necessarily finite because there is a finite number of charge carriers and their velocity is also finite. On the other hand, the conductivity is infinite which implies that $\boldsymbol{E} = 0$. Nothing can therefore be derived from (2.209) whose right-hand side of the the form $0 \cdot \infty$.

Similarly, the expression:

$$\nabla \times \boldsymbol{E} = -\frac{\partial \boldsymbol{B}}{\partial t} \tag{2.210}$$

implies, since $\boldsymbol{E} = 0$, the following condition:

$$\boldsymbol{B} = \text{constant} \tag{2.211}$$

This result does not contradict the Meissner effect but it does not imply this effect.

The *London equations* complete the Maxwell relationships for the description of superconductors.

2.8.9 The London Equations

London's first equation replaces (2.209) for the calculation of the current density due to superconducting electrons. It is established in the following manner. In the case of superconducting electrons, the equation for the motion of the electrons (2.15) is reduced to the following expression:

$$m_{\mathrm{n}} \frac{\mathrm{d}v_{\mathrm{d}}}{\mathrm{d}t} = -e\boldsymbol{E} \tag{2.212}$$

The current density associated with them is always given by (2.18) which here takes the form:

$$J_{\mathrm{sc}} = -eN_{\mathrm{sc}}v_{\mathrm{d}} \tag{2.213}$$

By combining (2.212) and (2.213), we obtain the first London equation:

$$\frac{\mathrm{d}J_{\mathrm{sc}}}{\mathrm{d}t} = \frac{1}{\mu\lambda^2}\boldsymbol{E} \tag{2.214}$$

with

$$\lambda^2 = \frac{m_{\mathrm{n}}}{e^2 N_{\mathrm{sc}} \mu} \tag{2.215}$$

The physical meaning of λ will appear in the next section; μ is the absolute permeability of the material. In order to avoid any ambiguity, it should be noted that we are dealing here with the true permeability of the material and not a fictitious permeability as introduced in Section 2.8.7 to describe the Meissner effect in terms of diamagnetism. Since superconductors are not ferromagnetic, we have $\mu = \mu_0$ in practice.

London's first equation shows that the electric field only differs from zero when the current density varies with time. This is a partial explanation of the losses produced in superconductors subjected to varying currents. In fact, the electric field accelerates both the superconducting electrons and the normal electrons. It is the collisions of the normal electrons with the lattice that are responsible for the Joule effect. However, these losses remain small with respect to those that appear in an ordinary conductor because the electric field in a superconductor remains very small, as the following example shows. In tin ($\lambda = 50$ nm), a current density with an amplitude of 10^4 A/m^2 at a frequency of 1 MHz gives, because of (2.214), $E = 0.2 \cdot 10^{-9}$ V/m. The same current density in tin in the normal state, at 20 K, corresponds to $E = 0.11 \cdot 10^{-4}$ V/m.

The simultaneous existence of superconducting and normal electrons makes it possible to represent a superconductor section by means of the equivalent circuit of Fig. 2.55.

The top branch of the circuit corresponds to the path of the superconducting electrons and the bottom branch to that of the normal electrons.

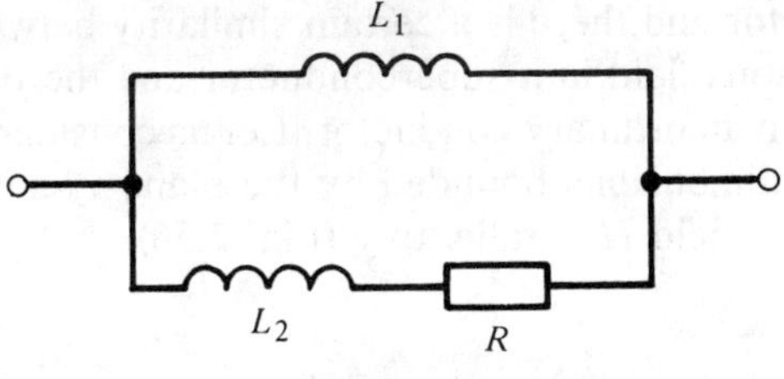

Fig. 2.55

The second London equation relates the magnetic field to the current density. It is derived from the first London equation by the following calculation. By taking the rotational of both sides of (2.214), we obtain

$$\nabla \times \frac{\mathrm{d}\boldsymbol{J}_{\mathrm{sc}}}{\mathrm{d}t} = \frac{e^2 N_{\mathrm{sc}}}{m_{\mathrm{n}}} \nabla \times \boldsymbol{E} \tag{2.216}$$

By replacing $\nabla \times \boldsymbol{E}$ in (2.216) with its value obtained from one of Maxwell's equations, we have

$$\nabla \times \frac{\mathrm{d}\boldsymbol{J}_{\mathrm{sc}}}{\mathrm{d}t} = -\frac{e^2 N_{\mathrm{sc}}}{m_{\mathrm{n}}} \mu \frac{\partial \boldsymbol{H}}{\partial t} \tag{2.217}$$

from which, by integrating over time:

$$\nabla \times \boldsymbol{J}_{\mathrm{sc}} = -\frac{e^2 N_{\mathrm{sc}}}{m_{\mathrm{n}}} \mu (\boldsymbol{H} - \boldsymbol{H}_0) \tag{2.218}$$

In order to include the Meissner effect in this expression, it is sufficient to assume that the integration constant H_0 is equal to zero. We thus obtain London's second equation:

$$\nabla \times \boldsymbol{J}_{\mathrm{sc}} = -\frac{e^2 N_{\mathrm{sc}}}{m_{\mathrm{n}}} \mu \boldsymbol{H} = -\frac{1}{\lambda^2} \boldsymbol{H} \tag{2.219}$$

2.8.10 Penetration Depth of the Current under Steady-State Conditions

The eddy currents, invoked to explain the Meissner effect, flow in a layer with small but necessarily finite thickness underneath the surface of the superconductor. The same applies to the current imposed by an external generator. The magnetic field is therefore not *totally* excluded from the superconductor and there is a certain similarity between the penetration of a continuous field in a superconductor and the penetration of an alternating field in an ordinary conductor. Let us consider a superconductor with infinite dimensions bounded by the plane xy and subjected to a constant magnetic field H parallel to y (Fig. 2.56).

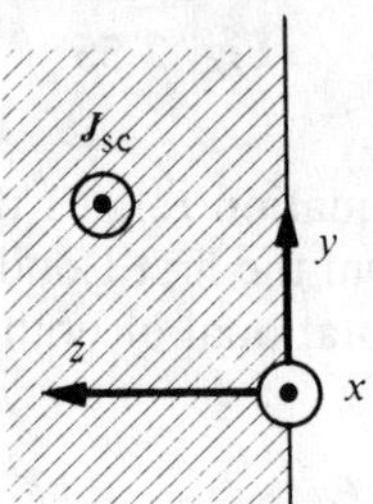

Fig. 2.56

The variation of H as a function of z is given by London's second equation. Let us take the rotational of (2.219):

$$\nabla \times \nabla \times \boldsymbol{J}_{sc} = \nabla(\nabla \cdot \boldsymbol{J}_{sc}) - \nabla^2 \boldsymbol{J}_{sc} = -\frac{e^2 N_{sc}}{m_n} \mu \nabla \times \boldsymbol{H} \qquad (2.220)$$

Now,

$$\nabla \cdot \boldsymbol{J}_{sc} = 0 \qquad (2.221)$$

On the other hand, under steady-state conditions $\boldsymbol{E} = 0$ (2.214); therefore, $\boldsymbol{J} = \boldsymbol{J}_{sc}$. Consequently,

$$\nabla \times \boldsymbol{H} = \boldsymbol{J}_{sc} \qquad (2.222)$$

In the case of Fig. 2.56, (2.220) therefore becomes

$$\frac{d^2 \boldsymbol{J}_{sc}}{dz^2} = \frac{e^2 N_{sc}}{m_n} \mu \boldsymbol{J}_{sc} = \frac{1}{\lambda^2} \boldsymbol{J}_{sc} \qquad (2.223)$$

The general solution of (2.223) has the form:

$$\boldsymbol{J}_{sc} = \boldsymbol{J}_0 \exp(-z/\lambda) + \boldsymbol{J}_1 \exp z/\lambda \qquad (2.224)$$

where $\boldsymbol{J}_0$ and $\boldsymbol{J}_1$ are the integration constants. It is obvious that $\boldsymbol{J}_{sc}$ cannot tend to infinity even if the plate is infinitely thick so that (2.224) reduces to

$$\boldsymbol{J}_{sc} = \boldsymbol{J}_0 \exp(-z/\lambda) \tag{2.225}$$

The quantity λ (Fig. 2.57) therefore represents the penetration depth of the direct current in a superconductor. On the order of 50 nm at $T \approx 0$ K in most pure metals, λ does not depend on H. On the other hand, the penetration depth varies as a function of temperature, mainly in the neighborhood of the critical temperature (Fig. 2.58). A law of the type:

$$\lambda = \frac{\lambda_0}{\sqrt{1 - \left(\frac{T_c}{T_{c0}}\right)^4}} \tag{2.226}$$

may be accepted.

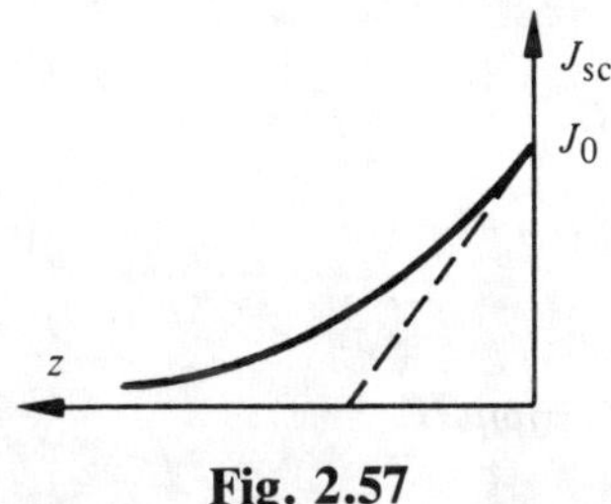

Fig. 2.57

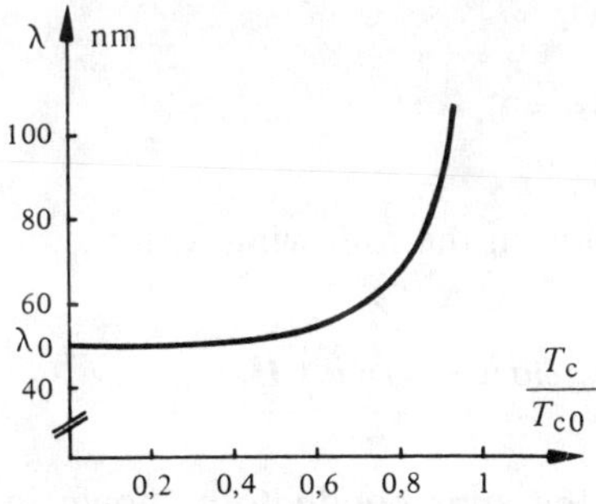

Fig. 2.58 The Function (2.226) for Tin

2.8.11 Penetration Depth of the Magnetic Field under Steady-State Conditions

In this section and the next, we will write the useful quantities in the phasor notation. As usual, the angular velocity is denoted by ω, the permittivity of the material by ε, and its conductivity by σ.

Under dynamic conditions, the current has three components. The first two are due to the superconducting and normal electrons, the last being the displacement current. Maxwell's second equation therefore takes the form:

$$\nabla \times \underline{\boldsymbol{H}} = \underline{\boldsymbol{J}}_{sc} + \underline{\boldsymbol{J}}_{no} + j\omega\varepsilon\underline{\boldsymbol{E}} \tag{2.227}$$

The rotational of (2.227) makes it possible to use the second London equation. This rotational can be written

$$\nabla \times \nabla \times \underline{\boldsymbol{H}} = \nabla \times \underline{\boldsymbol{J}}_{sc} + \nabla \times \underline{\boldsymbol{J}}_{no} + j\omega\varepsilon\nabla \times \underline{\boldsymbol{E}} \tag{2.228}$$

In this expression:

$$\nabla \times \underline{\boldsymbol{J}}_{no} = \sigma\nabla \times \underline{\boldsymbol{E}} \tag{2.229}$$

and

$$\nabla \times \underline{\boldsymbol{E}} = -\frac{\partial \underline{\boldsymbol{B}}}{\partial t} = -j\omega\mu\underline{\boldsymbol{H}} \tag{2.230}$$

By expressing $\nabla \times \underline{\boldsymbol{J}}_{sc}$ by means of London's second equation and allowing for the fact that

$$\nabla \cdot \underline{\boldsymbol{H}} = \frac{1}{\mu}\nabla \cdot \underline{\boldsymbol{B}} = 0, \tag{2.231}$$

(2.228) is finally written in the following form:

$$\nabla^2 \cdot \underline{\boldsymbol{H}} = \left(\frac{1}{\lambda^2} + j\omega\sigma\mu - \omega^2\varepsilon\mu\right)\underline{\boldsymbol{H}} \tag{2.232}$$

In the case of the superconductors shown in Fig. 2.56 and for a planar wave propagating along a direction perpendicular to y, (2.232) becomes

$$\frac{\partial^2 \underline{\boldsymbol{H}}}{\partial x^2} + \frac{\partial^2 \underline{\boldsymbol{H}}}{\partial z^2} = \left(\frac{1}{\lambda^2} + \mathrm{j}\omega\sigma\mu - \omega^2\varepsilon\mu\right) \underline{\boldsymbol{H}} \tag{2.233}$$

This equation has a solution of the type:

$$\underline{\boldsymbol{H}} = \underline{\boldsymbol{H}}_0 \exp(\alpha x) \exp(\gamma z) \tag{2.234}$$

By introducing this expression in (2.233), we have

$$\alpha^2 + \gamma^2 = \frac{1}{\lambda^2} + \mathrm{j}\omega\sigma\mu - \omega^2\varepsilon\mu \tag{2.235}$$

An interesting case is when the plane xy serves to guide the wave. In this case, we must have: $\alpha^2 \ll \gamma^2$. By neglecting the displacement current, (2.234) reduces to

$$\underline{\boldsymbol{H}} = \underline{\boldsymbol{H}}_y(z) = \underline{\boldsymbol{H}}_0 \exp\left[-\left(\sqrt{\frac{1}{\lambda^2} + \mathrm{j}\omega\sigma\mu}\right) z\right] \tag{2.236}$$

It can be easily verified that for $\omega = 0$, (2.236) agrees with (2.225) and that

$$\underline{\boldsymbol{J}}_0 = \frac{\underline{\boldsymbol{H}}_0}{\lambda} \tag{2.237}$$

2.8.12 Surface Impedance of a Superconductor

An example of application of the above considerations is the calculation of the surface impedance $\underline{Z}_s$ of a superconductor. Let us also consider the case of a superconductor with infinite dimensions bounded by the plane xy (Fig. 2.59).

The surface impedance of a section of length L traversed by a current $\underline{I}$ per unit length along x is equal to

$$\underline{Z}_s = \frac{E_x(z = 0)}{\underline{I}} \tag{2.238}$$

This is an important parameter in the study of the properties of microstrip lines (Vol. XIII). In order to evaluate (2.238), it is sufficient to express the numerator and the denominator as a function of $\underline{\boldsymbol{H}}$.

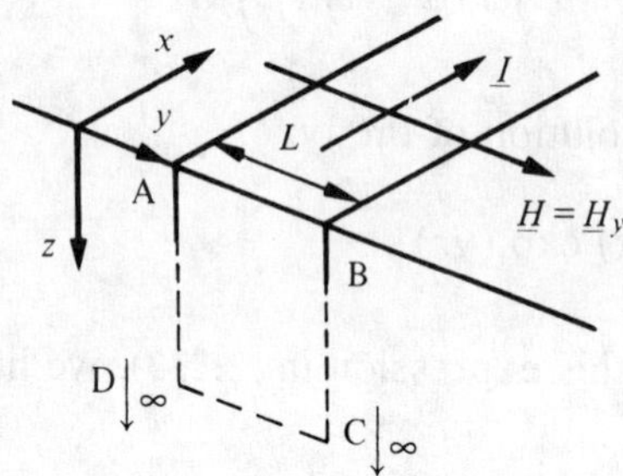

Fig. 2.59

By Ampere's theorem, over the path ABCDA, we have

$$\int_0^L \underline{\boldsymbol{H}}_y(z = 0)\mathrm{d}y + \int_L^0 \underline{\boldsymbol{H}}_y(z = \infty)\mathrm{d}y = \underline{\boldsymbol{I}} \tag{2.239}$$

from which, due to (2.236), we obtain

$$\underline{\boldsymbol{H}}_0 L = \underline{\boldsymbol{I}} \tag{2.240}$$

The integrals over sections BC and DA are zero because $\underline{H}$ is perpendicular to the displacement.

Because of (2.227), still neglecting the displacement current and the attenuation of $\underline{H}$ along x compared to the attenuation of $\underline{H}$ along z, we obtain

$$\nabla \times \underline{\boldsymbol{H}} = -\frac{\partial \underline{\boldsymbol{H}}_y}{\partial z} = \underline{\boldsymbol{J}}_{\mathrm{no}} + \underline{\boldsymbol{J}}_{\mathrm{sc}} \tag{2.241}$$

By replacing $\underline{\boldsymbol{J}}_{\mathrm{no}}$ and $\underline{\boldsymbol{J}}_{\mathrm{sc}}$ in this expression with their values given by Ohm's law and London's first equation, respectively, we have

$$-\frac{\partial \underline{\boldsymbol{H}}_y}{\partial z} = \left(\sigma + \frac{1}{\mathrm{j}\omega\mu\lambda^2}\right) \underline{\boldsymbol{E}}_x \tag{2.242}$$

Using (2.236) to calculate the first term of this equation, we have

$$\underline{E}_x(z = 0) = \frac{\sqrt{\dfrac{1}{\lambda^2} + \mathrm{j}\omega\sigma\mu}}{\sigma + \dfrac{1}{\mathrm{j}\omega\mu\lambda^2}} \underline{H}_0 = \frac{\mathrm{j}\omega\mu}{\sqrt{\dfrac{1}{\lambda^2} + \mathrm{j}\omega\sigma\mu}} \underline{H}_0 \tag{2.243}$$

from which we have

$$\underline{Z}_s = \frac{1}{L} \frac{j\omega\mu}{\sqrt{\frac{1}{\lambda^2} + j\omega\sigma\mu}} \tag{2.244}$$

The penetration depth δ of a normal current (Vol. III, Sec. 6.1.2) is given by

$$\delta = \sqrt{\frac{2}{\omega\sigma\mu}} \tag{2.245}$$

By introducing δ into (2.244), we put this equation in a more symmetric form:

$$\underline{Z}_s = \frac{1}{L} \frac{j\omega\mu}{\sqrt{\frac{1}{\lambda^2} + j\,\frac{2}{\delta^2}}} \tag{2.246}$$

in which the role of the two penetration depths δ and λ are clearly revealed. The depth δ varies as a function of frequency whereas, in practice, λ only depends on the temperature. At moderate frequency, except in the neighborhood of the critical temperature, $\lambda \ll \delta$ and (2.246) reduces to

$$\underline{Z}_s \cong \frac{1}{L} j\omega\mu\lambda \tag{2.247}$$

$\underline{Z}_s$ is therefore a pure inductance. There is, therefore, no dissipation of energy at the surface of the superconductor. At high frequency or close to the critical temperature, λ becomes much greater than δ, so that (2.246) can be written

$$Z_s \cong \frac{1}{L} \frac{j\omega\mu}{\sqrt{j\,\frac{2}{\delta^2}}} = \frac{1}{\sqrt{2}L} \sqrt{\frac{\omega\mu}{\sigma}}\,(1 + j) \tag{2.248}$$

2.8.13 The BCS Theory

The BCS theory has proved the simultaneous existence of normal and superconducting electrons and provides a quantum description of the behavior of the latter. Presented for the first time in two articles, now famous [17], this theory has been reproduced in several works (see, for example, [18]). We shall limit ourselves here to presenting the most important aspects of this theory.

The parameter that distinguishes normal electrons from superconducting electrons is obviously the energy. Below T_{c0}, a forbidden band of width ΔW is centered on the Fermi energy (Fig. 2.60).

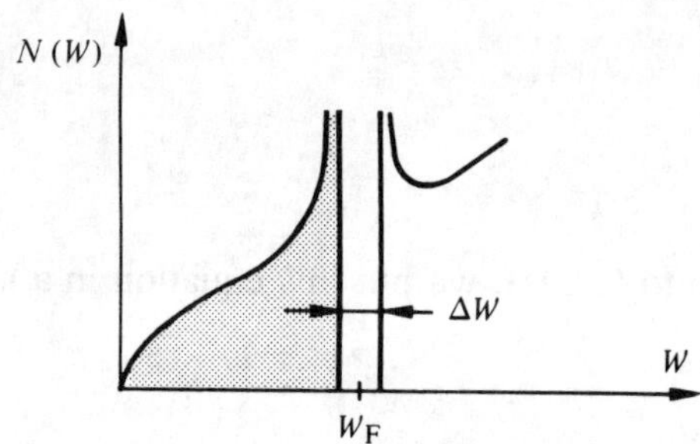

Fig. 2.60 Occupation of Energy Levels in a Superconductor at 0 K

The superconducting electrons occupy the states for which $W < W_F - \Delta W/2$, and the normal electrons occupy the states where $W > W_F + \Delta W/2$. Although the appearance of the forbidden band modifies the form of $N(W)$, the number of states situated below W_F does not vary, so that at 0 K all the electrons can be superconducting electrons. At a finite temperature, a fraction of the electrons are excited and pass to the normal state by crossing the forbidden band. The width of this band varies continuously as a function of temperature. It is zero at the critical temperature and increases when the temperature drops below T_{c0} (Fig. 2.61). It reaches $\alpha k_B T_{c0}$ at 0 K. According to the theory, $\alpha = 3.52$. Experimental values are close to this number, as shown by Table 2.62.

The appearance of the forbidden band and the lowering of energy which accompanies it for a fraction of the electrons corresponds to a particular type of electron-phonon interaction. A picture of this interaction can be given as follows.

At $T < T_c$, the vibrational movements of the lattice are small enough that the deformation of the lattice produced by a collision from an electron e_1 can be experienced by a second electron e_2 relatively far from e_1. These coupling between e_1 and e_2 produced in this way lowers the energy of these two electrons, which, because of this, become bound to each other. They form what is called a *Cooper pair*. The electrons of the pair have opposite spins. Their momenta are equal and opposite, being equal to $\hbar \boldsymbol{k}$ and $-\hbar \boldsymbol{k}$, respectively, where $\boldsymbol{k}$ is the wave vector. When a current

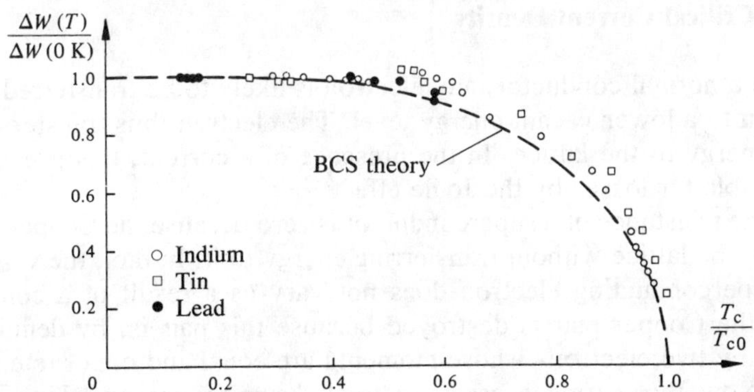

Fig. 2.61 Normalized Width of the Forbidden Band *versus* Temperature. Experimental Results for In, Sn, and Pb, Compared to the BCS Theory (after [19]).

Table 2.62

Element	ΔW *at 0 K* 10^{-3} *eV*	$\alpha = \frac{\Delta W\ (0\ K)}{k_B T_{c0}}$	*Element*	ΔW *at 0 K* 10^{-3} *eV*	$\alpha = \frac{\Delta W\ (0\ K)}{k_B T_{c0}}$
Al	0.34	3.3	Pb	2.73	4.38
In	1.05	3.6	V	1.6	3.4
Sn	1.15	3.5	Nb	3.05	3.8

corresponding to a drift velocity $\boldsymbol{v}_d$ exists, the Cooper pairs are formed from electrons whose momenta are $m_n\boldsymbol{v}_d + \hbar\boldsymbol{k}$ and $m_n\boldsymbol{v}_d - \hbar\boldsymbol{k}$, respectively.

The *coherence length* ξ is the distance over which the attractive force binding the two electrons of the pair is exerted. The value of ξ is surprisingly large: on the order of 10^{-8} to 10^{-7} m which corresponds to about 100 interatomic distances.

The role of the lattice in the formation of Cooper pairs is confirmed by the fact that for a given element, the critical temperature is a function of the atomic weight A, i.e., of the isotopic composition of the specimen:

$$T_c \sim A^{-a}, \tag{2.249}$$

where a is close to 0.5 for elements other than the transition metals.

2.8.14 Critical Current Density

In a normal conductor, any electron is likely to be transferred by a collision to a lower vacant energy level. The electron thus transfers part of its energy to the lattice. In the presence of a current, this process is responsible for losses by the Joule effect.

The resistance of a superconductor is zero because the Cooper pairs move in the lattice without transferring energy to it. In fact, the velocity of a superconducting electron does not vary as a result of a collision unless the Cooper pair is destroyed because this pair is, by definition, formed by two electrons whose momenta are equal and opposite except for $\boldsymbol{v}_d$. Now, the energy necessary to break the pair is not available as long as the drift velocity $\boldsymbol{v}_d$ remains less than a critical velocity $\boldsymbol{v}_c$.

We shall calculate $\boldsymbol{v}_c$ at 0 K for the case of a strip with thickness appreciably less than the penetration depth λ so that the current can be considered uniformly distributed over the entire thickness. Let xy be the plane of the strip and v_x and v_y be the components of the electron velocity along the axes.

In the absence of current, the representative points of all the electrons in the plane v_x, v_y lie inside a circumference whose radius v_F corresponds to the velocity of the electrons at the Fermi energy (Fig. 2.63) or, more precisely, the energy $W_F - \Delta W/2 \approx W_F$ (ΔW is approximately 1000 times smaller than W_F).

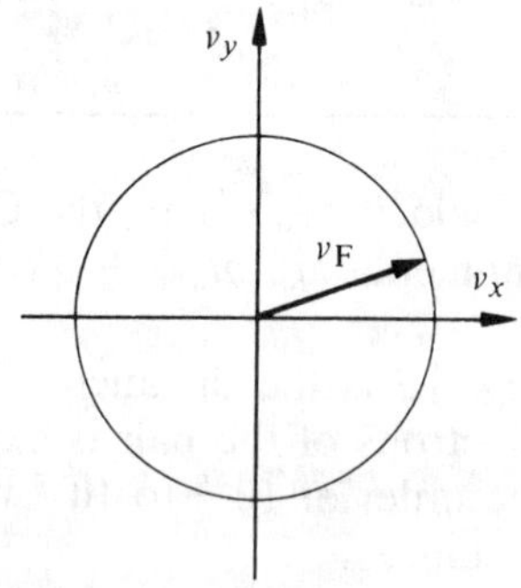

Fig. 2.63

In the presence of a current density $\boldsymbol{J}$ oriented along x, the circumference is translated by a quantity v_d along the x-axis (Fig. 2.64). The maximum energy that the lattice can receive under the collision of a *normal electron* corresponds in Fig. 2.64 to the transfer of an electron

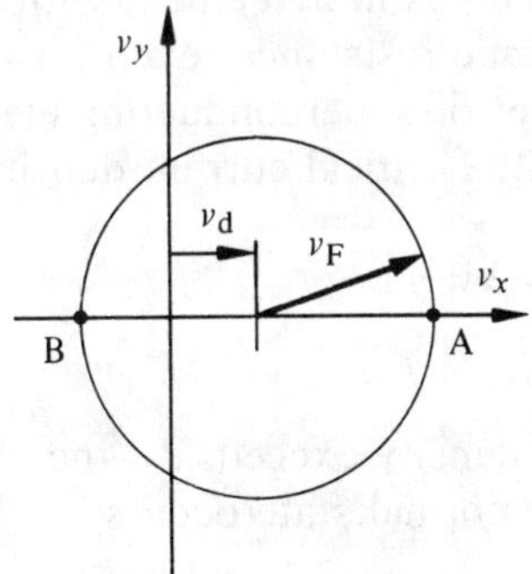

Fig. 2.64

from point A to point B. During this collision, the variation in energy δW_n of the electron is equal to

$$\delta W_n = \frac{1}{2} m_n(v_F - v_d)^2 - \frac{1}{2} m_n(v_F + v_d)^2 = -2m_n v_F v_d \tag{2.250}$$

Let us now assume that the electron corresponding to the point A is a *superconducting electron*. It therefore forms a Cooper pair with the electron corresponding to point B. The variation in energy δW_s of the superconducting electron corresponding to a transfer from point A to point B is

$$\delta W_s = \Delta W - 2m_n v_F v_d \tag{2.251}$$

because the Cooper pair would be destroyed. Now, δW_s cannot be negative because the electron considered has the maximum velocity. By choosing the point B as representative of the electron after collision, we have already obtained the minimum of δW_s. For $\delta W_s < 0$, it is therefore necessary in all cases that

$$v_d \geqslant \frac{\Delta W}{2m_n v_F} = v_c \tag{2.252}$$

In conclusion, as long as the drift velocity of the Cooper pairs remains less than a limiting velocity v_c, the Cooper pairs are displaced without transferring energy to the lattice. This situation corresponds to the superconducting state. As soon as v_d exceeds v_c, the Cooper pairs are destroyed and the transition from the superconducting state to the normal state occurs. Above the critical temperature, ΔW is equal to zero and consequently $v_c = 0$. This is in agreement with the fact that the normal conductors have an ohmic resistance, even for the smallest currents.

If the total number of superconducting electrons n_{sc} is known, we can deduce from (2.252) a critical current density J_c at $T = 0$ K:

$$J_c = n_{sc} e v_c = \frac{n_{sc} e \Delta W}{2 m_n v_F} \tag{2.253}$$

As soon as the current density exceeds J_c, the transition from the superconducting state to the normal state occurs.

2.8.15 Relationship among Critical Field, Critical Current, and Critical Temperature

It can be concluded from the preceding section that the critical magnetic field is nothing other than the field that produces eddy currents with a density that exceeds J_c. For a given material, the critical field varies according to the form of the specimen and its orientation with respect to the field because the intensity of the eddy currents depends on these two parameters.

In the presence of a transport current (i.e., a current due to an external generator), it is the sum of the eddy current density and transport current density that determines the initiation of the transition to the normal state.

A superconductor with a given form (for example, a rectilinear wire) subjected to a field with given orientation may therefore be described, in terms of the transition, by a diagram of the type shown in Fig. 2.65, which is a generalization of Fig. 2.50.

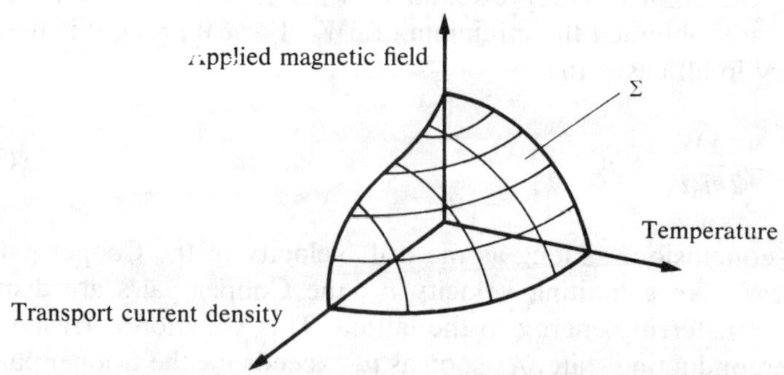

Fig. 2.65 The Points Situated Between Σ and the Coordinate Planes Correspond to the Superconducting State

2.8.16 Spatial Variation of the Free Enthalpy

Under steady-state conditions, the transition zone separating a normal region from a superconducting region is stationary. The free energy G is consequently the same on each side of this zone. In fact, if G were different, the region with lower G would have a tendency to increase to the detriment of the region with higher G, in this way creating a movement of the transition zone. Let G_0 be the free enthalpy outside this zone.

Inside the transition zone, the free enthalpy is likely to be higher or, on the contrary, lower than G_0.

Since the temperature is uniform, only two components of G vary inside the transition zone. These are

- the magnetic energy w_{mag}, given by (2.190);
- the mean energy of the electrons w_n, which has not been taken into account in the simplified thermodynamic model of Section 2.8.7 but which plays a very important role in the intermediate state.

In the normal region, the magnetic field freely penetrates, the magnetic polarization is zero and with it w_{mag}. The mean energy of the electrons is a maximum because there are no Cooper pairs.

When the transition zone is traveled through from the normal region to the superconducting region, the magnetic energy increases because the magnetic polarization gradually increases until it is equal to $-\mu_0 H$, in this way cancelling the induction B. This variation of w_{mag} essentially takes place over the penetration depth λ. In parallel, the mean energy of the electrons reduces as the number of paired electrons increases. This variation is essentially produced over the coherence length ξ.

Two different situations are observed depending on whether the ratio $K = \lambda/\xi$, called the Ginsburg-Landau factor [20], is large or small. In Fig. 2.66, the penetration depth λ of the magnetic field is less than the coherence length ξ. The result of this is an increase in the free enthalpy in the transition zone. On the contrary, in Fig. 2.67, $\lambda > \xi$, and the free enthalpy decreases in the transition zone.

In this case, the material will tend to evolve to a state with less energy by multiplying the transition zones. Thus we find it finely divided into normal and superconducting regions (Fig. 2.68). The normal regions in which the magnetic induction is different from zero have the form of filaments. They are called *fluxoids* or *vortices* because they are surrounded by eddy currents.

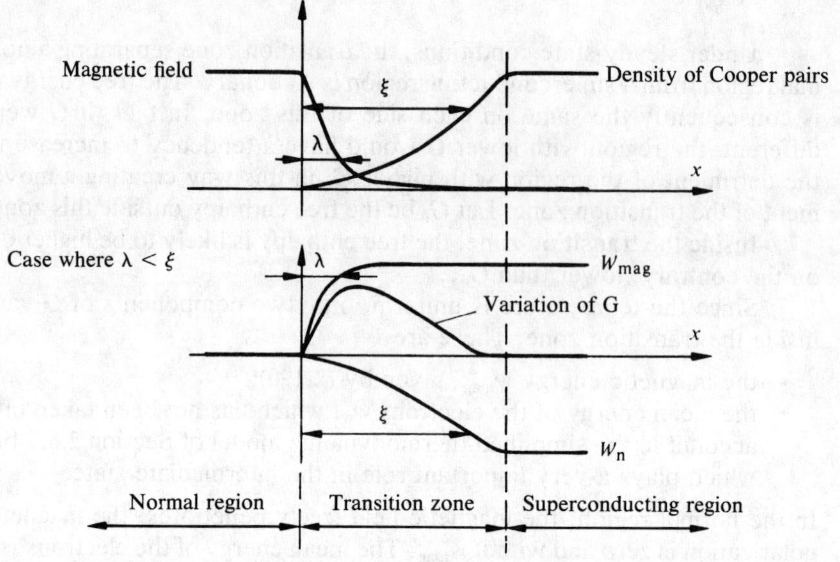

Fig. 2.66

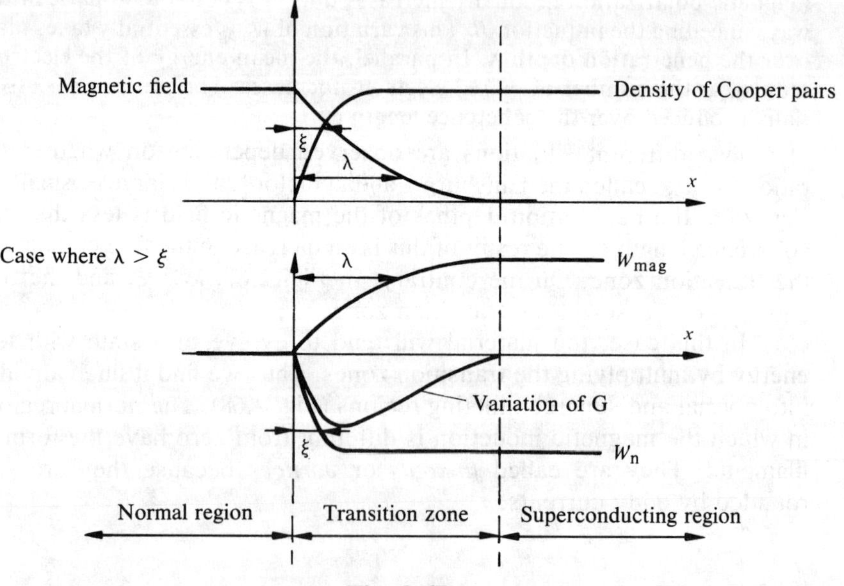

Fig. 2.67

Fig. 2.68 Fluxoids and Eddy Currents. The Zones in Which B = 0 are White and the Zones in Which the Applied Field Freely Penetrates are in Black (After [21].)

2.8.17 Fluxoid Structure

The central region of the fluxoid is always in the normal phase. The total number of Cooper pairs is zero on the fluxoid axis and increases with the radius as shown by Fig. 2.69. The fluxoid radius is approximately equal to the coherence length.

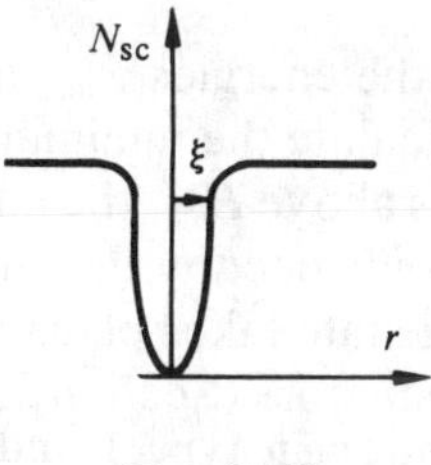

Fig. 2.69

The eddy currents appear above this radius and their density has the variation shown diagrammatically in Fig. 2.70. The effect of these currents, due exclusively to the movement of the Cooper pairs, is to progres-

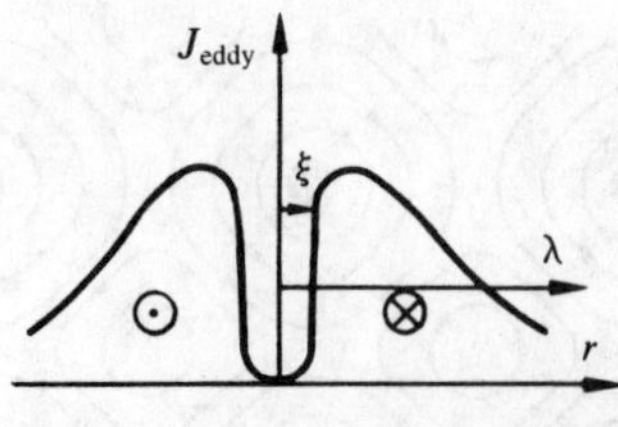

Fig. 2.70

sively cancel B whose penetration depth will therefore be limited to λ (Fig. 2.71).

It can be shown [21] that the flux associated with each vortex is the same. This flux quantum Φ_q is equal to

$$\Phi_q = \frac{h}{2e} = 2.07 \cdot 10^{-15} \text{ Wb} \tag{2.254}$$

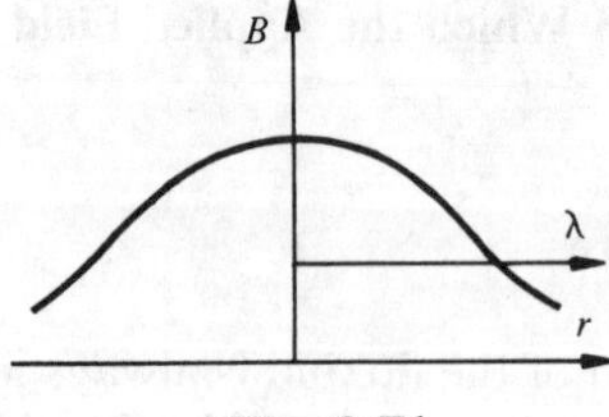

Fig. 2.71

2.8.18 Intermediate State

A detailed analysis of the energies w_{mag} and w_n involved in the fluxoids makes it possible to calculate the minimum field H_{c1} needed for them to appear. When H increases above H_{c1}, the number of fluxoids increases.

For $H = H_{c2}$, the fluxoids occupy the entire volume of the material and the return to the normal state takes place. Here we can recognize the behavior of type 2 superconductors (Section 2.8.6). As a mean value over space, the magnetic polarization in type 1 and type 2 superconductors, *versus* the applied field, has the form shown in Fig. 2.72.

The Ginsburg-Landau theory predicts that the values of the ratio λ/ξ are less than $1/\sqrt{2}$ for type 1 and greater than $1/\sqrt{2}$ for type 2.

Different methods allow normal and superconducting regions to be revealed in a specimen in the intermediate state. Fig. 2.73 shows the

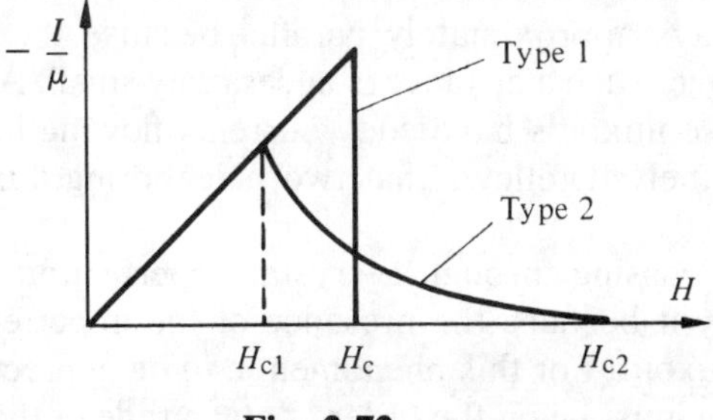

Fig. 2.72

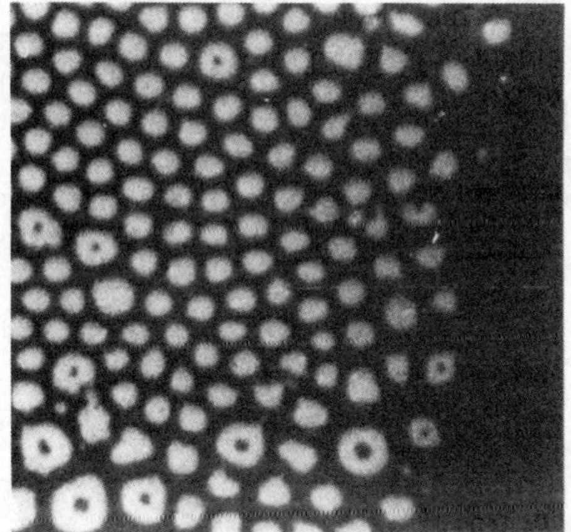

Fig. 2.73

surface of a piece of indium observed under the microscope (magnification 30×) using a magneto-optic technique [22]. The superconducting regions appear dark and the normal regions appear light. The magnetic field is perpendicular to the plane of the image. The normal regions with approximately circular shape each contain between 2000 and 5000 vortices.

2.8.19 Forces Acting on the Fluxoids

Different forces can act on the fluxoids. Some of them make them move; others, on the contrary, have the effect of confining them to certain points in the crystal. In this case, one talks of *pinning forces*. Three typical situations are described below.

Two neighboring fluxoids, i.e., separated by a distance of up to a few λ, are always approximately parallel because the variation in the magnetic field over such a distance is necessarily small. Along their line of "contact," these fluxoids have eddy currents flowing in opposite directions to each other. It follows that two neighboring fluxoids repel each other.

A fluxoid passing through a crystal imperfection tends to remain pinned down by it because the presence of the imperfection lowers the energy of the fluxoid. For this phenomenon to be appreciable, the size of the imperfection must be on the order of magnitude of the diameter of the fluxoid, i.e., approximately ξ. This means precipitates, grain boundaries, and dislocation clusters rather than point imperfections or isolated dislocations.

A fluxoid is subject to a Lorentz force if a transport current flows in its vicinity. By way of illustration, let us consider a superconducting strip subjected to a field $\boldsymbol{H}$ along z and traversed by a transport current along y (Fig. 2.74).

In this strip, each fluxoid is subjected to a force acting along x.

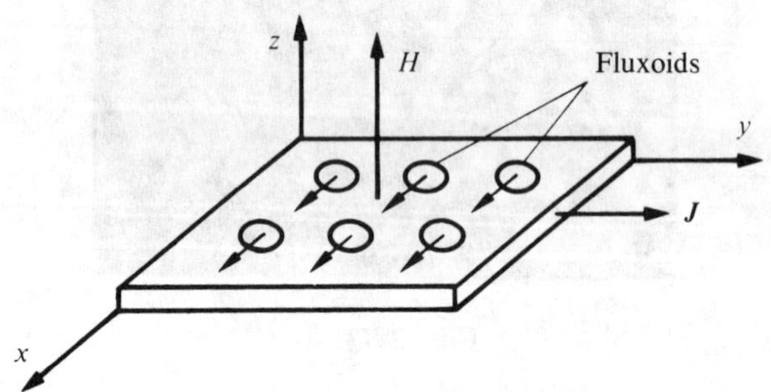

Fig. 2.74

2.8.20 Movement of Fluxoids and Dissipation of Energy

The fluxoids mainly move under the action of the Lorentz force. Movements take place for a variation in the transport current, in the applied magnetic field, or in both these variables at the same time. Under steady-state conditions, the fluxoids find an equilibrium position (at constant temperature) resulting from their mutual interactions and their interactions with the crystal imperfections.

It is crucial to control these movements because they are accompanied by a dissipation of energy which can be explained phenomenologically by the existence of a viscosity force impeding the movement of the fluxoids. It is therefore desirable to provide superconductors used under dynamic conditions with the most efficient possible pinning points.

Treatment making it possible to obtain such points (precipitation, cold-working, *et cetera*) makes the material hard and fragile. This is the reason why type 2 superconductors, with a structure that blocks the fluxoids, are called *hard superconductors*. For the same reason, magnetic materials with a structure that blocks the Bloch walls (Section 3.8.6) are called hard magnetic materials.

2.8.21 Stabilization of a Superconductor

Since fluxoids cannot be completely stopped from moving, measures must be taken for efficiently removing the energy dissipated by these movements. In fact, this energy appears in the material in the form of heat. If this heat is not extracted sufficiently rapidly, the temperature of the material can reach T_c. If it is not controlled, the return of the superconductor to the normal state can have very serious consequences due to the sudden appearance of very considerable losses by the Joule effect.

Stabilization of a superconductor is the name given to all the measures designed to avoid such an accidental return to the normal state.

The principle of stabilization involves placing the superconductor in the form of filaments into a mass of pure metal called the *matrix*. The matrix is nearly always made of copper. Two examples of stabilized superconductors are shown in Figs. 2.75 and 2.76.

Performed in this way, the stabilization allows small portions of a few filaments to return temporarily to the normal state without any adverse consequences. Stabilization essentially acts in two ways.

First, the presence of the matrix and the division of the superconductor into fine filaments considerably improve the transfer of heat between the superconductor and the cooling fluid. In fact, the thermal conductivity of the superconductor is very small compared to that of the matrix, as seen in Table 2.77.

Second, the matrix reduces the energy dissipated by the Joule effect in the vicinity of the superconducting mass, temporarily in the normal state, by offering the current an alternative path with low ohmic resistance. The resistivity of a superconducting alloy in the normal state is much higher than that of copper, (Table 2.77). The effect of stabilization is shown diagrammatically in Fig. 2.78.

The study of the optimum diameter of the filaments and their best distribution in the matrix may be carried out using different stability criteria, a review of which is given in [24].

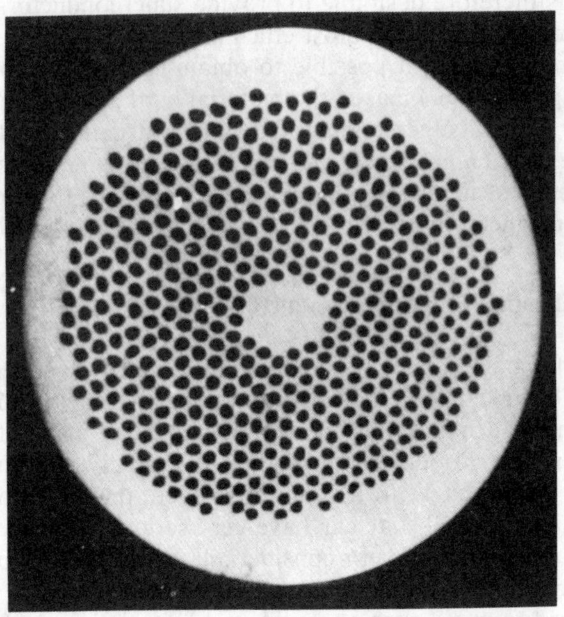

Fig. 2.75 Superconducting Wire Consisting of 480 Filaments of NbTi Each With a Diameter of 22 μm, Immersed in a Copper Matrix with a Diameter of 1 mm [23].

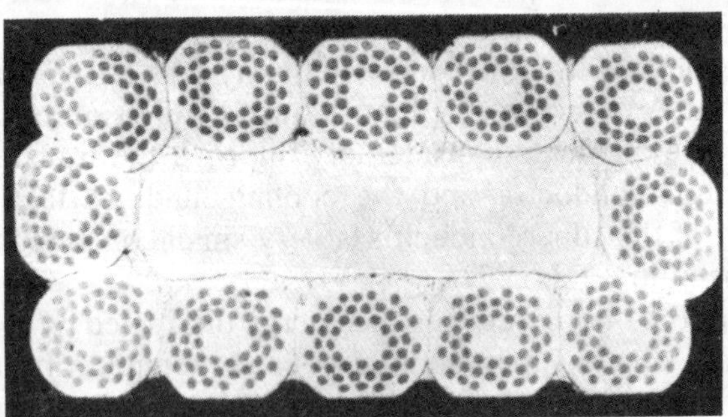

Fig. 2.76 Example of a Superconducting Cable Formed from Filament Elements Assembled by Tinning. Critical Current: 2070 A at 4.2 K and Under 5 T. Dimensions: 1.9 × 3.8 mm [23].

Table 2.77

	Superconducting alloy (in the normal state)	*Pure metal (e.g., electrolytic Cu)*
Thermal conductivity W/m K	**0.1–1**	**100–1000**
Resistivity $\Omega \cdot m$	**10^{-7}–10^{-6}**	**10^{-10}**

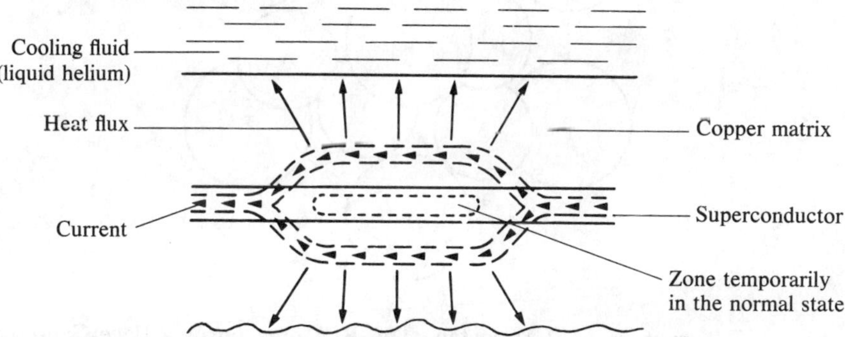

Fig. 2.78 Heat Flux and Circulation of the Current in the Vicinity of a Superconducting Filament, Part of Which is in the Normal State

2.8.22 Superconductivity and the Periodic Table

The BCS theory does not allow the prediction of the critical temperature of a superconductor nor of which elements or materials are superconductors. In this area, experiment is the main approach. The following facts have been observed [25].

No alkali metal has been found to be a superconductor, even at temperatures below 0.1 K. The same applies to the noble metals such as gold, silver, and copper.

The ferromagnetic metals are not superconductors. On the other hand, they are capable of considerably lowering the critical temperature of superconductors that contain them as impurities in very low concentrations. The effect of nonmagnetic impurities is much less marked.

Compounds with the highest critical temperatures contain a transition element, vanadium or niobium, and at normal temperatures have the crystal structure of β-tungsten (Fig. 2.79), corresponding to a chemical formula of the type A_3B.

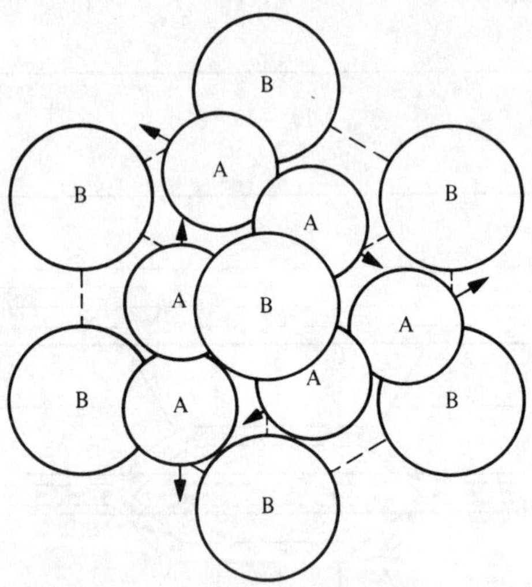

Fig. 2.79 Structure of β-Tungsten. The B Atoms Form a Face-Centered Cubic Lattice. The Atoms A are on the Faces of the Cube

2.8.23 The Josephson Effect

A system of two superconductors (SC1 and SC2) separated from each other by a thin insulating layer with a thickness δ on the order of 1–3 nm (Fig. 2.80) is called a *Josephson junction.*

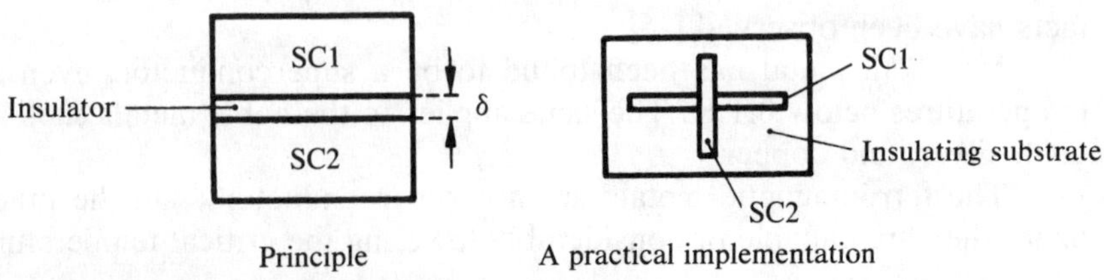

Fig. 2.80 The Josephson Junction

Such a junction is often obtained in the following manner : a first superconducting layer is deposited on an insulating substrate such as glass by evaporation. By letting this first deposit undergo surface oxidation, an insulating layer is obtained on which a second superconductor is deposited.

Above the critical temperature, a low current flows in the junction by the ordinary tunnel effect and an approximately linear current-voltage characteristic is observed. The junction behaves as a resistance.

Below T_c, the behavior of the junction changes completely. Because the coherence length is larger than the thickness of the insulator, the interaction forming the Cooper pairs takes place across it. The result is that a direct current, less than a limiting value I_c, can cross the junction without a potential difference appearing between the faces of the insulator. This phenomenon constitutes the *dc Josephson effect*.

When the current exceeds I_c, a potential difference U appears abruptly at the terminals of the junction (Fig. 2.81). Josephson [26] has shown that this voltage is related to the current I flowing in the junction by the expressions:

$$I = I_c \sin \varphi \tag{2.255}$$

and

$$U = \frac{\hbar}{2e} \frac{d\varphi}{dt} \tag{2.256}$$

where φ represents the phase difference between the wave functions on each side of the junction. This behavior of the junction is known as the *ac Josephson effect*. From (2.255) and (2.256), we derive

$$I = I_c \sin\left(\frac{2e}{\hbar} \int U \, dt\right) \tag{2.257}$$

which shows that if the junction is polarized by a direct voltage, the current is purely sinusoidal:

$$I = I_c \sin(\omega_0 t) \tag{2.258}$$

with

$$f_0 = \frac{\omega_0}{2\pi} = \frac{2e}{h} U; \qquad \frac{2e}{h} = 483.6 \text{ MHz}/\mu\text{V} \tag{2.259}$$

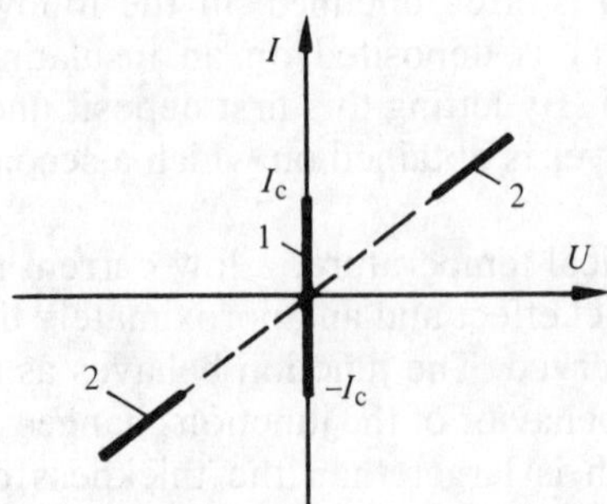

Fig. 2.81 1. DC Josephson Effect. 2. AC Josephson Effect

2.8.24 Applications of Superconductivity

The main obstacle to the development of the applications of superconductivity was the need, until now, to use liquid helium (4.2 K) as a cooling fluid. Despite this drawback, the unique properties of superconductors have given impetus to their use in a large number of applications.

The oldest of these is the production of very high magnetic fields in a volume that is sometimes quite considerable. The large bubble chamber of CERN, for example, is provided with a superconducting coil (NbTi), 4.7 m in diameter, producing a field of 3.5 tesla (T). Today, it is common to manufacture coils with smaller dimensions with fields reaching 20 T and above.

Prototypes of high-power dc motors and generators are being studied in several countries. The unipolar machine where only the stator is a superconductor, the rotor operating at room temperature, seems to be an interesting solution. Superconducting cables would be competitive with current conventional cables for transported powers above a few gigawatts but the problems associated with the production of a cryogenic enclosure of a few kilometres are not simple to solve.

The new superconductors that are cooled simply with liquid nitrogen, which is a cooling fluid used more often today than liquid helium, will considerably extend the applications of superconductivity in the near future. The fields of electronics, instrumentation, and communication will be the first ones affected.

Not all superconductivity applications involve high powers. The Josephson junction in particular allows the construction of devices with unique properties. We can give as examples:

- the SQUIDS (Superconducting Quantum Interference Devices) formed from one or two junctions placed in a superconducting ring.

They constitute ultrasensitive magnetometers capable of achieving a resolution of 10^{-15} T;

- voltage standards directly using the relationship (2.259) with precision several orders of magnitude greater than that of the classical Weston element;
- switching elements for logic circuits in computers. These components, with two states corresponding to the dc and ac Josephson effects, are expected to have computation speeds appreciably higher than those achieved with semiconductors.

2.9 PROBLEMS

2.9.1 Calculate the average velocity of electrons belonging to an ensemble in which the velocities are assumed to follow the Maxwell distribution.

2.9.2 Establish the expression for the Hall constant R_H, in the case where the conduction results from a simultaneous movement of electrons and holes. If R_H is negative, can one deduce from this that the total number of electrons is larger than the total number of holes?

2.9.3 Consider a conducting plate with thickness $2a$. Establish the field produced in this plate by a space charge varying linearly with x. Deduce from this that it is always possible to consider the space charge as constant inside a conductor.

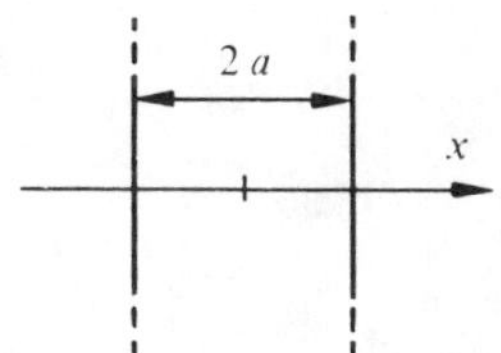

Fig. 2.82

2.9.4 Measurements on an aluminum specimen have given the following results: resistivity, 28 nΩm; Hall constant, $-0.43 \cdot 10^{-10}$ Vm^{3}A^{-1}Wb^{-1}. The density of aluminum is 2700 kg/m^3. Calculate the total number of electrons and their mobility. How many electrons per atom are involved in the conduction?

2.9.5 A semiconductor has the following characteristics: the mobility of the electrons is three times higher than that of the holes, the resistivity is 1.17 Ω · m, the total number of electrons is 10^{19}m^{-3} and of holes is 10^{20}m^{-3}. Calculate the mobility of the electrons and the holes.

2.9.6 A silicon specimen contains one donor atom per 10^6 silicon atoms. The acceptor concentration is one percent less than that of the donor concentration. Calculate the total number of electrons and holes (in order to simplify, take $m_n^* = m_p^* = m_n$).

Chapter 3
Magnetic Properties

3.1 HISTORICAL SURVEY

Magnetism has concerned mankind since the earliest time. There is even a legend relating to magnetic materials that would date their discovery by the Chinese as 2500 BC. Magnetic stone (magnetite, Fe_3O_4) is supposed to have been used at this time for the manufacture of rudimentary compasses guiding travelers in the desert.

If we limit ourselves to historical evidence, the first observation of magnetism dates from the sixth century BC. At this time, Thales of Milet notes that certain iron minerals from Magnesia (Asia Minor) attract iron. From the application point of view, the oldest text mentioning a magnetic needle is Chinese and dates from the end of the eleventh century. It took another century in order to find any similar mention in a European text. It is certain that the West inherited the compass from China in an already advanced form.

There have been many theories of magnetism. Lucretius (first century BC) had predicted in *De Rerum Natura* that the road that would lead to an understanding of the phenomena observed with magnetic stone would be very long. This prediction was to be highly relevant. In fact, the end is not in sight even today!

The letter on magnetic stone (1269) by Peregrinus of Maricourt can be considered as the first scientific contribution to knowledge on magnetism due to the experimental approach it detailed. Despite this very precise document, animistic theories continued to be developed for a long time. For example, in the sixteenth century, Francis Bacon was still writing: "for we see that there exists in almost all natural bodies a manifest force of discernment and even a kind of choice [whereby] they unite with friendly substances and flee enemy substances."

In 1895, Pierre Curie opened up the way to a modern theory of magnetism by distinguishing diamagnetism from paramagnetism and ferromagnetism. He also observed the transition from ferromagnetism to paramagnetism with an increase in temperature.

Although the compass truly revolutionized navigation in the middle ages, it is difficult to overestimate the importance of magnetic materials in contemporary society. One simply has to think of the production of electricity, which would be totally impossible on the scale we know it without the existence of ferromagnetic metals in the generators!

3.2 REMINDER OF DEFINITIONS AND NOTATION

3.2.1 Magnetic Induction and Magnetic Field in Vacuum

In vacuum, there is a property called the magnetic field or magnetic induction. It can be demonstrated at any point P by one of the following experiments (Figs. 3.1 and 3.2).

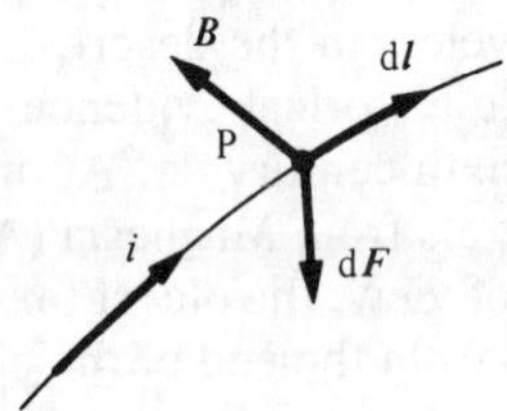

Fig. 3.1 Experiment No. 1

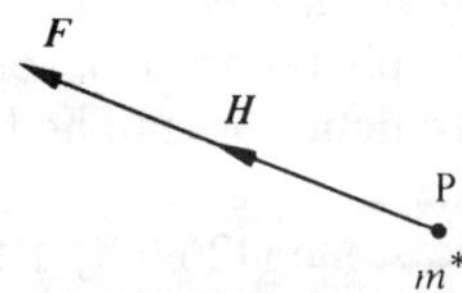

Fig. 3.2 Experiment No. 2

In experiment No. 1, a conductor carrying a current i is made to pass through P. The element of length d$\boldsymbol{l}$ of this conductor around P is subject to a force d$\boldsymbol{F}$ given by the Laplace law (Vol. I, Sec. 2.4):

$$d\boldsymbol{F} = i d\boldsymbol{l} \times \boldsymbol{B} \tag{3.1}$$

The force $d\boldsymbol{F}$ is a direct measure of the *magnetic induction* $\boldsymbol{B}$ with a unit called the tesla (T).

In experiment No. 2, a magnetic mass m^* is placed at P and this is subject to a force $\boldsymbol{F}$ given by Coulomb's law:

$$\boldsymbol{F} = m^*\boldsymbol{H} \tag{3.2}$$

The force $\boldsymbol{F}$ is a direct measure of the *magnetic field* $\boldsymbol{H}$, with a unit of ampere per meter (A/m).

The question of knowing whether magnetic masses exist in nature and whether it is possible to place one of them at P is not relevant here. It is sufficient to consider these masses as objects that behave in a magnetic field identically to the way that an ordinary mass behaves in a gravitational field, or rather to the way that a charge behaves in an electric field (Vol. I, Sec. 2.2). The imaginary nature of experiment No. 2 does not diminish its value in any way.

The effects observed at P may originate from

- the presence of electric currents in the vicinity of P;
- the presence of magnetic masses in the vicinity of P;
- a combination of these causes.

Experiments No. 1 and No. 2 do not give any means of determining which of these causes is responsible for the measured force.

In conclusion, $\boldsymbol{B}$ and $\boldsymbol{H}$ *reflect the same property in vacuum*. This is seen in a different light depending on whether a magnetic mass or a current is present. It is natural to make $\boldsymbol{B}$ and $\boldsymbol{H}$ proportional to each other:

$$B = \mu_0 H \tag{3.3}$$

and even

$$\boldsymbol{B} = \mu_0 \boldsymbol{H} \tag{3.4}$$

due to (3.1) and (3.2) and the fact that the forces observed in the two experiments are always perpendicular to each other.

The units chosen for B (Wb/m^2) and H (A/m) determine the constant of proportionality μ_0:

$$\mu_0 = 4\pi \cdot 10^{-7} \text{H/m}$$

We call μ_0 the *magnetic permeability of vacuum*.

3.2.2 Magnetic Materials: Definition

When subjected to magnetic induction, certain substances begin to produce a magnetic induction themselves, in the volume that they occupy and to the outside. It is said that they are *magnetized* or that they are *polarized magnetically*. This is a general property of matter. However, this property is only exhibited very visibly in certain materials called *magnetic materials*.

The spin of the electrons and, to a small extent, their orbital movement around the nucleus are responsible for this phenomenon whose very essence can only be studied by quantum physics. However, from the point of view of the engineer, the detailed behavior of the electrons in the process of magnetization is more usefully described by the concept of the atomic magnetic moment than by that of the wave function. We will therefore limit ourselves to state some of the results of quantum theory, most often in a qualitative manner. In fact, although a few elementary quantum calculations are sufficient to give a good picture of semiconductors, it is unfortunately not the same with magnetic materials, for which the calculations—when they exist—are always highly complex.

3.2.3 Atomic Magnetic Moment: Tool for Studying Magnetic Materials

The magnetic moment of an atom is a property that can be represented by two models:

- The magnetic dipole (Fig. 3.3);
- The Amperian current (Fig. 3.4).

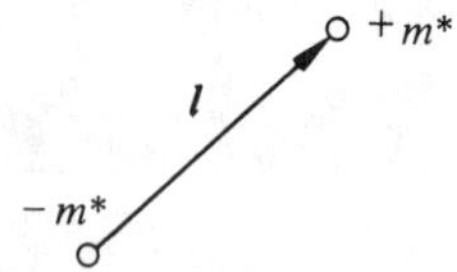

Fig. 3.3

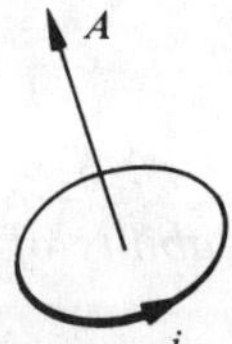

Fig. 3.4

The magnetic dipole consists of two magnetic masses with opposite sign separated by a distance l. The magnetic moment associated with this model is called the *dipole magnetic moment* $\boldsymbol{m}_d$. It is defined by

$$m_d = m^* l \qquad \text{Wb} \cdot \text{m} \tag{3.5}$$

The Amperian current is a small circular current, considered to exist on the atomic scale, that could represent the resultant of the movements of the electrons around the nucleus. The magnetic moment associated with this model is called the *Amperian magnetic moment* $\boldsymbol{m}_A$. It is defined by (3.6) in which $\boldsymbol{A}$ is the vector representing the surface bounded by the circular current i:

$$\boldsymbol{m}_A = i\boldsymbol{A} \qquad \text{A} \cdot \text{m}^2 \tag{3.6}$$

It is essential to bear in mind that there are no Amperian currents just as there are no magnetic dipoles in matter. These are just useful concepts for representing the properties of spin that are of interest to us.

In diagram form, the complete study of magnetism consists of stages (a) to (d) of Fig. 3.5. A quantum theory, only existing today in a fragmentary form, establishes the idea of the atomic magnetic moment. This moment, represented by the Amperian current model or the magnetic dipole model, serves as the basis for the science of materials.

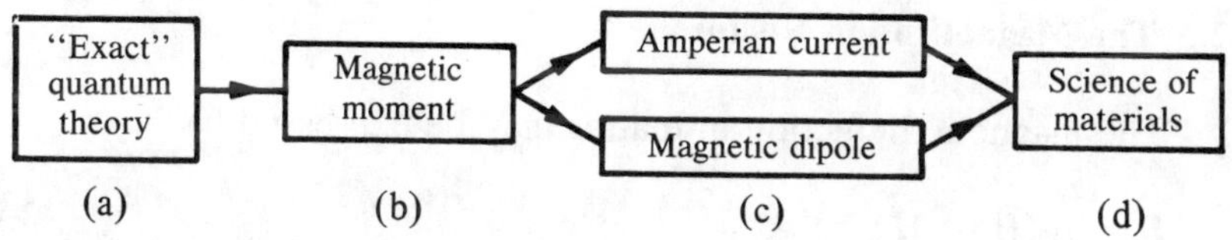

Fig. 3.5 Stages in a Complete Study of Magnetism

3.2.4 Magnetic Induction and the Magnetic Field in Matter

The experiments described in Figs. 3.1 and 3.2 may theoretically be carried out in matter. It will be assumed that the hypothetical presence of the magnetic mass m^* or the Amperian current i between the atoms does not modify the magnetic polarization of the specimen. The forces acting on m^* or i may have several origins, which can be divided into two classes:

- In the first, only the magnetic polarization of the specimen itself is involved, resulting from the presence in it of magnetic dipoles or Amperian currents.

- The second contains all the other factors likely to act on m^* and i such as the currents flowing outside or even inside the specimen, e.g., eddy currents, magnets situated in the vicinity, *et cetera*.

By using the subscripts 1 and 2 to indicate the contributions of the factors belonging to the corresponding classes, we can write, by generalizing (3.1) and (3.2):

$$\mathrm{d}\boldsymbol{F} = i\mathrm{d}\boldsymbol{l} \times (\boldsymbol{B}_1 + \boldsymbol{B}_2) \tag{3.7}$$

and

$$\boldsymbol{F} = m^*(\boldsymbol{H}_1 + \boldsymbol{H}_2) \tag{3.8}$$

Equation (3.3) would still be valid if we were to set $\boldsymbol{B} = \boldsymbol{B}_1 + \boldsymbol{B}_2$ and $\boldsymbol{H} = \boldsymbol{H}_1 + \boldsymbol{H}_2$. The magnetic induction and the magnetic field would then have totally comparable roles in vacuum as well as in matter. Now,

- the magnetic induction $\boldsymbol{B}$ is, *by definition,* a measure of the factors belonging to classes 1 *and* 2;
- the magnetic field $\boldsymbol{H}$ is, *by definition,* a measure of the factors *only* belonging to class 2.

It follows that (3.1) is valid everywhere, whereas equations (3.2) and (3.3) are only valid in vacuum.

3.2.5 The Magnetization Vector

The magnetic induction in matter may be expressed by

$$\boldsymbol{B} = \mu_0(\boldsymbol{H} + \boldsymbol{M}) \tag{3.9}$$

This equation defines the *magnetization vector* $\boldsymbol{M}$, a measure of the density of the amperian magnetic moment. The unit for $\boldsymbol{M}$ is the ampere per meter.

3.2.6 Magnetic Polarization Vector

The magnetic induction in matter may also be expressed by

$$\boldsymbol{B} = \mu_0\boldsymbol{H} + \boldsymbol{I} \tag{3.10}$$

This equation defines the *magnetic polarization vector* $\boldsymbol{I}$, a measure of the density of the dipole magnetic moment. The unit for $\boldsymbol{I}$ is the tesla (T).

$\boldsymbol{I}$ and $\boldsymbol{M}$ account for the same phenomenon; consequently, we will mainly use $\boldsymbol{I}$, as it is the one used in the study of ferromagnetism.

3.2.7 Magnetic Susceptibility

The *absolute magnetic susceptibility* χ is the name given to the ratio:

$$\chi = \boldsymbol{I}/\boldsymbol{H} \qquad \text{H/m} \tag{3.11}$$

The *relative magnetic susceptibility* χ_r is defined by

$$\chi_r = \boldsymbol{M}/\boldsymbol{H} \qquad 1 \tag{3.12}$$

It is assumed in (3.11) and (3.12) that $\boldsymbol{I}$, $\boldsymbol{M}$, and $\boldsymbol{H}$ are colinear. Provided this is the case, verified in isotropic materials (at least on the macroscopic scale), the susceptibility is a scalar. In anisotropic materials, the susceptibility is a tensor.

3.2.8 Magnetic Permeability

The *absolute magnetic permeability* μ is the name given to the quantity

$$\mu = \mu_0 + \chi = \mu_0(1 + \chi_r) \qquad \text{H/m} \tag{3.13}$$

The *relative magnetic permeability* μ_r is defined by

$$\mu_r = \mu/\mu_0 \qquad 1 \tag{3.14}$$

3.2.9 The Bohr Magneton

Theory (Section 7.6) shows that a magnetic moment is always an integral multiple of a unit magnetic moment called the *Bohr magneton* m_B. In the Amperian model:

$$m_B = 9{,}273 \cdot 10^{-24}\ \text{A} \cdot \text{m}^2 \tag{3.15}$$

In the dipole model:

$$m_B = 1{,}165 \cdot 10^{-29}\ \text{Wb} \cdot \text{m} \tag{3.16}$$

3.3 CLASSIFICATION OF THE TYPES OF MAGNETISM

3.3.1 Diamagnetism

This type of magnetism is characterized by a negative relative susceptibility of low amplitude. Diamagnetism is due to the orbital movement of the electrons produced by an applied magnetic field. This movement may be considered as a microscopic current the behavior of which is comparable to that of a current induced in a solenoid. By virtue of Lenz's law, the induced current is opposed to the field that produces it, which is in agreement with the fact that χ_r is negative. The noble gases, certain metals, most metalloids, and a large number of organic compounds are diamagnetic. Their relative susceptibilities, on the order of 10^{-5} to 10^{-6}, are of little interest to the engineer (Table 3.6).

Table 3.6

Material	χ_r	*Material*	χ_r
Si	$-1.2 \cdot 10^{-6}$	Se	$-4.0 \cdot 10^{-6}$
Cu	$-1.08 \cdot 10^{-6}$	Ag	$-2.4 \cdot 10^{-6}$
Zn	$-1.9 \cdot 10^{-6}$	Pb	$-1.4 \cdot 10^{-6}$
Ge	$-1.5 \cdot 10^{-6}$	Al_2O_3	$-3.5 \cdot 10^{-6}$

3.3.2 Paramagnetism

Paramagnetism is characterized by a positive relative susceptibility of low amplitude, i.e., between 10^{-6} and 10^{-3}. It is found in substances with atoms that have a *permanent* magnetic moment when these moments are not coupled with each other. These moments tend to align themselves under the action of a magnetic field. However, the resulting polarization remains very small because the effect of thermal motion, which gives a random orientation to the magnetic moments, remains predominant.

Apart from a few exceptions such as uranium and titanium, the relative susceptibility obeys Curie's law, i.e., it varies inversely with the absolute temperature.

Most gases, certain metals, in particular the alkali metals, some salts, and ferromagnetic materials and ferrimagnetic materials when they are heated above their Curie temperature (Section 3.3.5) are paramagnetic.

Table 3.7 gives some values for the susceptibility at room temperature and for the two substances marked with an asterisk at 1000°C.

Table 3.7

Material	χ_r	*Material*	χ_r
Na	$8.6 \cdot 10^{-6}$	Pt	$1.2 \cdot 10^{-5}$
Al	$7.7 \cdot 10^{-6}$	U	$3.3 \cdot 10^{-5}$
Mn	$1.2 \cdot 10^{-4}$	CoO	$0.75 \cdot 10^{-3}$
Ta	$1.1 \cdot 10^{-6}$	Fe_3C*	$3.7 \cdot 10^{-5}$
W	$3.5 \cdot 10^{-6}$	Fe_γ*	$2.5 \cdot 10^{-5}$

3.3.3 Antiferromagnetism

Antiferromagnetism is distinguished by a variation in the susceptibility as a function of temperature with a very particular form (Fig. 3.8).

As with paramagnetic materials, the atoms have a permanent magnetic moment but these moments are not independent of each other; on the contrary, they are strongly coupled. This interaction, which is called *antiferromagnetic coupling,* gives rise to an antiparallel arrangement of the moments shown diagrammatically in Fig. 3.9.

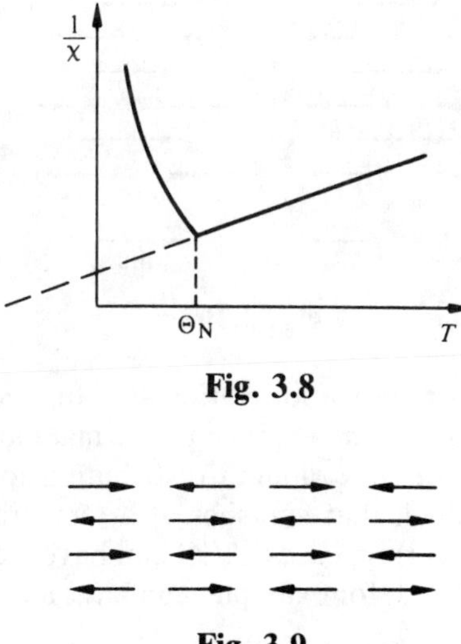

Fig. 3.8

Fig. 3.9

When the temperature increases, this arrangement breaks down. The concomitant reduction in the effect of the alignment forces makes the action of an external field more significant. This explains that $1/\chi_r$ decreases when the temperature increases, up to a temperature Θ_N, called the Neel temperature, at which the antiferromagnetic coupling disappears. Above Θ_N, the behavior of the antiferromagnetic materials becomes comparable to that of paramagnetic materials, but in general the extrapolation of $1/\chi(T)$ does not pass through the origin. Quite a large number of oxides, chlorides, and other compounds of the transition metals are antiferromagnetic.

3.3.4 Ferrimagnetism

Ferrimagnetism is the magnetism of a class of oxides called ferrites. Two families of sites A and B occupied by ions with magnetic moments m_A and m_B, respectively, can be distinguished in the crystal structure of these materials. The number of A sites differs from the number of B sites and most often $m_A \neq m_B$. The strong antiferromagnetic coupling between sites A and B therefore produces a spontaneous polarization $\boldsymbol{I}_s$, i.e., a polarization that exists *in the absence of an applied magnetic field* (Fig. 3.10).

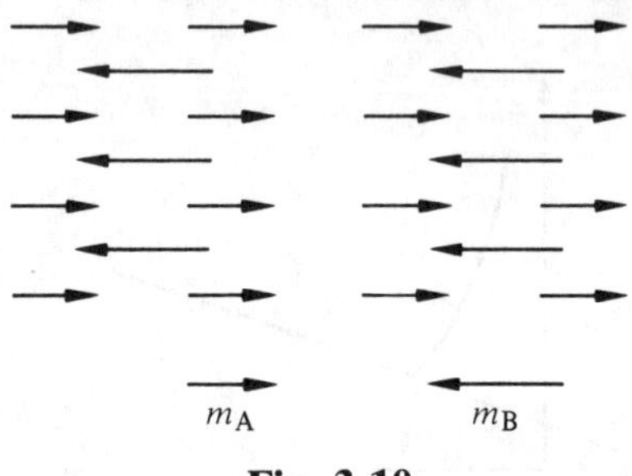

Fig. 3.10

This spontaneous polarization varies as a function of the temperature as shown in Fig. 3.11. The increasing thermal motion with temperature has a tendency to give a random distribution to $\boldsymbol{m}_A$ and $\boldsymbol{m}_B$. This results in a reduction in I_s that vanishes at a temperature Θ, called the Curie temperature. (In certain cases, $I_s(T)$ can have a different form from that shown in Fig. 3.11 and, for example, can have a maximum value for a temperature between 0 and Θ.) Above the Curie temperature, we progressively return to paramagnetic behavior. The asymptote of $1/\chi$ generally intersects the T axis on its negative side. Ferrimagnetic materials of great importance to the engineer are studied in detail in Sections 3.6, 3.9.6, and 3.10.3.

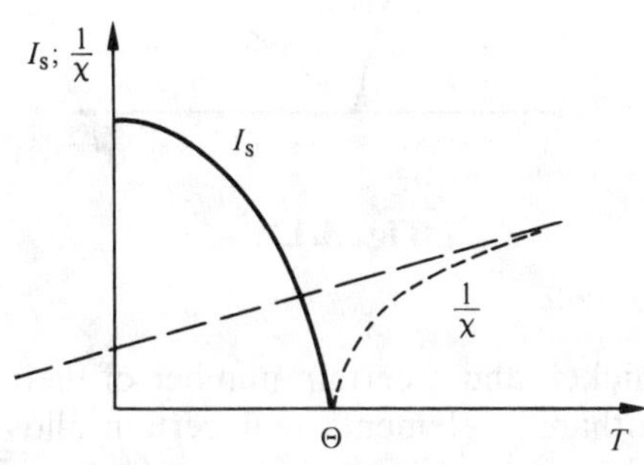

Fig. 3.11

3.3.5 Ferromagnetism

Ferromagnetism is a type of magnetism resulting from the alignment of permanent magnetic moments. These moments are oriented parallel to each other by a mutual interaction called *ferromagnetic coupling* (Fig. 3.12). Therefore, ferromagnetic materials also have spontaneous polarization. What has been said for ferrimagnetic materials concerning the return to a random distribution of the magnetic moments under the effect of an increase in temperature, applies equally here (Fig. 3.13). Ferromagnetic materials also have a Curie temperature Θ, above which they become paramagnetic, their susceptibility then obeying the Curie-Weiss law (Section 3.4.5).

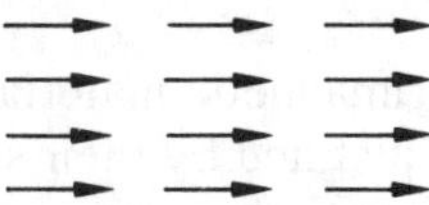

Fig. 3.12 Diagrammatic Representation of the Alignment of Magnetic Moments in a Ferromagnetic Material

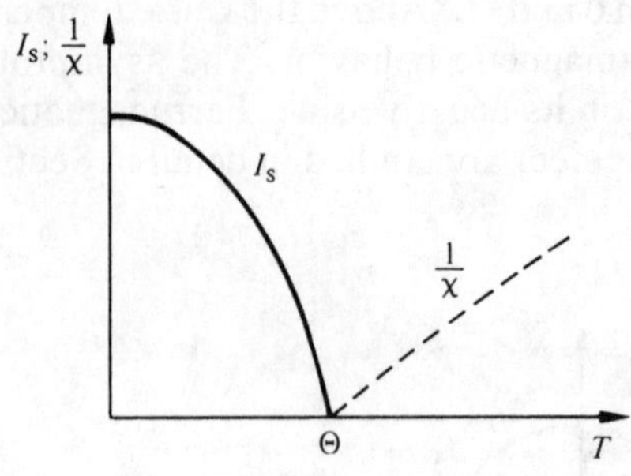

Fig. 3.13

Iron, cobalt, nickel, and a certain number of their alloys are ferromagnetic. Some lanthanide elements and certain alloys of magnesium with aluminum and copper are also ferromagnetic. Ferromagnetism, whose technical significance is well known, is studied in detail in Section 3.4.

3.3.6 Comments

The phenomenon of the orientation of magnetic moments inside ferromagnetic and ferrimagnetic materials takes place in a localized manner. A specimen of macroscopic size will therefore be divided into a large number of small volumes, called magnetic domains (Section 3.7) each with a spontaneous polarization oriented differently from those of its neighbors.

3.4 FERROMAGNETISM

3.4.1 Introduction

Ferromagnetic and ferrimagnetic materials are often called together magnetic materials which is justified by their susceptibility 10^3–10^{11} times higher than that resulting from other types of magnetism. In magnetic materials, stage (d) of Fig. 3.5 is divided into two parts.

The first consists of studying the alignment of dipoles on the scale of a magnetic domain. It is discussed in Section 3.4.4 for ferromagnetism and in Section 3.6.5 for ferrimagnetism.

The second part concerns the calculation of the polarization on the scale of the specimen resulting from the superimposition of the polarizations belonging to each magnetic domain. This is discussed in Sections 3.7 and 3.8.

The theory of ferromagnetism due to Weiss is based on the formal approach developed by Langevin in the case of paramagnetism.

3.4.2 Paramagnetism: Langevin's Theory

This theory gives an answer to the following questions:

- To what extent is a magnetic field capable of orienting permanent atomic magnetic moments that are independent of each other?
- What role does the temperature play?

Consider a medium containing N atoms per unit volume, each carrying a dipole magnetic moment $\boldsymbol{m}$. When subjected to an external field $\boldsymbol{H}$, these moments are subject to a torque $\boldsymbol{C}$.

$$\boldsymbol{C} - \boldsymbol{m} \times \boldsymbol{H} \tag{3.17}$$

The angle γ between $\boldsymbol{m}$ and $\boldsymbol{H}$ will be measured starting from $\boldsymbol{H}$ the orientation of which will serve as a reference. We have

$$C = mH \sin \gamma \tag{3.18}$$

The potential energy W_{pot} associated with C is equal to

$$W_{pot} = -\int_0^{\gamma} - mH \sin \gamma \mathrm{d}\gamma = mH(1 - \cos \gamma) + \text{constant} \tag{3.19}$$

This expression is simplified by setting $W_{pot} = 0$ for $\gamma = \pi/2$, so that we have

$$W_{pot} = -mH \cos \gamma \tag{3.20}$$

Since the moments are independent of each other, it is reasonable to assume that their energy distribution is given by Boltzmann's law (Section 7.2). Let $n(\gamma)\mathrm{d}\gamma$ be the number of moments per unit volume making an angle lying between γ and $\gamma + \mathrm{d}\gamma$ with $\boldsymbol{H}$ and let $\mathrm{d}\Omega$ be the solid angle that they occupy.

Allowing for (7.32), we have

$$n(\gamma)\mathrm{d}\gamma = n_0 \exp\left(-\frac{W_{\mathrm{pot}}}{k_{\mathrm{B}}T}\right)\mathrm{d}\Omega \tag{3.21}$$

from which

$$n(\gamma)\mathrm{d}\gamma = n_0\, 2\pi \sin\gamma \exp\left(\frac{mH\cos\gamma}{k_{\mathrm{B}}T}\right)\mathrm{d}\gamma, \tag{3.22}$$

where n_0 is a normalization constant defined by the condition:

$$\int_0^{\pi} n(\gamma)\mathrm{d}\gamma = N \tag{3.23}$$

The polarization I is obtained by summing the projections of all the moments on $\boldsymbol{H}$:

$$I = \int_0^{\pi} n(\gamma) m \cos\gamma\, \mathrm{d}\gamma \tag{3.24}$$

This expression can easily be integrated provided that it is rewritten in the following form:

$$I = Nm \frac{\displaystyle\int_0^{\pi} \cos\gamma \sin\gamma \exp\left(\frac{mH\cos\gamma}{k_{\mathrm{B}}T}\right)\mathrm{d}\gamma}{\displaystyle\int_0^{\pi} \sin\gamma \exp\left(\frac{mH\cos\gamma}{k_{\mathrm{B}}T}\right)\mathrm{d}\gamma} \tag{3.25}$$

By setting

$$\alpha = \frac{mH}{k_{\mathrm{B}}T} \tag{3.26}$$

and

$$\cos\gamma = x \tag{3.27}$$

we have

$$I = Nm\frac{\int_{1}^{-1} x\exp(\alpha x)\mathrm{d}x}{\int_{1}^{-1} \exp(\alpha x)\mathrm{d}x} = Nm\left(\coth\alpha - \frac{1}{\alpha}\right) = Nm\mathrm{L}(\alpha) \qquad (3.28)$$

In this expression, L(α) denotes the Langevin function (Fig. 3.15).

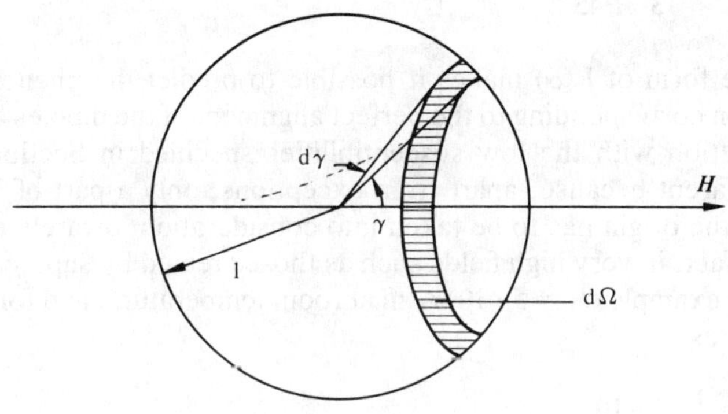

Fig. 3.14

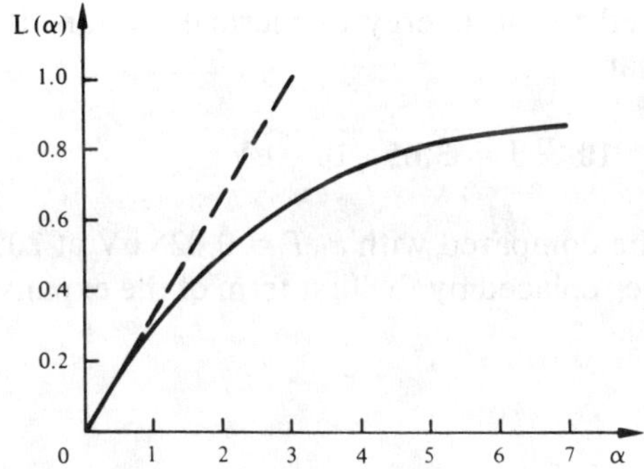

Fig. 3.15 Langevin Function

This function has a horizontal asymptote:

$$\lim_{\alpha \to \infty} \mathrm{L}(\alpha) = 1 \tag{3.29}$$

In the neighborhood of the origin, L(α) can be represented by the expansion:

$$\mathrm{L}(\alpha) = \frac{\alpha}{3} - \frac{\alpha^3}{45} + \cdots \tag{3.30}$$

The form of L(α) makes it possible to predict the phenomenon of saturation corresponding to the perfect alignment of the dipoles on $\boldsymbol{H}$. The contradiction with the low susceptibilities specified in Section 3.3.2 is only apparent because, apart from exceptions, only a part of L(α) very close to the origin has to be taken into consideration, α rarely exceeding 10^{-2}. In fact, in very high fields such as those created by superconducting coils, for example, $H = 5 \cdot 10^6$ A/m at room temperature, and for $m = m_B$, (3.26) gives

$$\alpha = 1.45 \cdot 10^{-2} \tag{3.31}$$

The potential energy communicated by $\boldsymbol{H}$ to the dipoles therefore remains very small compared to the energy of thermal motion. It can easily be seen from (3.20) that

$$|W_{pot}| \leqslant 5.85 \cdot 10^{-23} \text{ J} \triangleq 3.65 \cdot 10^{-4} \text{ eV} \tag{3.32}$$

This value should be compared with $k_B T = 0.025$ eV at 20°C. Therefore, L(α) in (3.28) can be replaced by the first term of the expansion (3.30). We have

$$I = \frac{Nm^2H}{3k_B T} \tag{3.33}$$

from which we have

$$\chi = \frac{Nm^2}{3k_B T} \tag{3.34}$$

Relationship (3.34) expresses Curie's law: the susceptibility of paramagnetic materials varies in inverse proportion to the absolute temperature.

In the neighborhood of absolute zero, fields on the order of 10^6 A/m produce a saturation polarization in certain paramagnetic substances, corresponding to values of α between 5 and 10, as shown by the curves of Fig. 3.16.

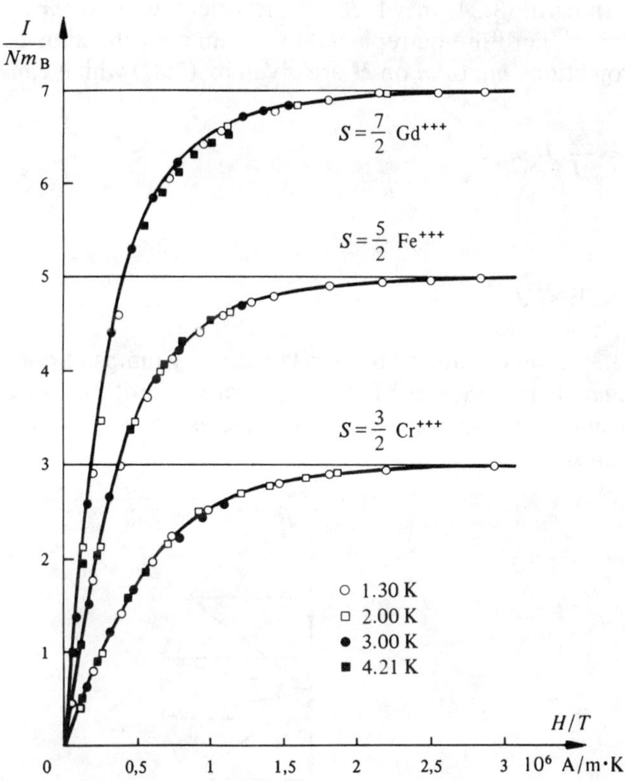

Fig. 3.16 Apparent Atomic Magnetic Moment *I*/*N* in Units of m_B at Low Temperatures For the Bivalent Ions of Gadolinium, Iron, and Chromium. (After [27].) Measurements Made on Paramagnetic Complex Salts. The Brillouin Functions (Section 3.4.3) are Shown as Solid Lines. *S*: Spin of the Ions (Section 7.4)

3.4.3 Corrections Made by Quantum Theory

Quantum theory shows (Section 7.4) that the modulus of the atomic magnetic moment varies by discrete values and that its orientation in space referred to an external field $\boldsymbol{H}$ can only correspond to certain specific angles γ. We have already considered the first result but not the second which implies a new quantization of the projections of the moment on $\boldsymbol{H}$.

The integral (3.24) in which all the orientations of $\boldsymbol{m}$ in space are allowed, must therefore be replaced by a sum over the allowed values of γ. The projections m_{H} of $\boldsymbol{m}$ on $\boldsymbol{H}$ are given by (7.84) which can be written

$$m_{\mathrm{H}} = \frac{M}{J} Jgm_{\mathrm{B}} \tag{3.35}$$

where

$$-J \leqslant M \leqslant J \tag{3.36}$$

In (3.36), the quantum number M varies by jumps of one unit, and J is an integer or half-integer. The maximum value of m_{H} is Jgm_{B} and in (3.35) the quotient M/J plays the same role as $\cos\gamma$ in Section 3.4.2, as Fig. 3.17 shows.

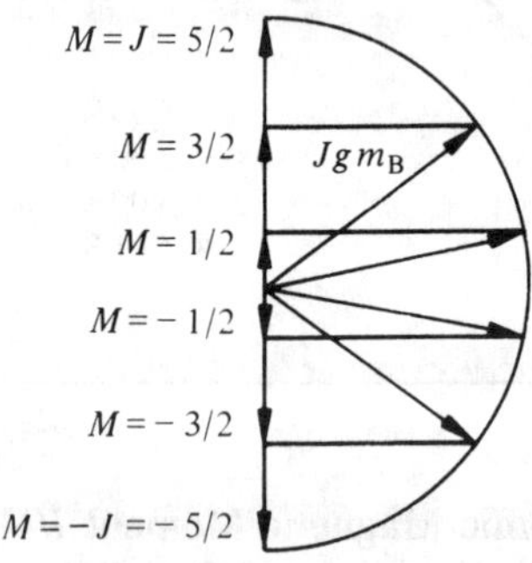

Fig. 3.17 Quantization of the Projection of $\boldsymbol{m}$ on $\boldsymbol{H}$

Expression (3.25) therefore takes the form:

$$I = NJgm_B \frac{\sum_{M=-J}^{M=+J} \frac{M}{J} \exp\left(\frac{\frac{M}{J} Jgm_B H}{k_B T}\right)}{\sum_{M=-J}^{M=+J} \exp\left(\frac{\frac{M}{J} Jgm_B H}{k_B T}\right)}$$

$$= NJgm_B \frac{\sum_{M=-J}^{M=+J} \frac{M}{J} \exp\left(\alpha \frac{M}{J}\right)}{\sum_{M=-J}^{M=+J} \exp\left(\alpha \frac{M}{J}\right)} \tag{3.37}$$

The quotient of these sums is known under the name of the Brillouin function $B_J(\alpha)$, from which we have

$$I = NJgm_B B_J(\alpha) \tag{3.38}$$

where

$$B_J(\alpha) = \frac{2J+1}{2J} \coth \frac{2J+1}{2J} \alpha - \frac{1}{2J} \coth \frac{1}{2J} \alpha \tag{3.39}$$

Examining (3.28) and (3.38), it can be seen that the quantization of the orientation of $\boldsymbol{m}$ has the effect of replacing the Langevin function by a Brillouin function in the expression for the polarization. Fig. 3.18 makes it possible to compare these functions and demonstrates that in reality, saturation occurs more rapidly than is predicted by the classical theory of Langevin.

For $\alpha \ll 1$, the Brillouin functions may be represented by

$$B_J(\alpha) = \frac{J+1}{3J} \alpha - \frac{[(J+1)^2 + J^2](J+1)}{90J^3} \alpha^3 + \cdots \tag{3.40}$$

By retaining only the first term of this expansion, (3.38) leads to a new expression for Curie's law:

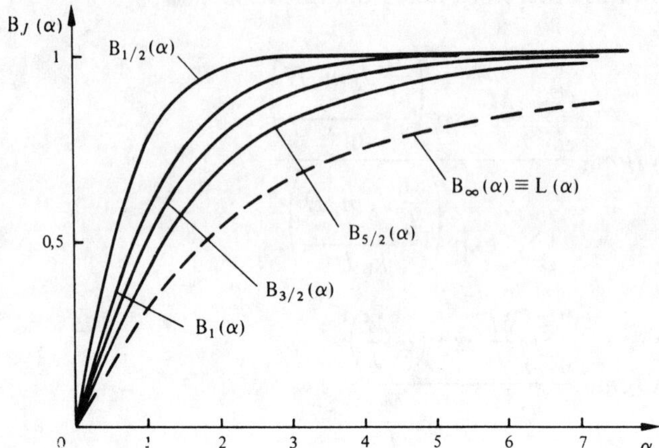

Fig. 3.18 Brillouin Functions

$$\chi = \frac{NJ(J + 1)g^2 m_B^2}{3k_B T}, \tag{3.41}$$

to be compared with (3.34).

3.4.4 Basic Principles of the Weiss Theory

Why do magnetic moments align themselves parallel to each other in ferromagnetic materials? The explanation obviously depends on the model chosen to describe the phenomenon. Quantum theory invokes ideas of exchange energy between neighboring magnetic moments and of overlapping of crystal orbits. The Weiss theory is a much simpler phenomenological approach with general application. It gives a first valid approximation for the variation in spontaneous polarization as a function of temperature.

Weiss attributes the alignment of the magnetic moments to the existence of a field H_m, called the molecular field. This field is very intense. By referring to Fig. 3.18, it can be seen that it is necessary at least that $\alpha \geqslant 5$ in order to reasonably approach saturation. At room temperature and for $m = m_B$, we obtain from (3.26):

$$H_m \geqslant 1.8 \cdot 10^9 \text{ A/m} \tag{3.42}$$

The molecular field is therefore approximately 100 to 1000 times higher than the highest fields produced by engineers.

By assumption, H_m is related to the polarization by

$$H_m = wI \tag{3.43}$$

The constant of proportionality w is independent of the temperature. The coupling of a magnetic moment with its neighbors is represented, due to (3.43), in the form of a coupling with $\boldsymbol{H}_m$. Each dipole therefore appears as independent from the others but subjected to a potential. Under these conditions, Boltzmann's distribution still applies as well as the calculation of the polarization based on Langevin's theory. It is sufficient to assume that the effective field $\boldsymbol{H}_{eff}$ acting on the magnetic moments is equal to

$$\boldsymbol{H}_{eff} = \boldsymbol{H} + w\boldsymbol{I} \tag{3.44}$$

Formally, the polarization is always given by (3.28), but the argument α is now equal to

$$\alpha = \frac{m(H + wI)}{k_B T} \tag{3.45}$$

from which

$$I = \alpha \frac{k_B T}{mw} - \frac{H}{w} \tag{3.46}$$

3.4.5 The Weiss Theory: Discussion of Results

It is instructive to derive I from the graphic solution of the simultaneous equations (3.28) and (3.46) shown in Fig. 3.19.

At temperature T_1 and for $H = 0$, the straight line (3.46) intersects the Langevin function (3.28) at O and P. The point O corresponds to an unstable solution: if all the coupled magnetic moments were randomly oriented, we would definitely have $I = 0$. But if two neighboring moments are aligned somewhere at a given point in time, a local polarization appears.

By means of the molecular field that it produces, this polarization produces the immediate alignment of all the dipoles, which corresponds to a transition to the stable state represented by P.

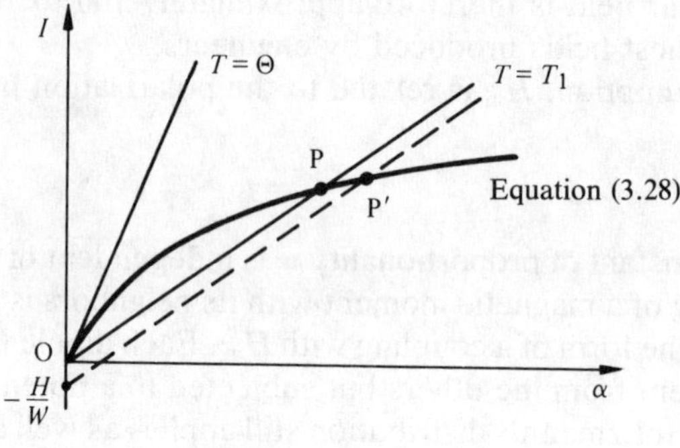

Fig. 3.19

The variation in spontaneous polarization as a function of temperature can easily be derived from the displacements of P.

If the temperature tends to 0 K, P is displaced to infinity which corresponds to saturation:

$$I = Nm = I_{\text{sat}} \tag{3.47}$$

If the temperature increases, the slope of the straight line OP increases and the spontaneous polarization decreases. When the Curie temperature Θ is reached, the spontaneous polarization vanishes, and the straight line has the same slope as the tangent to the Langevin function at the origin. We obtain the Curie temperature by expressing the equality of these slopes:

$$\Theta = \frac{Nm^2w}{3k_B} \tag{3.48}$$

The relative value of the spontaneous polarization can be written in a simple form as a function of the reduced variable T/Θ. By introducing (3.47) and (3.48) in (3.28), we have

$$\frac{I(T)}{I_{\text{sat}}} = \mathrm{L}\left(3\,\frac{I(T)/I_{\text{sat}}}{T/\Theta}\right) \tag{3.49}$$

If we take into account the corrections introduced in Section 3.4.3, equation (3.49) takes the form:

$$\frac{I(T)}{I_{sat}} = B_J\left(\frac{3J}{J+1}\,\frac{I(T)/I_{sat}}{T/\Theta}\right) \tag{3.50}$$

Solutions (3.49) and (3.50) are compared to the experimental results in Fig. 3.20.

The agreement of the experimental points with (3.50), in particular for $\boldsymbol{J} = \frac{1}{2}$, demonstrates the value of the Weiss model, the phenomenological character of which must, however, not be lost from view. In fact, it would be difficult to give a theoretical justification of the fact that $\boldsymbol{J} = \frac{1}{2}$ for iron and nickel (Table 3.24).

Let us now consider the effect of a field H applied in the direction of $\boldsymbol{I}$. This field (3.46) moves the straight line of Fig. 3.19 downwards by an amount $\boldsymbol{H}/w$. For $\boldsymbol{I}_{sat} = 2\text{T}$, (indicative value for iron), (3.42) and (3.43) show that

$$w \geqslant 0.9 \cdot 10^9 \text{ m/H} \tag{3.51}$$

Even in the most intense fields, $\boldsymbol{H}/w$ can therefore not exceed a few mT. Consequently, the polarization is practically not influenced by $\boldsymbol{H}$.

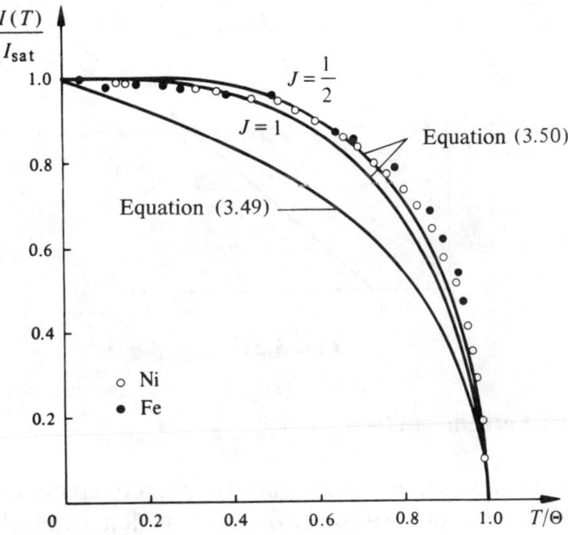

Fig. 3.20 (Experimental Points after [28])

There is no contradiction between this result and the experimental fact that a field of 1 A/m may be sufficient to polarize certain magnetic materials to saturation if we remember that the Weiss theory applies on the scale of a magnetic domain (Section 3.4.1).

We shall conclude this discussion by considering the case when T is just larger than Θ because it forms an exception in the sense that $\boldsymbol{H}$ has a nonnegligible effect on the polarization. Since $\boldsymbol{I}' \ll \boldsymbol{I}_{sat}$, we can represent $L(\alpha)$ in (3.28) by the first term of the expansion (3.30); we have

$$I = \frac{Nm^2}{3k_B T}(H + wI), \tag{3.52}$$

an expression from which we can immediately derive the susceptibility. By taking into account (3.48), we obtain

$$\chi = \frac{Nm^2}{3k_B(T - \Theta)} \tag{3.53}$$

This is the Curie-Weiss law, which makes it possible to determine Θ and m experimentally by plotting $1/\chi$ as a function of T (Fig. 3.21).

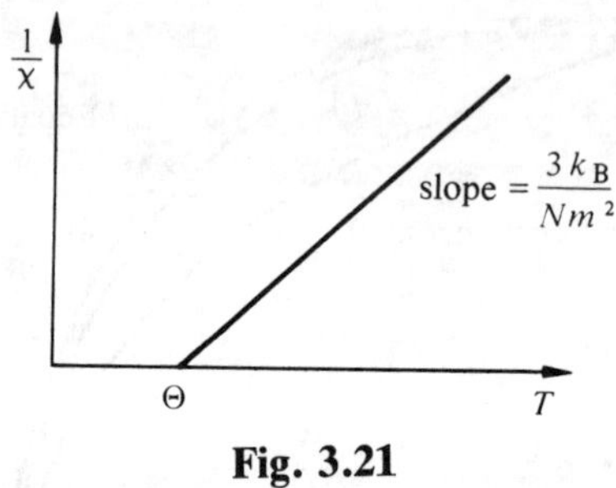

Fig. 3.21

3.4.6 Origin of Ferromagnetism

Experiment shows that ferromagnetism is a property connected with the characteristic electronic structure of the transition metals (first period) and the lanthanide elements. These materials have in common the characteristic of a not completely filled internal electron shell corresponding to the 3d levels for the transition metals and the 4s levels for the lanthanides. We will deal mainly with the case of the transition metals, to which iron, cobalt, and nickel belong. The case of the lanthanide elements is similar although more complex because of the larger number of electrons to be taken into account.

Measurement of the Lande factor (Section 7.6) shows that the contribution of the magnetic moment resulting from the spin of the electrons predominates over the contribution deriving from their orbital motion. The orbital motion reaches a maximum of 10% of the polarization and it will be neglected in what follows.

3.4.7 Structure of the 3d and 4s Bands and Ferromagnetic Coupling

The electron shells that are closer to the nucleus than the 3d shell are complete in the transition metals. They do not produce any magnetic moment.

Fig. 3.22 shows the typical form of the density of states in the 3d and 4s bands of the transition metals.

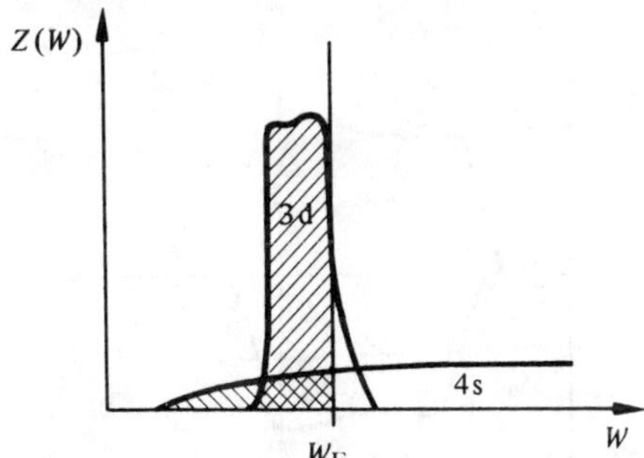

Fig. 3.22 The 3d and 4s Bands in Nickel

This figure calls for two remarks:

- The 3d and 4s bands overlap over a certain energy interval inside which they are both filled with electrons at the same time (Table 1.12).
- The 3d band is very narrow with respect to the 4s band.

The narrowness of the 3d band has two important consequences. First, the kinetic energy of the 3d electrons is small compared to that of the 4s electrons, for example. For confirmation of this, it is sufficient to refer to (2.127) and to Fig. 2.24: $\partial W/\partial k$ becomes smaller as the band becomes narrower. The 3d states are therefore closer to the states of a free atom than the 4s states. In other words, the 3d electrons belong more to "their" atom than the 4s electrons and therefore it makes sense to speak of the interaction between the 3d electrons of neighboring atoms.

The ferromagnetic coupling results from such an interaction between the 3d electrons belonging to two different atoms.

This interaction causes each level of the band to be divided into two close levels corresponding to a parallel and antiparallel alignment of the spins of the electrons concerned. We can therefore consider the 3d band as the superimposition of two half bands denoted $3d^+$ and $3d^-$ corresponding to the parallel and antiparallel alignment of the spins, respectively. These two half bands are shifted along the energy axis.

The second consequence of the narrowness of the 3d band is that the density of states is very high there. If the Fermi energy lies within this band, a very small shift in the $3d^+$ energy with respect to $3d^-$ energy is sufficient to create a considerable difference between the populations of these two half bands (Fig. 3.23), inducing a significant spontaneous polarization, which is what characterizes ferromagnetic materials.

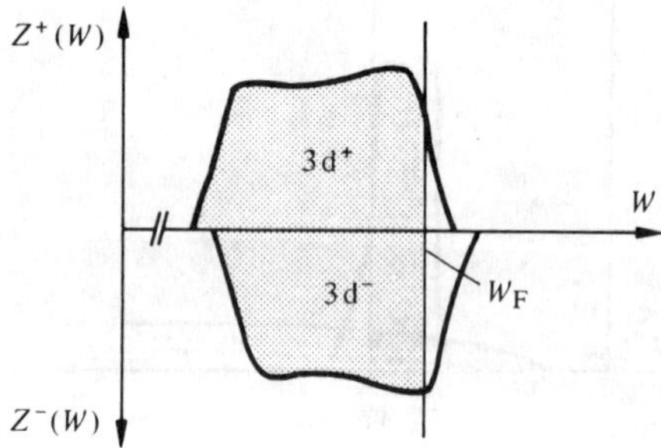

Fig. 3.23 Density of States and Occupation in the Two Half Bands $3d^+$ and $3d^-$

3.4.8 Experimental Methods

The great majority of problems encountered by engineers involving ferromagnetism cannot be solved by quantum theory. Experimental techniques therefore provide the most important source of information. Three methods given in order of increasing complexity in their implementation are currently used. These are

- the measurement of the polarization at saturation;
- x-ray diffraction;
- neutron diffraction.

Polarization at saturation may be derived from experiments based on the measurement of forces acting on a specimen (magnetic balances), the measurement of magnetic fields (magnetometers), or the measurement of the currents and voltages induced by the variation of a magnetic flux (integrator, oscilloscope).

X-ray diffraction (Section 1.6) is indicated for determining the mean number of electrons in the 3d shell as well as their distribution over space. The diffraction of polarized neutrons is carried out in the irradiation channel of a nuclear reactor. It makes it possible to obtain the orientation and the distribution of the spins in a crystal.

The information given in the four following sections has nearly all been obtained by one or another of these methods.

3.4.9 Hund's Rule

The 3d shell consists of five electrons with positive spin plus five electrons with negative spin (Section 1.2.9). In *free atoms,* this shell is filled in a particular manner. If we go through the series of the transition metals (first period) from the lightest to the heaviest, the states corresponding to the positive spins are occupied the first. There are never any negative spins while a state corresponding to a positive spin is still empty (Table 3.24). This process is known as *Hund's rule*. It has the consequence that the atomic magnetic moment resulting from all the electrons in the 3d shell is always a maximum.

3.4.10 Occupation of the 3d and 4s Bands in Pure Metals

Hund's rule ceases to be valid when the atoms are no longer in the free state but are bound by valence bonds, inside a metal, for example. Table 3.24 shows that the disparity between the $3d^+$ and $3d^-$ bands disappears completely in chromium, manganese, copper, and zinc. It remains in an attenuated form in iron, cobalt, and nickel.

Table 3.24
Electronic structure of the transition elements currently used in ferromagnetic alloys

Element	Number of electrons 3d + 4s	Spin	Free atom 3d	Free atom 4s	Metal 3d	Metal 4s	N_B
Cr	6	+	↑ ↑ ↑ ↑ ↑	↑			0
		−					
Mn	7	+	↑ ↑ ↑ ↑ ↑	↑			0
		−		↓			
Fe	8	+	↑ ↑ ↑ ↑ ↑	↑			2,22
		−	↓	↓			
Co	9	+	↑ ↑ ↑ ↑ ↑	↑			1,72
		−	↓ ↓	↓			
Ni	10	+	↑ ↑ ↑ ↑ ↑	↑			0,60
		−	↓ ↓ ↓	↓			
Cu	11	+	↑ ↑ ↑ ↑ ↑	↑			0
		−	↓ ↓ ↓ ↓ ↓				
Zn	12	+	↑ ↑ ↑ ↑ ↑	↓			0
		−	↓ ↓ ↓ ↓ ↓	↑			

In these metals, each atom has a magnetic moment equal to an integral number of Bohr magneton. However, in a given metal, all the atoms do not have an identical magnetic moment. This explains why the mean number N_B of Bohr magnetons per atom given in Table 3.24 is not an integer.

3.4.11 Remark

Certain types of valence perturb the electronic configuration of the free atom less than the metal valence. They are found in the ferrites (Section 3.6) and in certain salts, such as those shown in Fig. 3.16. In these salts, iron and chromium each give three electrons. By assuming that iron gives its two 4s electrons and then following Hund's rule, the 3d electron with negative spin, we obtain $N_B = 5$. Similarly, if chromium gives its 4s electron and then two 3d electrons by also obeying Hund's rule, we obtain $N_B = 3$. These two values of N_B are in agreement with the saturation polarizations shown in Fig. 3.16.

3.4.12 The Case of Alloys

Ferromagnetism also appears in the alloys of iron, cobalt, and nickel with other transition metals provided that the average number of electrons per atom in the 3d and 4s shells lies between approximately 6.5 and 10.5 (Fig. 3.25). The maximum value of N_B lies in the vicinity of 2.5 so we must accept that the difference in energy between $3d^+$ and $3d^-$ (Fig. 3.23) cannot lead to a difference in the populations of these bands exceeding 2.5 electrons per atom. Starting at 8.5 electrons per atom in 3d and 4s, all the alloys not containing metals heavier than iron are distributed along a straight line with slope −1. This is in agreement with the *variation* of electronic configurations for the free atoms given in Table 3.24. The existence of sections with positive slope has not yet received any entirely satisfactory explanation.

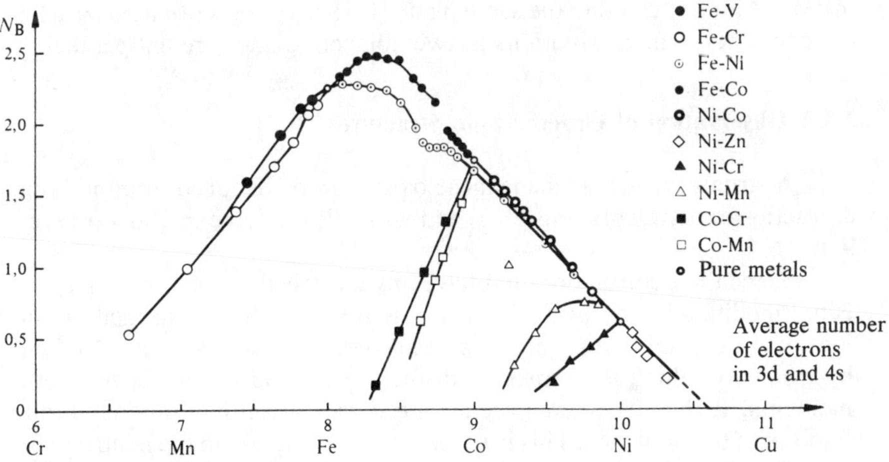

Fig. 3.25 Average Number of Electrons in 3d and 4s. (After Bozorth [29])

3.5 ANTIFERROMAGNETISM

3.5.1 Case of Manganese Oxide, MnO

Within the scope of this book, the study of antiferromagnetism is above all aimed at providing a base for the study of ferrimagnetism. The classical example of an antiferromagnetic material is manganese oxide, MnO, the structure of which is shown in Fig. 3.26.

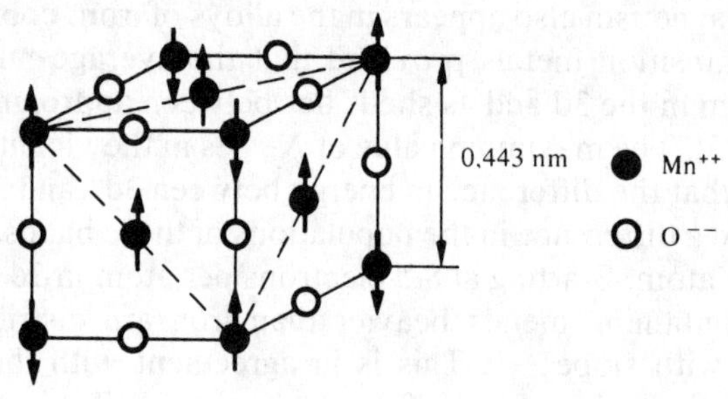

Fig. 3.26

It can be seen that the Mn^{++} ions form a face-centered cubic lattice, whereas the O^{--} ions occupy the midpoints of the edges of the cube. All the Mn^{++} ions situated in the same plane {111} have spins oriented parallel to each other. The orientations in two adjacent planes are antiparallel.

3.5.2 Observation of Ordered Spin Structures

A structure such as manganese oxide can be deduced from neutron diffraction experiments (Fig. 3.27) and x-ray diffraction experiments (Section 1.6).

Since the neutron does not have any electrical charge, its trajectory is not modified by the periodic distribution of the charges formed by the ions of the crystal. However, it has a magnetic moment of $1.04 \cdot 10^{-3}\ m_B$ by means of which it interacts with the atomic and molecular magnetic moments. If these moments are arranged regularly, they produce a diffraction of the neutrons. This is reflected by a variation in the neutron flux emerging from the specimen as a function of the angle α. The experimental results shown in Fig. 3.28 were obtained from monokinetic 0.074 eV neutrons and a powdered sample of manganese oxide. At room temperature, the antiferromagnetic coupling is destroyed by the thermal energy and the recorded maxima correspond to the nuclear magnetic moments

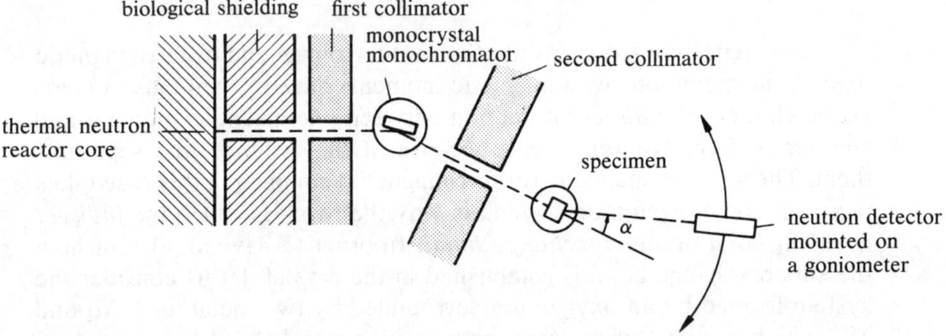

Fig. 3.27 Neutron Diffraction

and a cubic lattice with unit cell edge, a = 0.443 nm. Below 120 K, antiferromagnetic coupling appears and the diffractions corresponding to the planes (111) and (311) of a cubic lattice with unit cell edge, a = 0.885 nm, can be clearly seen. It is possible to show [30] that the only arrangement of magnetic moments corresponding to these observations is that of Fig. 3.26.

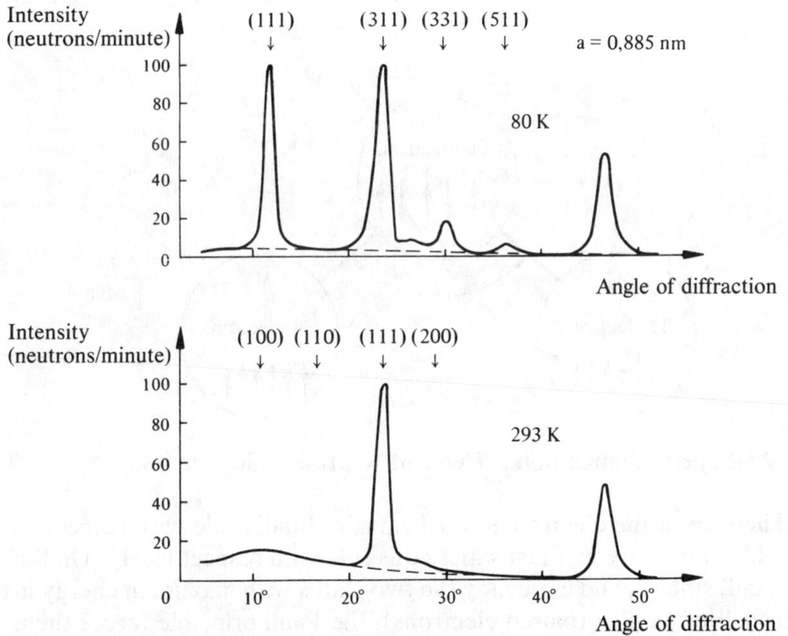

Fig. 3.28 Neutron Diffraction on MnO [30]

3.5.3 Superexchange Bond

In magnesium oxide, as in other antiferromagnetic or ferrimagnetic oxides, the metal ions with magnetic moments have oxygen ions as nearest neighbors. The *direct* interaction between one magnetic moment and another is therefore very weak because of the distance that separates them. The antiferromagnetic (or ferrimagnetic) coupling in this case takes place via the oxygen atom, which is why the result is called the *indirect covalent bond* or *superexchange bond*. In order to have an idea of how the superexchange bond is established in the crystal, let us consider the system formed by an oxygen ion surrounded by two metal ions, M_1 and M_2, which can be chosen from the transition metals lying between chromium and copper.

Oxygen lacks two electrons to have the configuration of a noble gas, neon ($1s^2\ 2s^2\ 2p^6$). Therefore, oxygen has the tendency to ionize twice, by capturing an electron e_1 from M_1 and an electron e_2 from M_2. It is known (Section 1.3.4) that it is very rare that a valency bond is purely of a specific type. Here, the ionic bond M_1–O–M_2 also has a covalent character.

It follows that e_1 is not rigidly transferred to the oxygen, but some of the time remains in the vicinity of M_1. The same applies to e_2 on M_2 (Fig. 3.29).

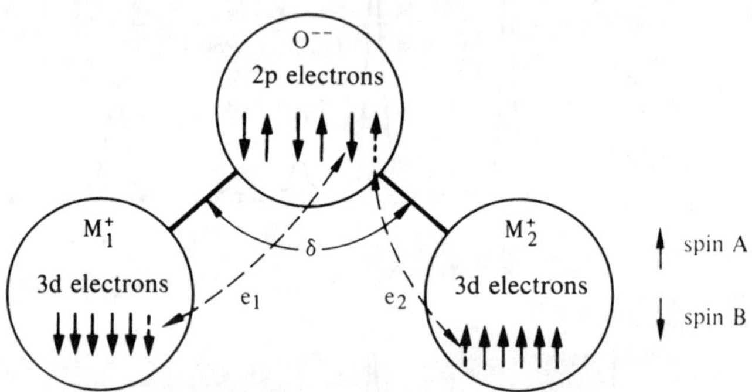

Fig. 3.29 Superexchange Bond. Here, M Represents Manganese

The spin of the electron e_1 must satisfy Hund's rule with respect to the ion M_1^+. Similarly, e_2 must satisfy this rule with respect to M_2^+. On the other hand, since e_1 and e_2 occupy the two states with maximum energy in the 2p shell of oxygen (paired electrons), the Pauli principle forces them to have opposite spins. The result is that the magnetic moments of M_1 and M_2 are oriented antiparallel. The same would apply to any pair where $M_1 = M_2$, chosen within the limits given above.

Fig. 3.30 2p Orbit

The form of the 2p orbit (Fig. 3.30) explains why the superexchange bond has its greatest intensity for $\delta = 180°$ and the smallest intensity for $\delta = 90°$: these are the angles for which there is a maximum and a minimum overlap with the orbits of M_1 and M_2, respectively.

Other electronegative elements may play the part of oxygen in the superexchange bond. These are, in particular, sulfur, tellurium, fluorine, and chlorine.

3.5.4 Neel's Theory

Adopting the concept of the molecular field introduced by Weiss, Neel and others have developed a phenomenological theory of antiferromagnetism.

Let us call A sites and B sites the positions in the crystal where spins of orientation A and B are found, respectively (Fig. 3.29). Let $\boldsymbol{I}_A$ and $\boldsymbol{I}_B$ be the polarization of the material resulting exclusively from the A sites and exclusively from the B sites, respectively. The molecular fields $\boldsymbol{H}_{mA}$ and $\boldsymbol{H}_{mB}$ present at the A and B sites are expressed by the equations:

$$\boldsymbol{H}_{\mathrm{mA}} = w_{\mathrm{AA}}\boldsymbol{I}_{\mathrm{A}} + w_{\mathrm{BA}}\boldsymbol{I}_{\mathrm{B}} \tag{3.54}$$

$$\boldsymbol{H}_{\mathrm{mB}} = w_{\mathrm{BB}}\boldsymbol{I}_{\mathrm{B}} + w_{\mathrm{AB}}\boldsymbol{I}_{\mathrm{A}} \tag{3.55}$$

which are direct generalizations of (3.43). The constants w_{AA}, w_{AB}, w_{BA}, and w_{BB} are independent of temperature. Because the A and B sites are perfectly equivalent, we set

$$w_{\mathrm{AA}} = w_{\mathrm{BB}} = w_1 \tag{3.56}$$

$$w_{\mathrm{BA}} = w_{\mathrm{AB}} = w_2\,, \tag{3.57}$$

where w_1 is positive and measures the ferromagnetic coupling between sites of the same name. The quantity w_2 is negative and measures the antiferromagnetic coupling between sites with different names. In the absence of an external field:

$$\boldsymbol{I}_{\mathrm{A}} = -\boldsymbol{I}_{\mathrm{B}} \tag{3.58}$$

Equations (3.54) and (3.55) therefore take the forms:

$$\boldsymbol{H}_{\mathrm{mA}} = (w_1 - w_2)\boldsymbol{I}_{\mathrm{A}} \tag{3.59}$$

$$\boldsymbol{H}_{\mathrm{mB}} = (w_1 - w_2)\boldsymbol{I}_{\mathrm{B}} \tag{3.60}$$

The orientation of the magnetic moments m present at sites A and B, subjected to the molecular fields (3.59) and (3.60), respectively, and to thermal motion is calculated by the procedure of Section 3.4.2. We have

$$\boldsymbol{I}_{\mathrm{A}} = \frac{N}{2}\, m\mathrm{L}\left(\frac{m(w_1 - w_2)\boldsymbol{I}_{\mathrm{A}}}{k_{\mathrm{B}}T}\right) \tag{3.61}$$

$$\boldsymbol{I}_{\mathrm{B}} = \frac{N}{2}\, m\mathrm{L}\left(\frac{m(w_1 - w_2)\boldsymbol{I}_{\mathrm{B}}}{k_{\mathrm{B}}T}\right) \tag{3.62}$$

where N is the total number of A and B sites. When we look at the graphical solution of these equations as was done in the case of ferromagnetism, it can easily be seen that the polarizations I_{A} and I_{B} vanish at a temperature called the *Neel temperature* Θ_{N} such that

$$\Theta_{\mathrm{N}} = \frac{C}{2}(w_1 - w_2) \tag{3.63}$$

where

$$C = \frac{Nm^2}{3k_{\mathrm{B}}} \tag{3.64}$$

C is called the *Curie constant.*

Below Θ_{N}, the variation of I_{A} and I_{B} as a function of temperature is of the same type as that shown in Fig. 3.20. The decrease in spontaneous polarization on approaching Θ_{N} is revealed in a particularly spectacular manner by neutron diffraction as shown by Fig. 3.31.

Just below Θ_{N}, it is easy to calculate an antiferromagnetic susceptibility, which is a function of the angle formed by $\boldsymbol{H}$ and $\boldsymbol{I}_{\mathrm{A}}$ (or $\boldsymbol{I}_{\mathrm{B}}$). Let us consider the case when $\boldsymbol{H}$ is parallel to $\boldsymbol{I}_{\mathrm{A}}$, corresponding to a susceptibility denoted $\chi_{/\!/}$. The magnetic moments of the two sites acquire different energies in $\boldsymbol{H}$ (3.20) so that $\boldsymbol{I}_{\mathrm{A}} \neq \boldsymbol{I}_{\mathrm{B}}$. By replacing the Langevin function by its expansion about the origin and by projecting all the vectors onto $\boldsymbol{H}$,

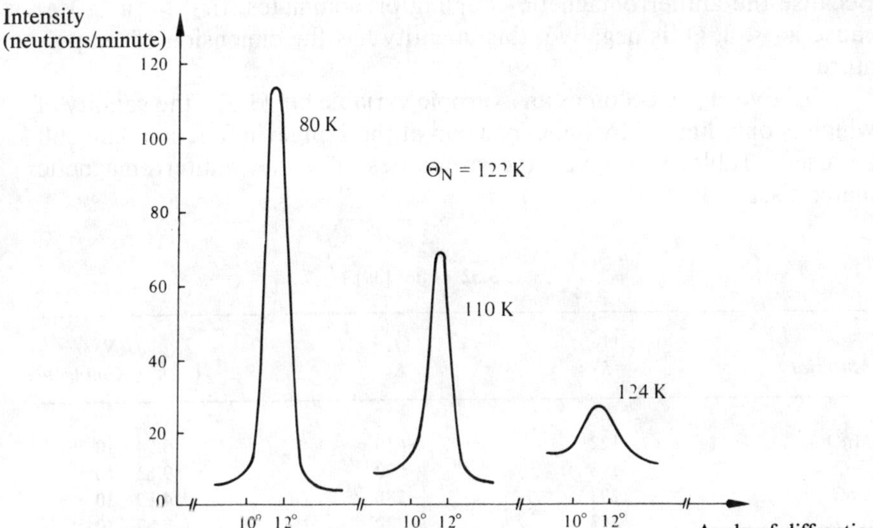

Fig. 3.31 Intensity of the Diffraction Due to the Plane (111) in MnO as a Function of Temperature. (After [4])

(3.61) and (3.62) can be written as

$$I_A = \frac{C}{2T}(H + w_1 I_A - w_2 I_B) \tag{3.65}$$

$$I_B = \frac{C}{2T}(-H + w_1 I_B - w_2 I_A) \tag{3.66}$$

from which we obtain

$$\chi_{\parallel} = \frac{I_A - I_B}{H} = \frac{C}{T - \Theta_a} \tag{3.67}$$

This expression has the same form as the Curie-Weiss law (3.53). The quantity Θ_a is called the *asymptotic temperature* and is equal to

$$\Theta_a = \frac{C}{2}(w_1 + w_2) \tag{3.68}$$

Because the antiferromagnetic coupling predominates, $|w_2| > |w_1|$. Because $w_2 < 0$, Θ_a is negative; this quantity has the dimension of temperature.

Above Θ_N, χ becomes an isotropic variable but (3.67), the validity of which is only limited by the expansion of the Langevin function, can still be used. Table 3.32 gives the properties of some antiferromagnetic materials.

Table 3.32 (After [39].)

Material	Θ_N *K*	Θ_a *K*	*C/N* $H \cdot m^2 \cdot K$*/molecule*
MnO	122	−610	$6.94 \cdot 10^{-11}$
FeO	198	−570	$9.83 \cdot 10^{-11}$
CoO	293	−280	$4.82 \cdot 10^{-11}$
V_2O_4	343	−720	$0.87 \cdot 10^{-11}$

3.6 FERRIMAGNETISM

3.6.1 Introduction

The magnetic stone of the ancients was a natural ferrimagnetic material. Microwave components using magnetic polarization employ artificial ferrimagnetic elements. Ferrimagnetism therefore is featured in the oldest and most recent applications of magnetism. Between these two extremes, ferrimagnetic materials were ignored until the studies of Snoek (1945) made these materials available on an industrial scale. The manufacture of magnetic alloys, benefitting from the knowledge gained in metallurgy, was able to start much earlier. The term *ferrite* is used to designate ferrimagnetic materials. It is also the name given to the α phase of iron in metallurgy.

In many respects, ferrites and magnetic alloys are comparable. In particular, they both have magnetic domains and exhibit the characteristic phenomena of saturation and hysteresis. However, for the engineer, each of these two classes of materials has a specific area of application resulting just as much from their mechanical and electrical characteristics as from their magnetic characteristics. The ferrites are ceramics and are therefore hard, brittle, and poor conductors of electricity as opposed to the metals which are good conductors and usually ductile. One result of this is that large components have to be manufactured as magnetic alloys whereas electromechanical components are an area of application for the ferrites.

Several properties of ferrites can be simply deduced from their crystal structure which we will study first. A phenomenological approach followed by a quantitative approach will then make it possible to examine the polarization mechanisms in more detail on the scale of a magnetic domain for different types of ferrites.

3.6.2 The Spinel Structure

The chemical formula for the ferrites can vary quite considerably. They have a cubic or hexagonal crystal structure. We will look in more detail at the spinel structure, which is the cubic structure of the most important class of the soft ferrites (Section 3.9.6).

The ferrites with spinel structure have the formula MFe_2O_4 in which iron is trivalent. The symbol M denotes a divalent metal chosen from the series Mg^{++}, Mn^{++}, Fe^{++}, Co^{++}, Ni^{++}, Cu^{++}, Zn^{++}, Cd^{++}. Of all these compounds, only $CoFe_2O_4$ is a hard ferrite (Section 3.10.3).

The spinel structure is formed from the alternation of two types of cubic subcells (Fig. 3.33) which are called octants. Each octant is formed from two concentric cubes. The inside cube always has four oxygen ions in tetrahedral sites and the outside cube has four metal ions also in tetrahedral sites. The similarity between the two octants stops here. The remaining ions are arranged in the following manner: four metal ions in tetrahedral sites in the inside cube for one octant, one metal ion at the centre of the cubes for the other (Figs. 3.34 and 3.35).

Examining Figs. 3.34 and 3.35, it can be seen that the metal ions are either at the center of a tetrahedron (tetrahedral site or site A) or at the center of an octahedron (octahedral site or site B). In the complete cell of Fig. 3.33, there are 64 A sites of which only 8 are occupied, and 32 B sites of which 16 are occupied. There are 32 oxygen ions. The complete cell therefore corresponds to the formula $8MFe_2O_4$. All the A sites constitute what is called the sublattice A and all the B sites, the sublattice B.

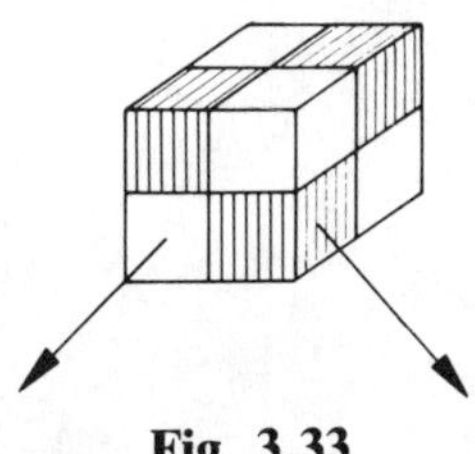

Fig. 3.33

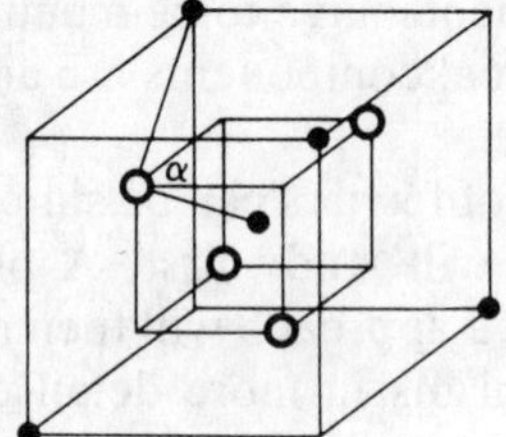

Fig. 3.34

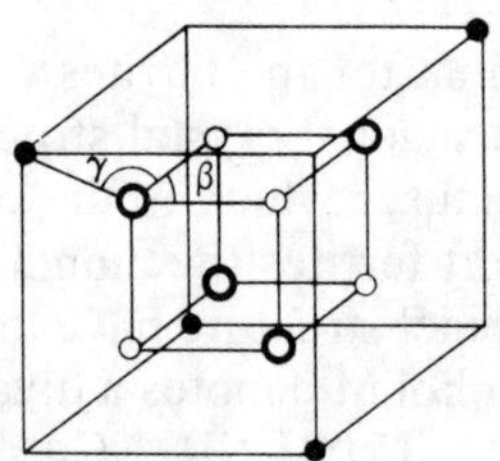

Fig. 3.35

○ oxygen
• metal ion occupying an A site
∘ metal ion occupying a B site

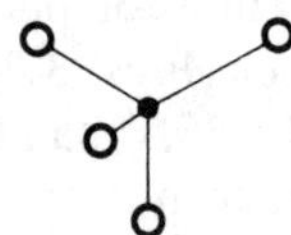

Fig. 3.36 Site A

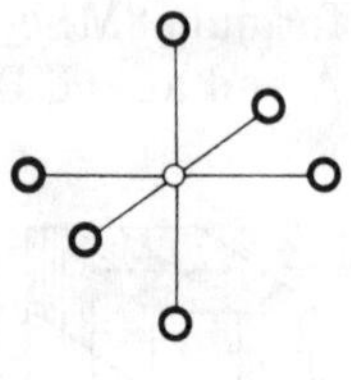

Fig. 3.37 Site B

At an A site, the metal has four nearest neighbors and at a B site it has six. It therefore appears normal that the divalent metal occupies the A sites and Fe^{+++} the B sites. This justifies the names of ferrite with *normal spinel* structure or *inverse spinel* structure used according to the repartition of metal ions between sites A and B, as shown in Table 3.38.

Table 3.38

Type of spinel	*Occupation of the A sites*	*Occupation of the B sites*
Normal	8 M^{++}	16 Fe^{+++}
Inverse	8 Fe^{+++}	8 M^{++} + 8 Fe^{+++}

In reality, a ferrite is always a mixture of the two types of spinels but very often there is an overall predominance of one type over another. The ratio of the different types of spinels may be modified to a certain extent by a suitable thermal treatment. This is particularly the case with $CuFe_2O_4$ and $MgFe_2O_4$. The following ferrites have a predominant inverse structure: $MgFe_2O_4$, Fe_3O_4, $CoFe_2O_4$, $NiFe_2O_4$, and $CuFe_2O_4$—they are all ferrimagnetic. $MnFe_2O_4$ is also ferrimagnetic with a dominant normal structure. On the other hand, $ZnFe_2O_4$ and $CdFe_2O_4$ with dominant normal structure are *not* ferrimagnetic but paramagnetic at room temperature and probably antiferromagnetic at low temperatures.

3.6.3 Phenomenology of Ferrimagnetism

The orientation of the spins in ferrimagnetic materials directly follows from the properties of the superexchange bond. Table 3.39 summarizes the six types of bonds to be considered and the ions involved for the case of a ferrite with inverse spinel structure.

Table 3.39

Sites	*Ions*	*Angle*	*Intensity of the Supercovalent Bond*
A-A	Fe^{+++}—O^{--}—Fe^{+++}	$\alpha \cong 80°$	**Weak**
B-B	M^{++}—O^{--}—M^{++}	$\beta = 90°$	**Very weak**
	M^{++}—O^{--}—Fe^{+++}		
	Fe^{+++}—O^{--}—Fe^{+++}		
A-B	M^{++}—O^{--}—Fe^{+++}	$\gamma \cong 120°$	**Strong**
	Fe^{+++}—O^{--}—Fe^{+++}		

For a given type of bond, the angle δ of Fig. 3.29 corresponds to one of the angles α, β, or γ of Figs. 3.34 and 3.35. The variation in the intensity of the superexchange bond as a function of δ produces a predominance of exchange between the A and B sites. The result is a particular configuration of the magnetic moments represented diagrammatically in Fig. 3.40.

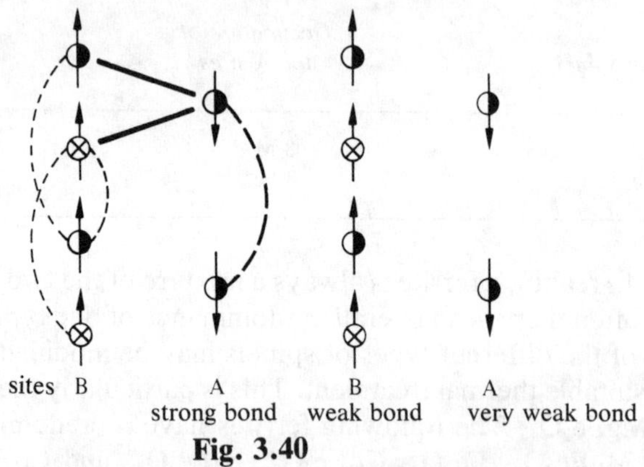

Fig. 3.40

The polarization due to the Fe^{+++} ion vanishes and macroscopically everything happens in a ferrite with inverse spinel as if only the M^{++} ion had a magnetic moment. This moment is always oriented in the direction of ⟨111⟩, except in $CoFe_2O_4$ where it is parallel to ⟨100⟩.

By assuming that the structure of the 3d shell for M^{++} is not modified in a ferrite with respect to that in the free atom (Table 3.24), it is possible to make a theoretical prediction of the saturation polarization of a ferrite with inverse spinel structure as a function of the nature of M^{++}. Figure 3.41 makes it possible to compare this prediction with observation.

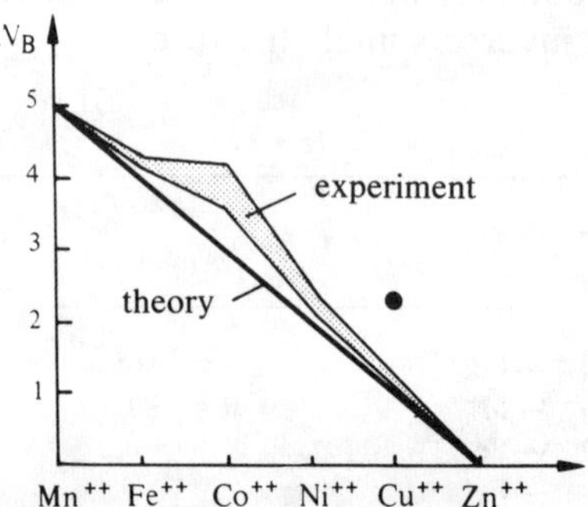

Fig. 3.41 N_B **= Number of Bohr Magnetons Per** MFe_2O_4 **Molecules. (After [31])**

The discrepancies between theory and experiment are relatively small and can be attributed to the presence of a certain quantity of normal spinel. Since Fe^{+++} has $5m_B$, the transfer of a fraction of these ions from the A sites to the B sites increases the saturation polarization except in the case of Mn^{++} where this transfer has no effect, Mn^{++} also having $5m_B$. The orbital magnetic moments can also produce a slight increase in the saturation polarization. This would be the case for the cobalt ferrite, in particular.

The point that clearly departs from the experimental area of Fig. 3.41 (case of copper) is an illustration of the modification of the relative amounts of normal and inverse spinel that can be obtained by thermal treatment.

3.6.4 Mixed Ferrites

It is possible to increase the saturation polarization of ferrites with inverse structure by adding a certain quantity of ferrites with normal structure. The compound $ZnFe_2O_4$ is often used for this purpose. One then speaks of a *mixed ferrite* with zinc in this particular case. The formula of such a ferrite can be written as

$$(1 - x)MFe_2O_4 + xZnFe_2O_4 \tag{3.69}$$

where $0 \leq x \leq 1$. The addition of spins is revealed by presenting this expression in the form:

$$(1 - x)[FeO \cdot MFeO_3] + x[ZnO \cdot Fe_2O_3] \tag{3.70}$$

with the convention that the metal ions occupying the A sites come before the point in the formula and the metal ions occupying the B sites come after the point. Expression (3.70) can also be written as

$$Fe_{1-x}\, Zn_x\, O \cdot Fe_{1+x}\, M_{1-x}\, O_3 \tag{3.71}$$

The magnetic moment in Bohr magneton units, N_B, of this molecule is

$$N_B = 5(1 + x) + n(1 - x) - 5(1 - x) = n + x(10 - n), \tag{3.72}$$

where n is the number of Bohr magnetons carried by M^{++}. (Fe^{+++} carries $5m_B$ and Zn^{++} does not carry any).

According to (3.72), regardless of n, for $x \rightarrow 1$, the molecule (3.71) would carry $10m_B$, which contradicts the fact that $ZnFe_2O_4$ is not magnetic. In reality, (3.72) is only valid for small values of x because the predominant type of superexchange varies as x increases, as can be seen from Table 3.42.

For $0 \leq x < 0.25$, the antiferromagnetic coupling between the A and B sites is predominant. At B, there is no modification of the magnitude of the magnetic moments present and no modification of their orientation. On the other hand, at A, the substitution of Fe^{+++} by Zn^{++} causes a reduction in the average magnetic moment of this site, and therefore an increase in N_B. The validity of (3.72) in this region of x appears in Fig. 3.43.

Table 3.42

Predominant Superexchange	x	*Occupation of Sites A*	*Occupation of Sites B*	N_B
AB	0	 ◑↑ ◑↑ ◑↑ ◑↑ ◑↑ ◑↑ ◑↑ ◑↑	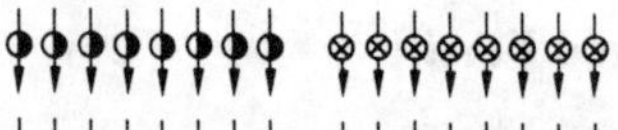 ◑↓ ◑↓ ◑↓ ◑↓ ◑↓ ◑↓ ◑↓ ◑↓ ⊗↓ ⊗↓ ⊗↓ ⊗↓ ⊗↓ ⊗↓ ⊗↓ ⊗↓	5
AB	0,25	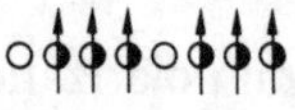 ○ ◑↑ ◑↑ ◑↑ ○ ◑↑ ◑↑ ◑↑	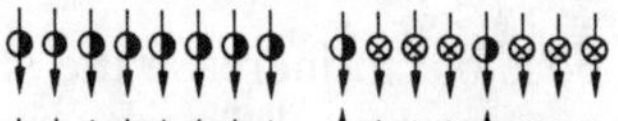 ◑↓ ◑↓ ◑↓ ◑↓ ◑↓ ◑↓ ◑↓ ◑↓ ◑↓ ⊗↓ ⊗↓ ⊗↓ ◑↓ ⊗↓ ⊗↓ ⊗↓	6,25
AB; BB	0,50	 ○ ○ ◑↑ ◑↑ ○ ○ ◑↑ ◑↑	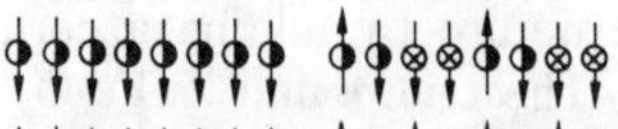 ◑↓ ◑↓ ◑↓ ◑↓ ◑↓ ◑↓ ◑↓ ◑↓ ◑↑ ◑↓ ⊗↓ ⊗↓ ◑↑ ◑↓ ⊗↓ ⊗↓	5
AB; BB	0,75	 ○ ○ ○ ◑↑ ○ ○ ○ ◑↑	 ◑↓ ◑↓ ◑↓ ◑↓ ◑↓ ◑↓ ◑↓ ◑↓ ◑↑ ◑↓ ◑↑ ⊗↓ ◑↑ ◑↓ ◑↑ ⊗↓	3,75
BB	1	○ ○ ○ ○ ○ ○ ○ ○	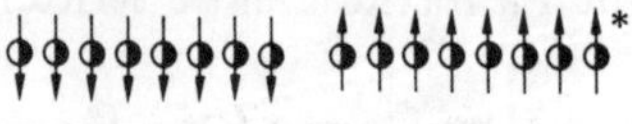 ◑↓ ◑↓ ◑↓ ◑↓ ◑↓ ◑↓ ◑↓ ◑↓ ◑↑ ◑↑ ◑↑ ◑↑ ◑↑ ◑↑ ◑↑ ◑↑*	0

◑, Fe^{+++}; ⊗, Mn^{++}; ○, Zn^{++} *** This structure is only produced at low temperatures, see text.**

For $0.25 \leq x < 0.5$, the reduction in the average magnetic moment at A weakens the couplings involved with this site, in particular the A–B coupling. The B–B coupling, until now completely masked by the A–B coupling, begins to emerge: some Fe^{+++} ions of B are oriented antiparallel to each other. The reduction in N_B that results from this is more or less counteracted by the increase of the polarization at A. If M^{++} has a strong magnetic moment, N_B increases further, otherwise it already decreases.

For $0.5 \leq x < 0.75$, the B–B coupling fairly covers the A–B coupling; in all cases, N_B decreases when x increases.

For $0.75 \leq x < 1$, the trend started in the interval $0.5 \leq x \leq 0.75$ is accentuated and finally the A–B coupling disappears, the A site no longer containing magnetic ions. Since the B–B coupling remains alone, $ZnO \cdot Fe_2O_3$ should be antiferromagnetic but without the addition of a certain amount of A–B coupling, the B–B coupling appears insufficient (at room temperature) to cause any alignment of magnetic moments. The material therefore becomes paramagnetic. This fact is confirmed by neutron diffraction.

Experimental results obtained with different mixed ferrites with $ZnO \cdot Fe_2O_3$ are shown in Fig. 3.43. They are in agreement with the mechanisms described above. The absence of experimental points for the highest values of x is due to the transition from ferrimagnetism to paramagnetism which has just been discussed.

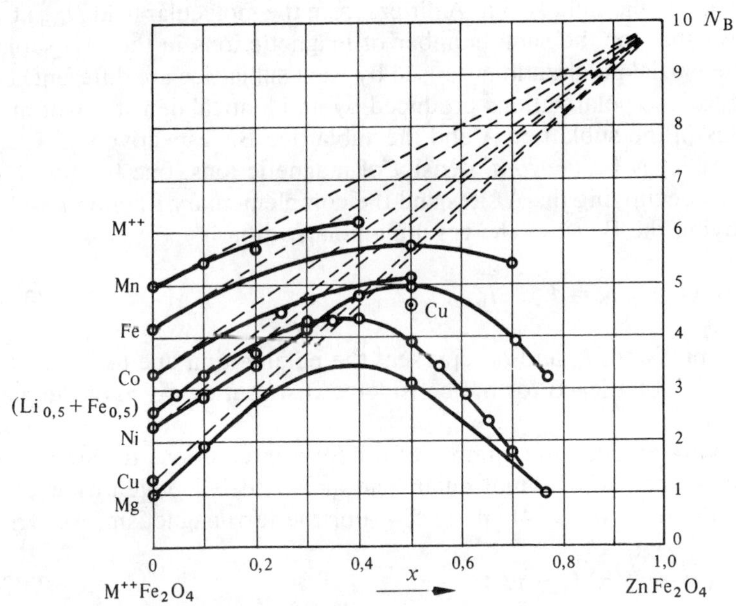

Fig. 3.43 (After [32])

3.6.5 Theory of Ferrimagnetism

Neel [33] has developed a phenomenological theory for ferrimagnetism similar to that given in Section 3.5.4 for antiferromagnetism. The basic principles are given below.

In the general case, there are a large number of different interactions due to the nature of the magnetic ions involved and due to their relative positions in space. The number of different interactions obviously depends on the crystal structure of the ferrites concerned.

In order to simplify this situation, Neel considers the ferrite as formed from a single type of magnetic ion, distributed unequally between the A and B sublattices. In this way, there only remain three types of interactions: *one* interaction of the A–A type, *one* interaction of the A–B type, and *one* interaction of the B–B type. Although this simplification appears rather radical, it should be noted that by proportioning the magnetic ions between the A and B sublattices, the theory gives a *global* representation of any polarization due to the A or B sites that could result from a more complex real situation.

The neighborhood of an A site is different from that of a B site so that the molecular field H_{mA} at A differs from the molecular field H_{mB} at B. It follows that for the same number of magnetic ions in the two sublattices, the partial polarizations caused by each sublattice are different. Let I_a and I_b be the polarizations produced by an identical density N of magnetic ions in the sublattice A and the sublattice B, respectively.

Now let N be the *total* density of magnetic ions, one fraction λ of these ions occupying the A sites and the complementary fraction $\nu = 1 - \lambda$ occupying the B sites. The resulting polarization is

$$I = \lambda I_a + \nu I_b = I_A + I_B \tag{3.73}$$

In this expression, I_A and I_B represent the polarization due to the sublattice A and sublattice B for the considered distribution (λ, ν) of the magnetic ions.

Having reduced the number of different interactions to three, as in antiferromagnetism, the molecular field at A and B is given by expressions of the form of (3.54) and (3.55). For the ferrimagnetism, we write

$$H_{mA} = w_{AB}(\alpha\lambda I_a - \nu I_b) \tag{3.74}$$

$$H_{mB} = w_{AB}(\beta\nu I_b - \lambda I_a) \tag{3.75}$$

with

$$\alpha = \frac{w_{AA}}{w_{AB}} \qquad \beta = \frac{w_{BB}}{w_{AB}} \tag{3.76}$$

Considering the magnetic moments at each site as subject to the molecular fields of the site and to a possible applied magnetic field H, the polarization produced by each sublattice is calculated by the Langevin theory for paramagnetism (Section 3.4.2). By successively substituting (3.74) and (3.75) in (3.28) we obtain, if $\boldsymbol{H}$ is parallel to $\boldsymbol{I}_b$:

$$I_a = Nm\mathrm{L}\left(\frac{mw_{AB}(\alpha\lambda I_a - \nu I_b) - mH}{k_B T}\right) \tag{3.77}$$

$$I_b = NmL\left(\frac{mw_{AB}(\beta\nu I_b - \lambda I_a) - mH}{k_B T}\right) \tag{3.78}$$

The solution of this system of equations is simple above the Curie temperature, the Langevin function being replaced by the first term of its expansion (3.30). On completion of a tedious calculation, we obtain

$$\frac{1}{\chi} = \frac{T}{C} + \frac{1}{\chi_0} - \frac{\sigma}{T - \Theta} \tag{3.79}$$

C is the Curie constant (3.64) and χ_0, σ, and Θ are constants depending on C, the fractions λ and ν, and the interaction coefficients w_{AB}, α and β.

Expression (3.79) is the equation of a hyperbola, whose part has a physical meaning and is shown in Fig. 3.44. For sufficiently large values of T, this hyperbola coincides with its asymptotes, the last term of (3.79) tending to zero. Then,

$$\chi = \frac{C}{T + C/\chi_0} \tag{3.80}$$

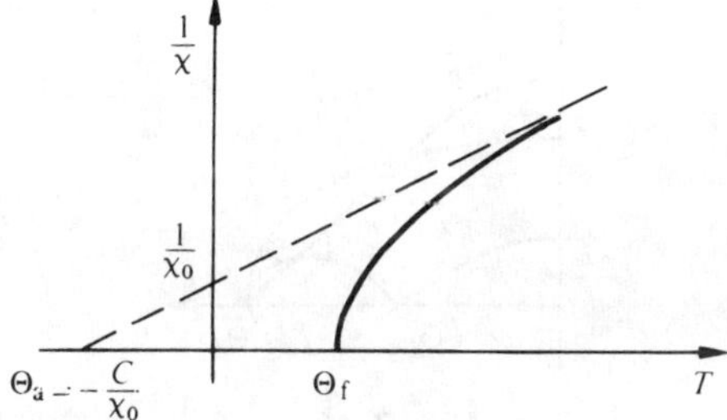

Fig. 3.44 Θ = the Curie Temperature

This is the Curie-Weiss law already encountered for ferromagnetism above the Curie temperature. As with antiferromagnetic materials, the asymptotic temperature Θ_a is negative.

Below the Curie temperature, each sublattice has a spontaneous polarization. The polarization on the macroscopic scale is still given by (3.73). To learn the value of I_a and I_b, it is necessary to solve the system of equations (3.77) and (3.78). The procedure used for the case of ferromagnetism (Fig. 3.19) can no longer be applied because these two equations are coupled. In fact, since the predominant superexchange bond is the one that binds the sublattice A to the sublattice B, the polarization due to the A sites essentially depends on that created by the B sites and *vice versa*.

3.6.6 Polarization versus Temperature

The solution of the system formed by the equations (3.77) and (3.78) shows [33] that the variation of the polarization versus temperature may take three different forms shown in Figs. 3.45 to 3.47.

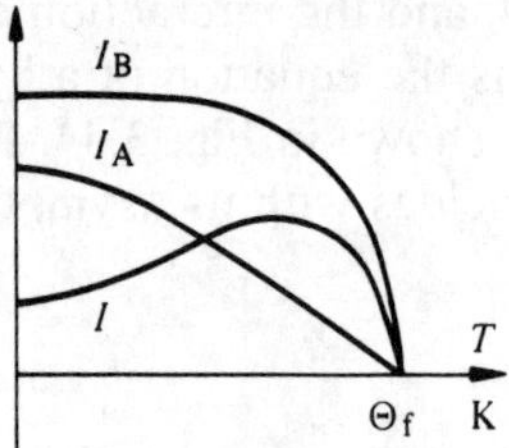

Fig. 3.45

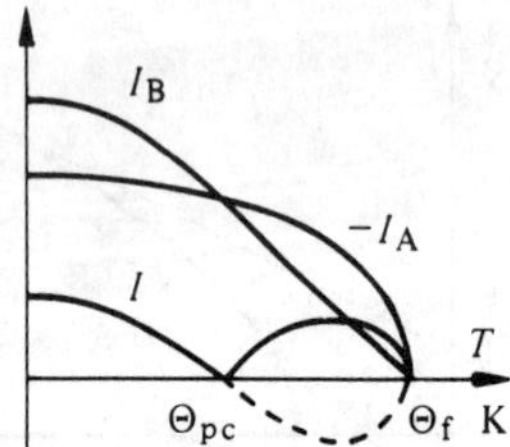

Fig. 3.46

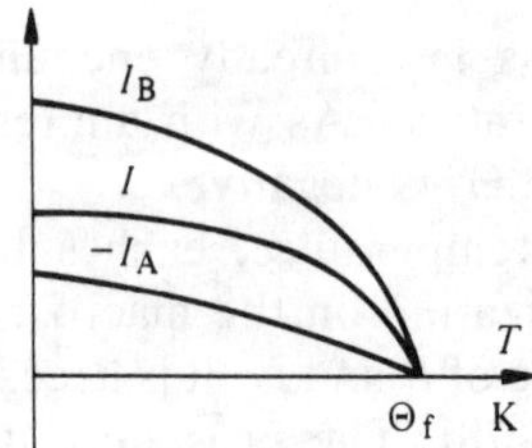

Fig. 3.47

The sublattices A and B always have the same Curie temperature because as soon as a sublattice no longer has spontaneous polarization, it can no longer align the magnetic moments of the other.

In Fig. 3.45, I_B varies little except in the vicinity of Θ, whereas I_A decreases regularly from 0 K to Θ. The maximum value of the spontaneous polarization is shifted from 0 K to approximately 0.5 Θ, for example, in the case of complex ferrites of the type $NiO \cdot Fe_{2-x}Al_xO_3$ with $x \approx 0.63$.

In Fig. 3.46, it is I_B that regularly decreases, I_A mainly varying in the vicinity of Θ. It follows that I_A and I_B intersect each other at a temperature called the *compensation point* Θ_{pc}. At this temperature, the ferrite is antiferromagnetic. Above Θ_{pc} and up to the Curie temperature, the polarization of the sublattice A overcomes that of sublattice B. Below Θ_{pc} it is the opposite, but in both cases the resulting spontaneous polarization is aligned with an applied field H (Fig. 3.46, solid line). On the other hand, in the absence of H, if the specimen has a remanent polarization, a change in the direction of this polarization can be seen on passing through Θ_{pc} (Fig. 3.46, dashed line). For example, the ferrite $Li_{0.5}FeCr_{1.5}O_4$, has a compensation point at approximately 0.65 Θ_f.

The variation in the polarization described by Fig. 3.47 is the most frequent. It is found in particular in all ferrites with spinel structure either simple or mixed with Zn (Fig. 3.48).

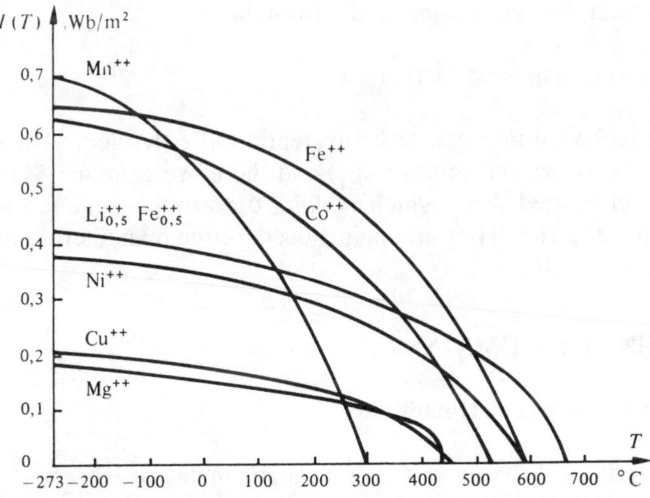

Fig. 3.48 Saturation Polarization *versus* Temperature for Some Simple Ferrites $M^{++}O \cdot Fe_2O_3$ with Spinel Structure. (After [34])

3.6.7 Other Ferrimagnetic Compounds

There are several families of ferrimagnetic compounds that do not have a spinel structure. We will just mention the most important of them. The superexchange bond always forms the basis for the mechanism of polarization of these ferrites but the crystal structures and the chemical formulae are more complex. There are sometimes three sublattices formed from particular magnetic sites as in ferrites with the garnet structure corresponding to the formula:

$$3M_2O_3 \cdot 2Fe_2O_3 \cdot 3Fe_2O_3 \tag{3.81}$$

The complete unit cell contains four molecules (3.81), i.e., 160 atoms. M is a metal chosen from the period of the lanthanide elements, or yttrium. In this case, the ferrite is known by the abbreviation YIG (yttrium iron garnet). All metal ions are trivalent. Most of these ferrites have a compensation point. Their value [35,36,37] essentially lies in a high electrical resistivity and very low magnetic losses at high frequency except at the position of a sharp maximum (resonance peak) situated in the microwave region. This peak is particularly narrow in YIG, which is in fact a material of prime choice for the manufacture of various components.

Of the ferrites with hexagonal structure, only the ferrites of barium and strontium, corresponding to the formula:

$$MFe_{12}O_{19} \text{ where } M = \text{Ba or Sr} \tag{3.82}$$

have any technical interest. M is divalent and Fe trivalent. The unit cell corresponds to two molecules (3.82) and therefore contains 64 atoms. It has a very elongated shape which explains the strong magnetic anisotropy of this type of ferrite. They are mainly used for the manufacture of permanent magnets [38].

3.7 MAGNETIC DOMAINS

3.7.1 Introduction and Definitions

The principal characteristic of magnetic materials is the spontaneous alignment of the atomic magnetic moments parallel to each other. Why is it, under these conditions, that any sample of magnetic material is not a permanent magnet in the absence of an applied field H? A first answer to this question was given in Section 3.3.6. It will be developed in detail in this and the next section.

The parallel orientation of magnetic moments is a *local* phenomenon. Appropriate observation techniques (Section 3.7.10) make it possible to show the distribution of magnetic polarization in matter. They show that a sample of macroscopic size is generally divided into a large number of polarized regions. Since the orientation of the polarization changes from one region to another, the overall polarization of the sample may quite likely be zero.

The *magnetic domain* or *Weiss domain* is the name given to each region in one block in which all the atomic magnetic moments are aligned parallel to each other. Magnetic domains have highly variable shapes. In their greatest extension, they can measure up to 1 mm and even more in wires. The modulus of the spontaneous polarization I_s has the same value in all the magnetic domains of a homogeneous material with a uniform temperature.

Each domain is separated from its neighbor by a transition zone in which the orientation of the magnetic moments passes progressively from the polarization direction in one of the domains to that in the other domain.

This transition zone, a sort of envelope marking the boundaries of the magnetic domain, is called the *Bloch wall* (Fig. 3.49).

The magnetic moment $\boldsymbol{M}$ of a sample of volume V has to lie between the following limits:

$$0 \leq \boldsymbol{M} = \int \boldsymbol{I}_s \, \mathrm{d}V \leq \boldsymbol{I}_s \, V \tag{3.83}$$

It can go from 0 to $\boldsymbol{I}_s V$ under the effect of an applied field $\boldsymbol{H}$, which implies a complete modification of the distribution of the magnetic domains, i.e., their number, their shape, and their distribution may vary radically.

All the magnetic properties of materials that most directly interest engineers, such as the permeability, hysteresis, *et cetera,* depend on the structure of the magnetic domain and especially on the more or less large variability of this structure under the effect of an external field.

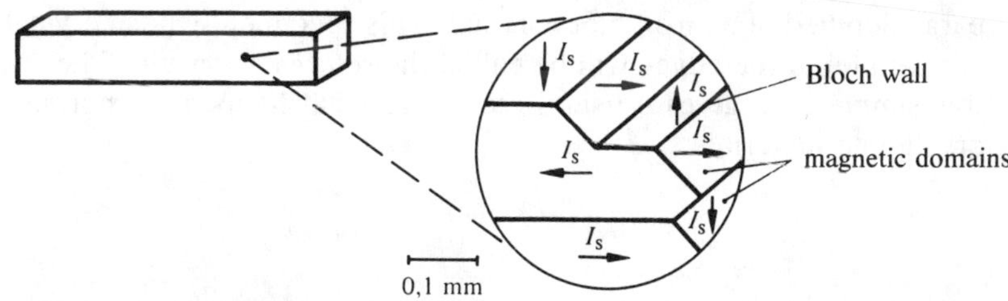

Fig. 3.49 Division of a Specimen Into Magnetic Domains

3.7.2 Structure of Magnetic Domains in a Monocrystal

The structure of magnetic domains in a monocrystal is defined by the following properties of magnetic domains expressed as statistical mean values for a given sample:

- the shape;
- the size;
- the orientation of $\boldsymbol{I}_s$.

Of all the possible structures, in practice, only those corresponding to *a* minimum of the internal energy of the sample can occur. A minimum and not *the* minimum because it can happen that the necessary energy ΔW to reach the minimum is lacking (Fig. 3.50).

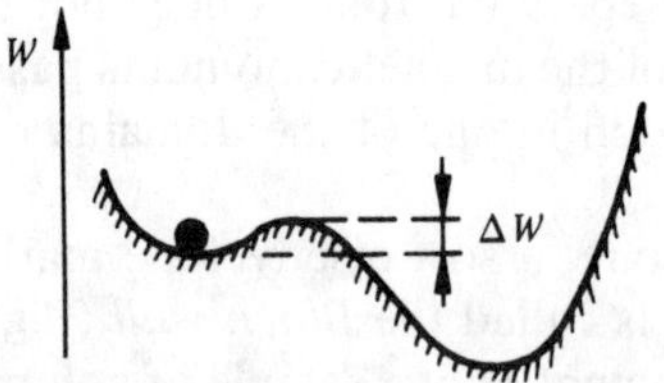

Fig. 3.50 Mechanical Analogy: the Structure of Magnetic Domains Can Correspond to a Metastable Equilibrium

The internal energy associated with the structure of the domains, W_{im} contains four terms:

$$W_{im} = W_{an} + W_{ms} + W_{ec} + W_{mt}, \tag{3.84}$$

where W_{an} is the magnetic anistropy energy, W_{ms} is the magnetostatic energy, W_{ec} is the exchange energy, and W_{mt} is the magnetostriction energy.

The effect of each of these terms will be examined in the case of a parallelepiped of an iron monocrystal. This specimen, whose crystal axes are parallel to the edges will be called the *reference crystal*. The study of this simple case gives a useful picture of what happens in normal poly-crystalline materials.

3.7.3 The Anisotropy Energy W_{an}

The intensity of interaction of atomic magnetic moments resulting in their parallel orientation is measured in terms of energy. A part of this energy varies as a function of the angle α of Fig. 3.51 and is called the *magnetocrystalline anisotropy energy*. There are other forms of magnetic anisotropies, for example, those produced in metals by rolling (Section 3.9.3). In the reference crystal, W_{an} is reduced to magnetocrystalline anisotropy.

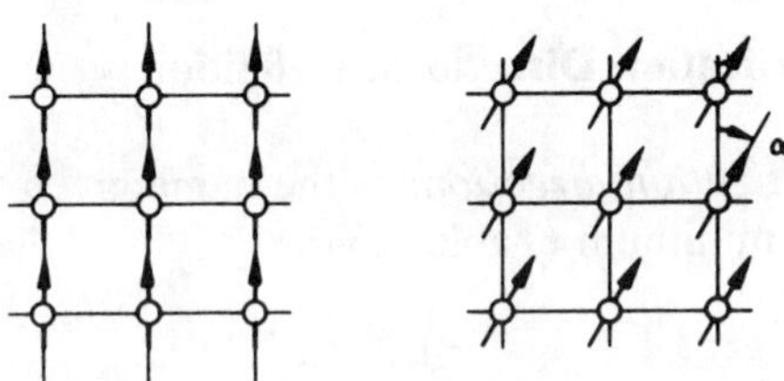

Fig. 3.51 Diagrammatic Representation of the Orientation of the Magnetic Moments with Respect to the Crystal Lattice

The function $W_{an}(\alpha)$ differs from one material to another. In three-dimensional space, this function is written as $W_{an}(\alpha_1, \alpha_2, \alpha_3)$ where the α_i are the direction cosines of $\boldsymbol{I}_s$ with respect to the crystal axes (Fig. 3.52). In iron, W_{an} is a minimum along the directions ⟨100⟩. Under the effect of the magnetocrystalline anisotropy energy *alone,* the spontaneous polarization will therefore tend to be aligned in parallel to the edges of the reference crystal (Fig. 3.53).

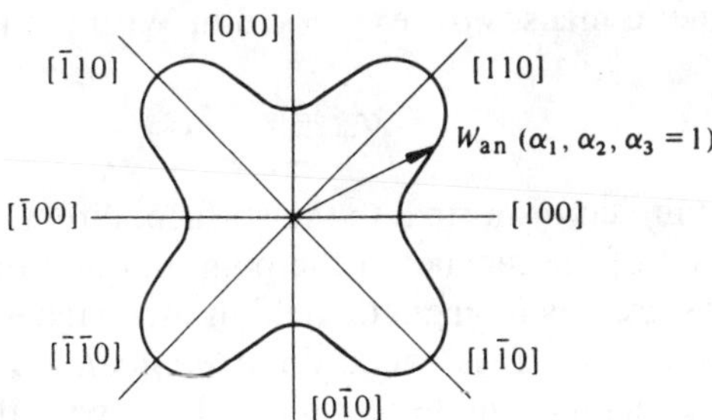

Fig. 3.52 Example of the Polar Diagram of W_{an} in the (001) Plane of a Cubic Crystal

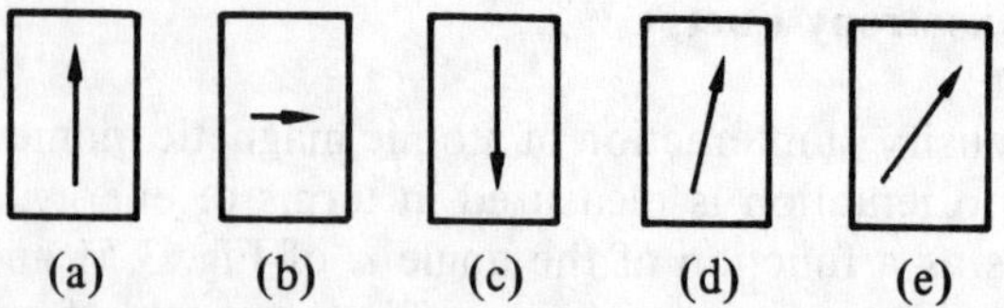

Fig. 3.53 Five Cases of Polarization of the Reference Crystal. W_{an} is a Minimum in Cases (a), (b), and (c), and Then Can Successively Increase in Cases (d) and (e)

3.7.4 Easy Magnetization Directions: Definition

Easy magnetization directions is the name given to the directions $\boldsymbol{d}$ in which W_{an} is a minimum (Table 3.54).

Table 3.54
Easy magnetization directions.

Element	*d*	*Crystal Structure*	*Type of Anisotropy*
Fe	⟨100⟩	CC	cubic
Co	axial	Hex	uniaxial
Ni	⟨111⟩	CFC	cubic

3.7.5 Magnetostatic Energy W_{ms}

At the surface of a permanent magnet placed in vacuum, there is a density of magnetic masses σ_m expressed in Wb/m² and given by

$$\sigma_m = \boldsymbol{n} \cdot \boldsymbol{I}, \tag{3.85}$$

where $\boldsymbol{n}$ is the unit vector normal to the surface, directed outwards, and $\boldsymbol{I}$ is the polarization of the magnet at the point in question. This density of magnetic masses creates a magnetic field in the magnet $\boldsymbol{H}_d$ with opposite orientation to that of $\boldsymbol{I}$. $\boldsymbol{H}_d$ is called the *demagnetizing field.*

In a homogeneous sample of simple shape with the same polarization $\boldsymbol{I}$ at every point, $\boldsymbol{H}_d$ is proportional to $\boldsymbol{I}$ and we set

$$\boldsymbol{H}_d = -\frac{1}{\mu_0} N\boldsymbol{I}, \tag{3.86}$$

where N is the *demagnetizing factor*. Its value is given for cylindrical bars and ellipsoids of all proportions in [29], (see Table 3.55).

Table 3.55
Demagnetizing Factor of a Cylinder Polarized Along Its Axis. (After [29].)

Length/Diameter	*0*	*1*	*2*	*5*	*10*
N	**1**	**0.27**	**0.14**	**0.040**	**0.0172**

The internal energy of the permanent magnet resulting from the fact that it is subjected to its own demagnetizing field is called the *magnetostatic energy* W_{ms} *of the magnet*. This energy is given by

$$W_{ms} = -\frac{1}{2}\int \boldsymbol{I} \cdot \boldsymbol{H}_d \, dV, \tag{3.87}$$

the integral applying to the entire volume of the magnet. The calculation of W_{ms} from (3.87) is always laborious except when N is known.

In the case of the parallelepiped shown in Fig. 3.56, expression (3.88) makes a good approximation [39] provided that l is small compared to the length of the sides:

$$W_{ms} \cong 1{,}1 \cdot 10^5 \frac{1}{n} I^2 b^2 c \tag{3.88}$$

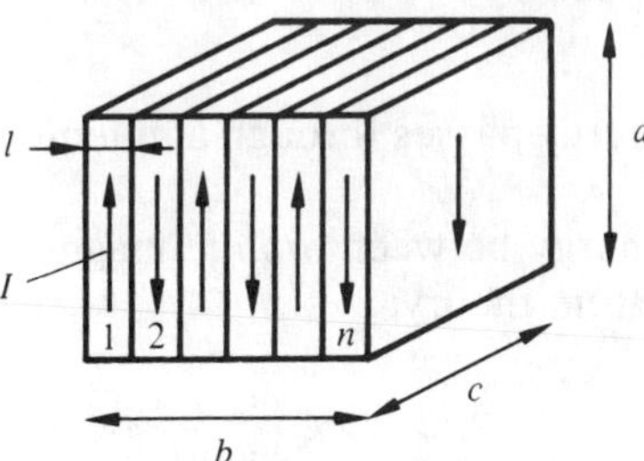

Fig. 3.56 Magnetic Domains in a Lamellar Structure

Thus, the magnetostatic energy divides up the reference crystal into as many lamellar domains as possible because W_{ms} is inversely proportional to the number n of lamellas.

When the magnetocrystalline anisotropy has a cubic symmetry, the configuration of Fig. 3.57(d) is possible and it cancels W_{ms}. The arrangement of the domains is such that the specimen does not produce any external field. On the diagonals, $\sigma_m = 0$ because the two adjacent domains produce equal and opposite magnetic mass densities. There is no demagnetizing field.

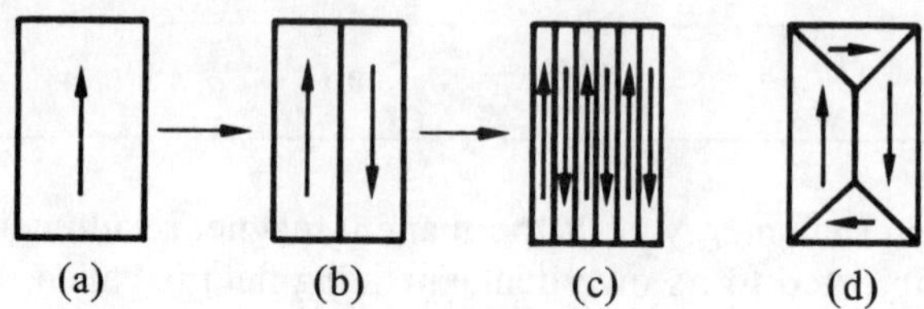

Fig. 3.57 From (a) to (c): Lamellar Structures with Decreasing W_{ms}

3.7.6 Exchange Energy W_{ex}

The *exchange energy* is the name given to the energy resulting from the interaction of two magnetic moments. This energy depends on the distance separating these moments as well as on their relative orientation. It can be easily calculated by classical theory, i.e., by representing the magnetic moments as small magnets. When the moments are arranged as shown in Fig. 3.58, we find [52]:

$$W_{ex} = -\frac{2m^2}{4\pi\mu_0 r^3} \cos\theta \tag{3.89}$$

The exchange energy passes through a minimum when the magnetic moments are parallel.

The exchange energy between *atomic* magnetic moments can only be calculated by quantum theory.

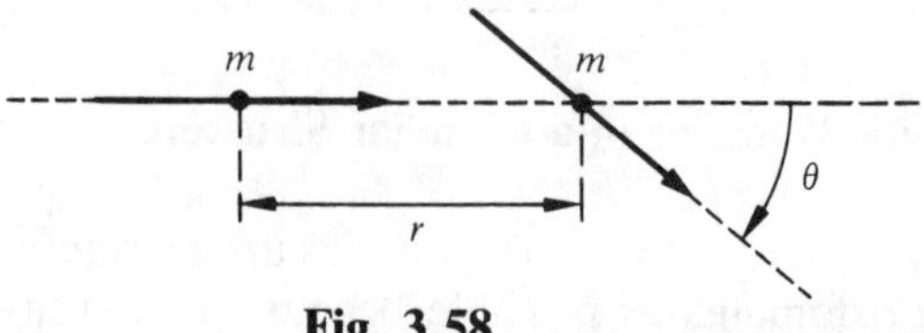

Fig. 3.58

One of the simplest quantum models, only taking into account the spin magnetic moments, results in a relationship that is formally similar to (3.89). Apart from an additive constant of no interest here, it can be written as

$$W_{ij} = -2J_{ij}S_iS_j \cos \theta \tag{3.90}$$

The subscripts i and j distinguish the two atoms with spin quantum numbers S_i and S_j. These numbers are multiples of ½ and are related to the magnetic moments of the atoms by $m = 2Sm_B$. The quantity J_{ij} is called the exchange integral; it has the dimension of an energy.

Often, the spins of the two atoms have the same value and then (3.90) is written in the form:

$$W_{ex} = -2JS^2 \cos \theta \tag{3.91}$$

The similarity between (3.89) and (3.91) should not make us forget the totally different approaches used to obtain these two relationships. For more information concerning (3.90) it is worth referring to [40]. The sign of J_{ij} determines the type of coupling. If $J_{ij} > 0$, W_{ij} is a minimum when the moments are parallel and the coupling is of the ferromagnetic type. When $J_{ij} < 0$, the coupling is antiferromagnetic.

The exchange energy and the magnetocrystalline anisotropy energy increase inside a Bloch wall. The sum of these two increases therefore represents an energy W_{SB} associated with the Bloch wall itself.

The existence of W_{SB} has the effect of limiting the total area of the Bloch walls. In particular, the division of the reference crystal into lamellar domains of increasing number (Figs. 3.57(a–c)) stops when the reduction in magnetostatic energy obtained in this way is compensated by the increase in energy in the Bloch walls.

3.7.7 Magnetostriction Energy W_{mt}

Magnetostriction is a dimensional variation related to the magnetic polarization. Magnetostriction is called *positive* when the specimen expands in the direction of $\boldsymbol{I}$, and *negative* if it contracts in the direction of $\boldsymbol{I}$ (Fig. 3.59). In relative values, the dimensional variation is on the order of 10^{-6} to 10^{-5}.

The *magnetostriction energy* W_{mt} is the elastic energy associated with the deformations and stresses originating from magnetostriction. In the reference crystal, this energy varies according to the structure of the magnetic domains (Fig. 3.60).

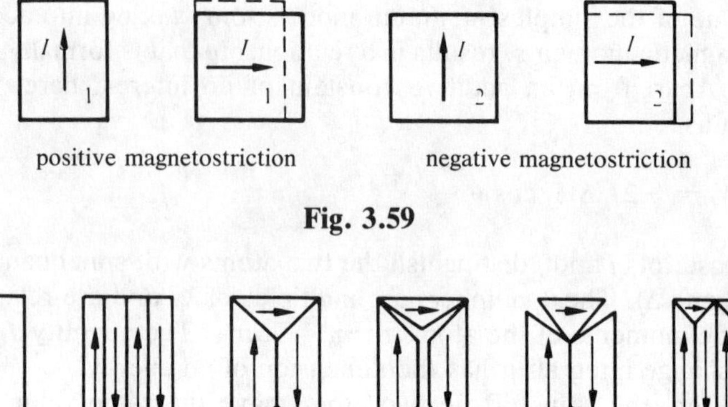

Fig. 3.59

(a) (b) (c) (d) (e) (f)

Fig. 3.60

Configurations (a) and (b) of Fig. 3.60 correspond to a low magnetostriction energy because it is only the result of the deformation of the crystal along the direction of polarization. A higher energy W_{mt} corresponds to the configuration (c). The increase in W_{mt} is due to the action of the stresses that the four domains exert on each other. It is easy to get an idea of these stresses by thinking of the crystal as partitioned by the Bloch walls. If the magnetostriction is positive, these segments will appear as in (d). Case (e) corresponds to a negative magnetostriction. The increase in magnetostriction energy represents the work that would be necessary to deform the segments in order to reassemble them. In case (f), W_{mt} is lower than in case (c).

3.7.8 Structure of Magnetic Domains in Polycrystalline Matter

The general principle according to which the structure of the domains corresponds to a minimum in the internal energy of the specimen remains valid. However, this minimum cannot be obtained by individually reducing the internal energy of each grain considered as isolated from all the others. It is obvious that the magnetostatic energy, for example, strongly depends on the environment of every grain. The division into domains within each grain will therefore only have a *tendency* to take place based on the principles that we have just been examining with regard to the reference crystal.

The effect of the shape of the specimen, predominant in the case of a monocrystal, disappears in the case of a polycrystal, where it is the structure of the grains that is of key importance. The structure of the domains is also influenced by the presence of imperfections that are point imperfections, relatively speaking, such as cavities, inclusions of nonmagnetic matter, or precipitates. These imperfections slow down the displacement of the Bloch walls by a process of pinning and give rise to secondary structures such as the blade structure (Fig. 3.61).

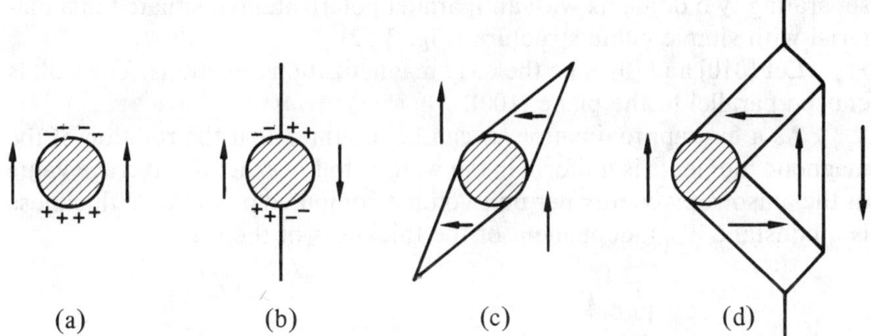

Fig. 3.61 Examples of the Possible Effects of an Inclusion on the Local Structure of Magnetic Domains

Inclusion produces a magnetostatic energy related to the existence of positive and negative magnetic masses (Fig. 3.61 (a)). This energy is reduced (Fig. 3.61 (b)) by the presence of a Bloch wall passing through the inclusion and causing a reduction in the average distance between the + and − masses. The Bloch wall therefore has a tendency to be retained by the inclusion which is called *pinning*. This effect is of great technological importance because of its direct effect on the form of the magnetization curve (Section 3.8). In the absence of preexisting Bloch walls in the vicinity of a point defect, the magnetostatic energy is often reduced by the creation of secondary blade-shaped domains (Fig. 3.61 (c)). The reduction in magnetostatic energy is sometimes obtained (Fig. 3.61 (d)) by an indirect pinning by one wall involving several secondary domains.

Mechanical treatment, cold rolling in particular; *thermal treatment,* annealing, melting, quenching, *et cetera*; and mixing of different elements for the *production of alloys* are the three main procedures making it possible to modify the structure of magnetic domains and the mobility of Bloch walls. These are the principal processes for manufacturing magnetic materials.

3.7.9 Thickness of a Bloch Wall

The thickness of a Bloch wall is fixed at a value δ so that the energy of the wall W_{SB} is a minimum. This condition can be written as

$$\frac{\partial W_{SB}(\delta)}{\partial \delta} = 0 \tag{3.92}$$

The calculation of δ is carried out for the case of a wall at 180° (i.e., separating two domains with antiparallel polarizations), situated in a material with simple cubic structure (Fig. 3.62).

Let [010] and [0$\bar{1}$0] be the easy magnetization directions. The wall is chosen parallel to the plane (100).

As a first approximation, it can be assumed that the rotation of the magnetic moments is uniform in the wall. It follows that the average value of the anisotropy energy per unit volume computed on the total thickness is a constant $\bar{W}_{an}$ independent of the thickness of the wall.

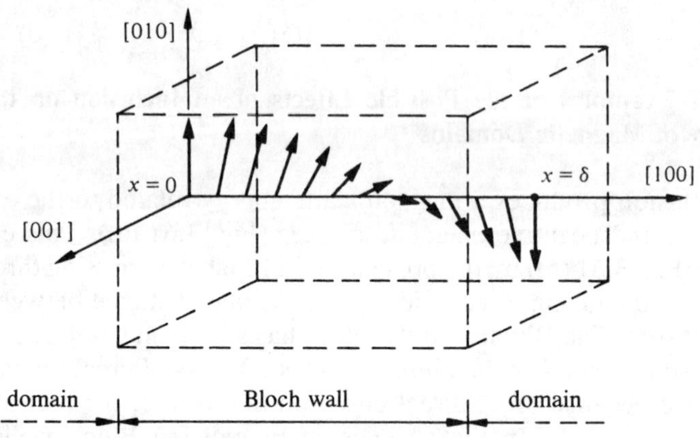

Fig. 3.62 Rotation of Magnetic Moments in a Bloch Wall at 180°

The number n of lattice planes contained in the thickness of the wall is sufficient for the angle θ between two adjacent magnetic moments to be small, which makes it possible to replace cos θ in (3.91) by the first two terms in the expansion of this function. By putting the exchange energy of two parallel magnetic moments equal to zero, we have

$$W_{ec} = JS^2 \, \theta^2 \tag{3.93}$$

The exchange energy per unit area of the wall, denoting the base vector of the lattice as a, can be written as

$$JS^2 \left(\frac{\pi}{n}\right)^2 n \frac{1}{a^2} \tag{3.94}$$

because each plane contains $1/a^2$ magnetic moments per unit area. Finally,

$$W_{\mathrm{SB}} = \overline{W}_{\mathrm{an}}\delta + \frac{JS^2\pi^2}{a\delta} \tag{3.95}$$

from which we obtain

$$\frac{\partial W_{\mathrm{SB}}}{\partial \delta} = 0 = \overline{W}_{\mathrm{an}} - \frac{JS^2\pi^2}{a\delta^2} \tag{3.96}$$

and

$$\delta = \sqrt{\frac{JS^2\pi^2}{a\overline{W}_{\mathrm{an}}}} \tag{3.97}$$

In the case of iron (with a structure that is centered cubic and not simple cubic, but here we are only looking for an order of magnitude of δ), $a = 0.287$ nm, $\overline{W}_{\mathrm{an}} = 4.2 \cdot 10^4\ \mathrm{J/m^3}$, $J = 2.16 \cdot 10^{-21}$ J if it is assumed that $S = 1$, from which we have

$$\delta = 0.042\ \mu\mathrm{m}$$

The thickness of a Bloch wall therefore corresponds to approximately 150 base vectors of the lattice. By substituting (3.97) in (3.95), we have

$$W_{\mathrm{SB}} = \overline{W}_{\mathrm{an}} \sqrt{\frac{JS^2\pi^2}{a\overline{W}_{\mathrm{an}}}} + \frac{JS^2\pi^2}{a} \sqrt{\frac{a\overline{W}_{\mathrm{an}}}{JS^2\pi^2}} = 2 \sqrt{\frac{JS^2\pi^2\overline{W}_{\mathrm{an}}}{a}} \tag{3.98}$$

It is found that the contribution of the anisotropy energy is equal to that of the exchange energy. With the above values for iron:

$$W_{\mathrm{SB}} = 3.5 \cdot 10^{-3}\ \mathrm{J/m^2} \tag{3.99}$$

Two advantages of the model we have just seen are its simplicity and the fact that it gives a good order of magnitude for δ. In a more sophisticated model, the assumption of uniform rotation of the magnetic moments, in particular, would have to be reexamined. In fact, the anisotropy energy is a maximum when the orientation of these moments is furthest away from an easy magnetization direction. At this point, we therefore observe a larger angle between adjacent magnetic moments, which reduces the number of magnetic moments with higher anisotropy energy at the cost of a certain increase in the exchange energy. In reality, the rotation of magnetic moments is not uniform but has the form given in Fig. 3.63. The thickness of the Bloch walls is defined by extrapolation.

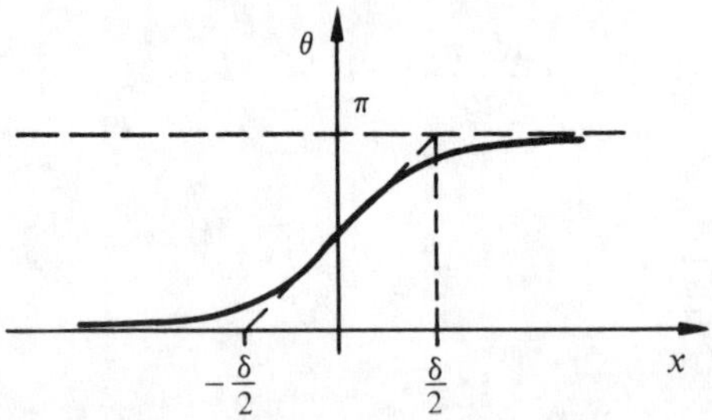

Fig. 3.63 Variation Across a Bloch Wall of the Angle α of the Magnetic Moments with Respect to an Easy Magnetization Direction

3.7.10 Methods for Observing Magnetic Domains

Evidence was provided for magnetic domains in an indirect manner for the first time in 1919 by the experiment of Barkhausen (Fig. 3.64).

A magnet rotates slowly, inducing a reversal of the polarization in a ferromagnetic rod following a periodic movement with a period of approximately one second. A winding captures the variations in magnetic flux in the rod, which are transmitted, after amplification, to a loudspeaker. Normally, no audible sound should be emitted. Now, each time the polarization is reversed, a noise reminiscent of the flow of small grains in a container can be heard. This phenomenon shows that under the effect of sufficiently large external fields, the polarization is established not in a continuous manner but in small successive jumps (Fig. 3.65). It is now known that these jumps correspond to the sudden change of pinning points on the Bloch walls.

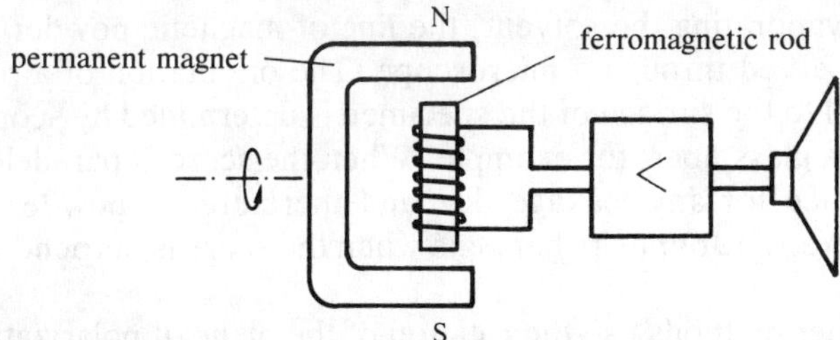

Fig. 3.64 Barkhausen's Experiment

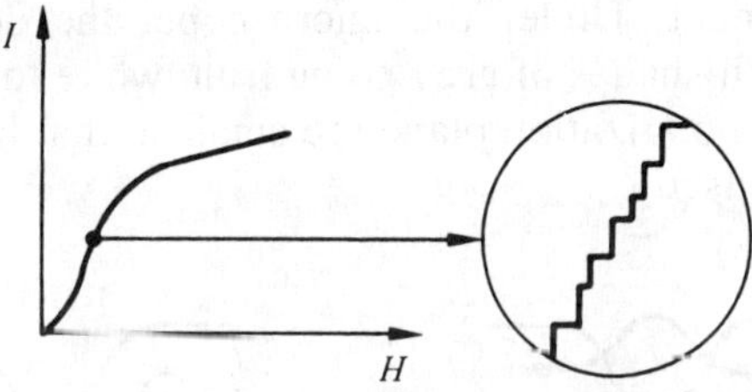

Fig. 3.65

Three methods of direct observation of magnetic domains are currently in use. The oldest of them, known as the Bitter method or the powder method, involves spreading a colloidal solution of magnetic particles on the surface of a specimen that has been previously electrolytically polished. The presence of magnetic masses at points where the Bloch walls intersect the surface of the specimen produces an accumulation of particles at these positions (Fig. 3.66).

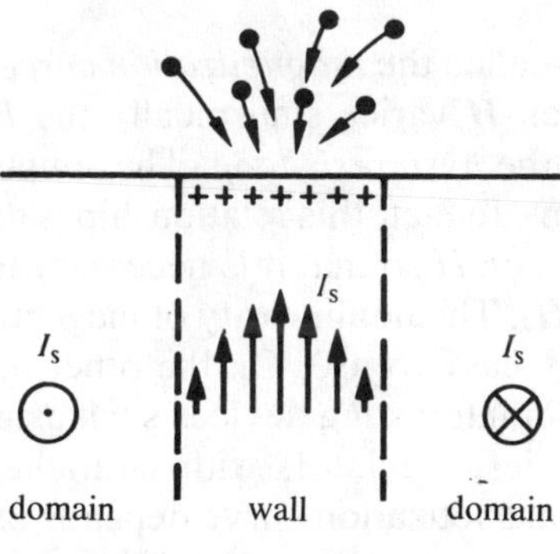

Fig. 3.66

After evaporating the solvent, the line of magnetic powder may be fixed and observed through a microscope. The orientation of a polarization I_s parallel to the surface of the specimen is determined by scoring this surface with a glass fiber, for example. When the score is parallel to I_s, it is not the place for any leakage flux and therefore no powder can be deposited there. The opposite happens when the score is perpendicular to I_s (Fig. 3.67).

Two other methods use the rotation of the plane of polarization of a linearly polarized light during transmission (the Faraday effect) or during reflection (the Kerr effect). Compared to the Bitter method, these techniques have the great advantage of allowing the Bloch walls to be observed in movement. Under the microscope, the domains appear as uniform areas with shades of grey going from white to black. The angles of rotation of the polarization plane are small so that high-quality optical systems must be used.

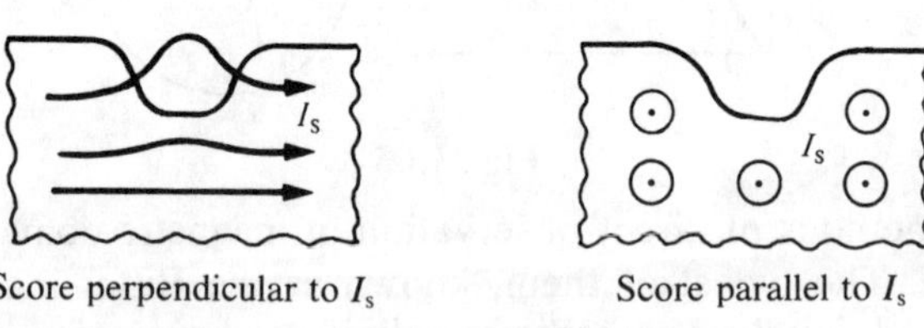

Fig. 3.67

3.8 MAGNETIZATION CURVE AND HYSTERESIS LOOP

3.8.1 Introduction

The *B-H* curve is called the *magnetization curve*. (See Section 3.2.1, Eqs. (3.3), (3.4).) When *H* varies sinusoidally the *B-H* curve takes the shape of a loop called the *hysteresis loop*. The simplicity of the equation $B = \mu H$ is only apparent. In fact, this relationship is not linear because the permeability μ depends on *H* so that it is necessary to have a representation of the function $B(H)$. The nonlinearity of magnetic materials does not make them particularly easy to use. On the other hand, it does make it possible to make certain interesting devices such as magnetic bubble registers, magnetic memories, regulators with saturable cores, *et cetera*.

The form of the magnetization curve depends on the mobility of the Bloch walls, the mobility being a function of the energies W_{an}, W_{ms}, W_{ex}, W_{mt} and of the applied field *H*. Depending on the case, a detailed mathematical model of the diagram is necessary, or on the contrary, knowing one or two representative parameters is sufficient.

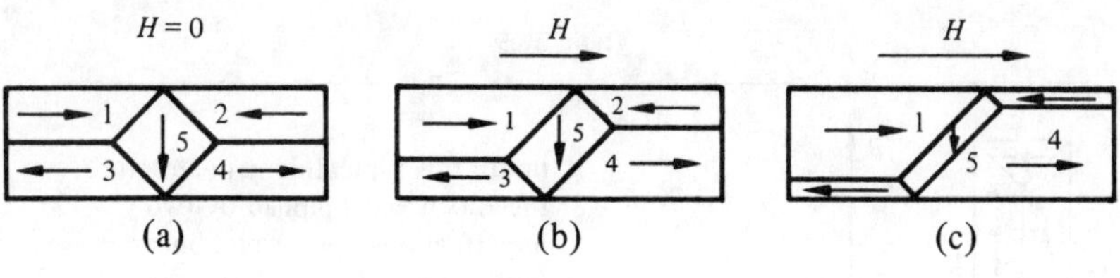

Fig. 3.68

3.8.2 Polarization Process and the Magnetization Curve

Consider a monocrystal of iron, practically without imperfections, in which the magnetic domains are as shown in Fig. 3.68(a).

As soon as a field $\boldsymbol{H}$ is applied as indicated, the energy of the magnetic moments is increased in domains 2 and 3 and to a smaller extent in domain 5. The condition of minimum energy in the specimen leads to a reduction in volume of these domains. In an increasing field $\boldsymbol{H}$, the structure of the domains will successively take the forms shown in Fig. 3.68(b) and then (c). Finally, a single domain oriented along $\boldsymbol{H}$ will remain. In this ideal example, the displacement of the Bloch walls is perfectly free. When $\boldsymbol{H}$ is removed, we therefore return to the initial state (Fig. 3.68(a)), the polarization process being reversible.

Crystal imperfections have a significant effect on the form of the hysteresis loop. The action of point imperfections on the displacement of the Bloch walls inside a monocrystal is illustrated in Table 3.69.

In a polycrystalline medium, a similar process to that described in Table 3.69 takes place in each grain. However, the situation is complicated by the magnetostatic and magnetostriction interactions occurring between neighboring grains. Again, the structure of the grains plays a very important part.

The form of the hysteresis loop may strongly vary from one magnetic material to another (Fig. 3.70). For a given material, it varies as a function of the conditions of use and more particularly as a function of the amplitude of the magnetic field (Fig. 3.71).

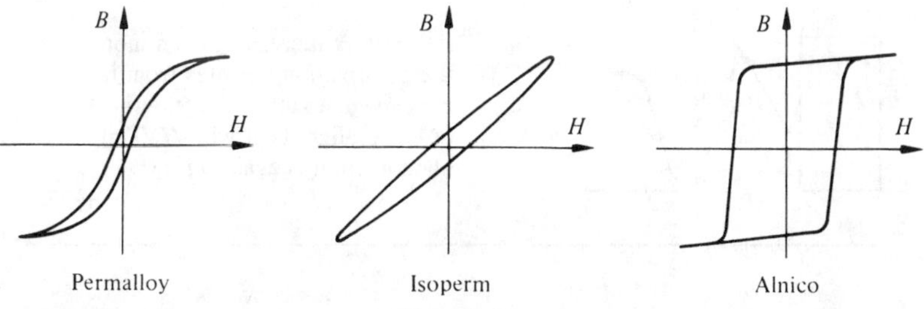

Fig. 3.70

Table 3.69

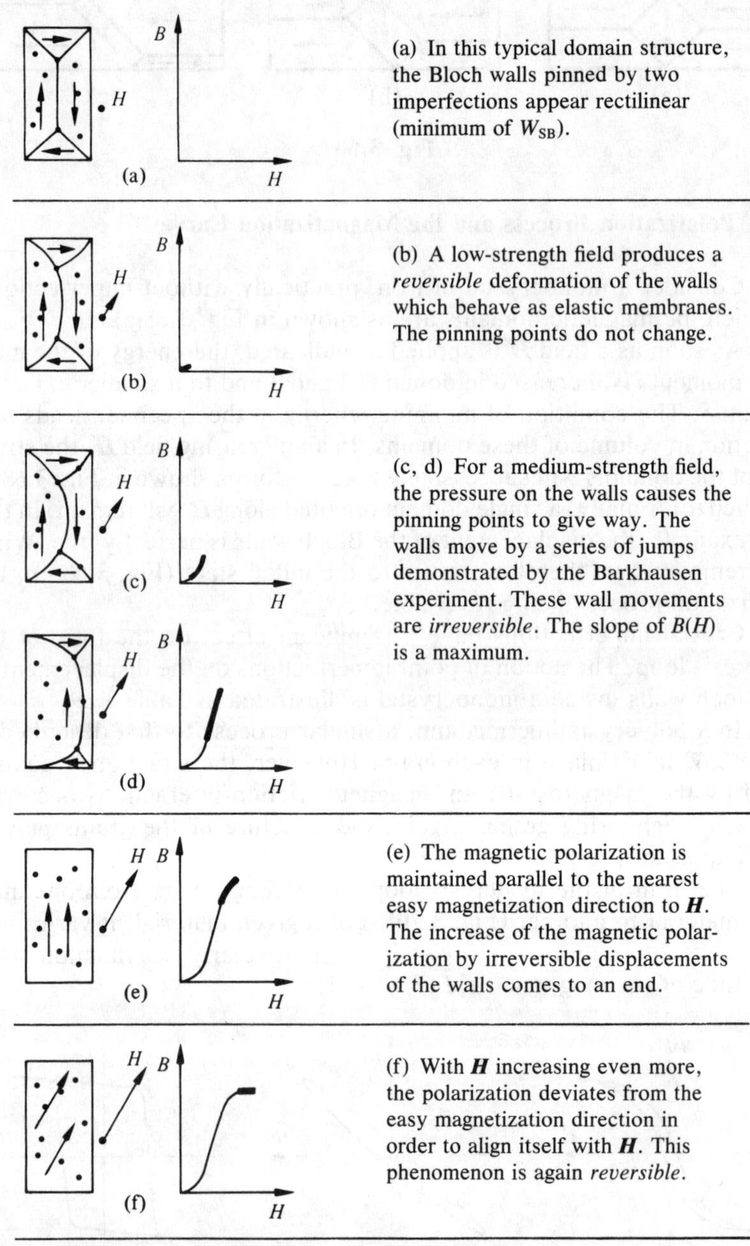

(a)	(a) In this typical domain structure, the Bloch walls pinned by two imperfections appear rectilinear (minimum of W_{SB}).
(b)	(b) A low-strength field produces a *reversible* deformation of the walls which behave as elastic membranes. The pinning points do not change.
(c), (d)	(c, d) For a medium-strength field, the pressure on the walls causes the pinning points to give way. The walls move by a series of jumps demonstrated by the Barkhausen experiment. These wall movements are *irreversible*. The slope of $B(H)$ is a maximum.
(e)	(e) The magnetic polarization is maintained parallel to the nearest easy magnetization direction to $\boldsymbol{H}$. The increase of the magnetic polarization by irreversible displacements of the walls comes to an end.
(f)	(f) With $\boldsymbol{H}$ increasing even more, the polarization deviates from the easy magnetization direction in order to align itself with $\boldsymbol{H}$. This phenomenon is again *reversible*.

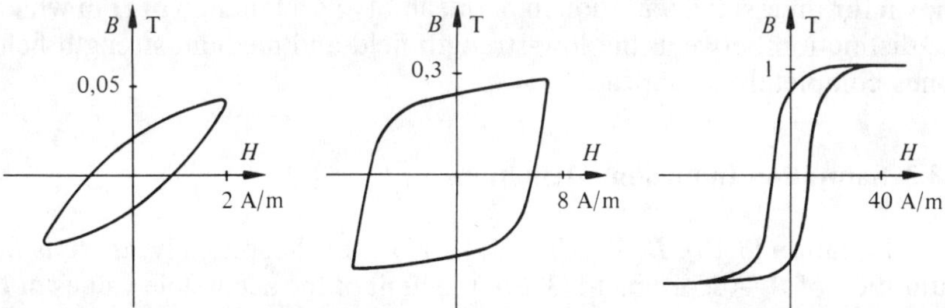

Fig. 3.71 Effects of the Field Amplitude on the Hysteresis Loop of an Iron-Nickel Alloy

The function $B(H)$ observed in an increasing field on a specimen that has never been polarized (on the macroscopic scale) is called the *initial magnetization curve*.

In a decreasing field, the function $B(H)$ deviates from the initial magnetization curve because of the irreversible nature of the polarization. The initial magnetization curve is always located within the hysteresis loop corresponding to the same extreme values of the applied field $\boldsymbol{H}$.

Only an increase in the temperature of the specimen above the Curie point makes it possible to travel through the initial magnetization curve again. In a manner of speaking, the magnetic matter keeps a memory of the polarizations that it has undergone.

The *major hysteresis loop* is the name given to the loop observed during a slow sinusoidal variation of H with sufficient amplitude to lead to saturation.

Conceptually, the magnetization curve can be divided into three zones:

- the *low-strength field zone* (Table 3.69(b)) characterized by an almost linear behavior of the matter as well as by moderate permeabilities;
- the *medium-strength field zone* (Table 3.69(c), (d), and (e)) in which the nonlinear effects are markedly strong, whereas the permeability passes through its maximum value;
- the *strong-field zone* (Table 3.69(f)), or *saturation zone*, where the permeability decreases and tends asymptotically to the permeability in a vacuum.

Except in certain monocrystals, the transition from one zone to another takes place progressively. Therefore, the three zones are not always clearly distinguished from each other. There are even certain alloys, known for their small variation in permeability as a function of H in which the distinction between the low-strength field and medium-strength field zones completely disappears.

3.8.3 Saturation Induction: Definitions

Equation (3.10), $\boldsymbol{B} = \mu_0\boldsymbol{H} + \boldsymbol{I}$, shows that theoretically, there is no saturation of $\boldsymbol{B}$. According to (3.10), the limit of the accessible values of $\boldsymbol{B}$ only results from the impossibility of creating fields $\boldsymbol{H}$ as large as desired. In fact, it is the magnetic polarization that reaches a saturation value on the scale of the magnetic domain if the temperature approaches 0 K or if $H \to \infty$. However, usage has adopted the expression saturation induction B_{sat}. In order to avoid any confusion, two values of B_{sat} are distinguished.

The *practical value* B_{sat} is the maximum value of B that the magnetic material is allowed to reach at the cost of a reasonable expenditure of magnetic potential. In order to be a consistent value, practical value B_{sat} must also contain the specification of the corresponding field H_{sat}.

The *theoretical value* B_{sat} corresponds to the saturation value of $\boldsymbol{I}$, in other words:

$$\text{Theoretical value } B_{sat} = \lim_{H\to\infty} (B - \mu_0 H) \tag{3.100}$$

In an Fe-Si alloy with 4% Si, for example, practical value B_{sat} = 1.80 T at H = 800 A/m.

3.8.4 Remanent Induction and Coercive Field: Definitions

Consider a specimen polarized to saturation, i.e., subjected to a magnetic field H higher than H_{sat}. The *remanent induction* B_r is the induction that remains in the specimen after H has been made to decrease to zero.

The *coercive field* H_c is the magnetic field necessary to cancel the remanent induction.

The remanent induction and the coercive field are specific properties of the material under consideration. They appear in the major hysteresis loop. It should be noted that B_r represents the maximum induction that can exist after a magnetic field has disappeared. However, it is obvious that the phenomenon of remanence also exists for maximum fields less than H_{sat}. Then, the induction that remains after the field has disappeared is less than B_r.

3.8.5 Definition of Four Relative Permeabilities

The nonlinearity of the function $B(H)$ has the effect that the relative permeability defined by (3.14) is not a constant but it varies with the magnetic field. It is not always indispensible to have a complete knowledge of this function $\mu_r(H)$. In many calculations, the permeability simply appears in the form of a number. We then distinguish four scalar permeabilities (Fig. 3.72).

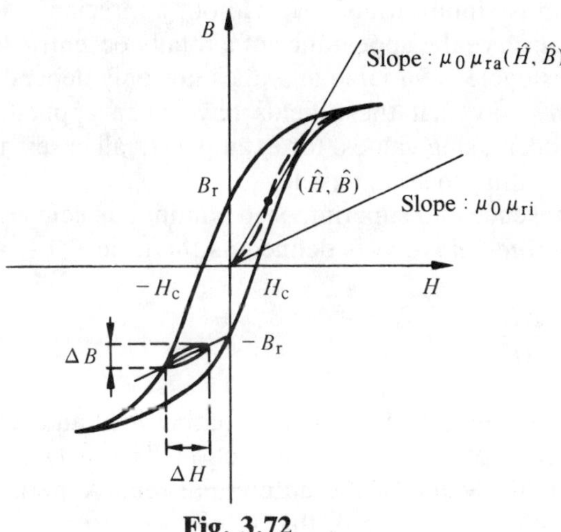

Fig. 3.72

The *initial relative permeability* μ_{ri} is defined as the quotient:

$$\mu_{ri} = \frac{1}{\mu_0} \lim_{H \to 0} \frac{B}{H} \tag{3.101}$$

determined for a specimen which has never been subject to irreversible polarization.

The initial relative permeability is a theoretical value which cannot be directly measured because it corresponds to a zero field. It therefore has to be determined by extrapolation. In practice, μ_{ri} is often given as the relative permeability measured in a weak field lying between 100 and 200 A/m, for example.

The *total relative permeability* is defined as the relative permeability in the region of medium-strength fields, defined by

$$\mu_{ra} = \frac{1}{\mu_0}\frac{\hat{B}}{\hat{H}} \tag{3.102}$$

the values of $\hat{B}$ and $\hat{H}$ being plotted on the initial magnetization curve. This is a good approximation of the relative permeability which would effectively be observed in a low-frequency alternating field with amplitude $\hat{H}$. A more precise value can only be obtained by measurement or calculation using a mathematical model of the hysteresis loop.

Let us now consider the case when an alternating field H_2 is superimposed on a continuous field H_1 acting parallel to H_2. If $\hat{H}_2 \gg H_1$, the hysteresis loop is simply translated without appreciable deformation. If $\hat{H}_2 \ll H_1$, we observe the appearance of a totally eccentric local loop. The position of this cycle is variable and does not only depend on H_1 and H_2 but also in the way that these fields have been applied: by means of increasing or decreasing values, for example. In all cases, the local loops are contained within the main cycle.

In the presence of a superimposed continuous field H_1, the *differential relative permeability* $\mu_{r\Delta}$ is defined as the ratio:

$$\mu_{r\Delta} = \frac{1}{\mu_0}\frac{\Delta B}{\Delta H} \tag{3.103}$$

where ΔH is the amplitude of the alternating field and ΔB is the corresponding variation of the magnetic induction. The *reversible relative permeability* μ_{rr} is the value of the differential relative permeability for an alternating field tending to zero, that is,

$$\mu_{rr} = \frac{1}{\mu_0}\lim_{\Delta H \to 0}\frac{\Delta B}{\Delta H} \tag{3.104}$$

In all the mathematical treatment of general application, the relative permeability is simply denoted by the symbol μ_r.

3.8.6 Hard and Soft Magnetic Materials: Definitions

Soft magnetic materials are those in which the coercive field is small (Tables 3.106, 3.107, and 3.108). The area of their major hysteresis loop is small. Permalloy and isoperm (Fig. 3.70) are two examples of such materials.

Hard magnetic materials are those in which the coercive field is large (Table 3.109). The area of their major hysteresis loop is large, as that of alnico shown in Fig. 3.70. In these materials, the Bloch walls are strongly pinned.

Impurities and other imperfections impeding the movement of the Bloch walls also hinder the displacement of the dislocations. This is why hard magnetic materials are also hard in the mechanical sense.

3.8.7 Magnetic Losses: General Considerations

Any variation of the magnetic induction in a magnetic material produces a dissipation of energy inside it. This energy most often appears in the form of heat, and in general cannot be recovered, thus the expression *magnetic losses* is used to describe this phenomenon. Three types of magnetic losses are distinguished:

- hysteresis loss;
- eddy current loss;
- after-effect loss.

Hysteresis loss is due to the work of the braking forces acting on the Bloch walls in motion. They are therefore maximum when the pinning forces are greatest, i.e., in hard magnetic materials. Loss due to hysteresis corresponds to the work that is necessary in order to slowly travel through the entire hysteresis loop.

Eddy current loss is due to the Joule effect resulting from currents created in any conducting matter (magnetic or not) by a flux varying over time. This loss may be considerable in materials with low electrical resistivity, such as magnetic alloys. The loss remains at a low level in ferrites.

After-effect is due to the delay in the change of the induction with respect to the change of the field, as shown in Fig. 3.73.

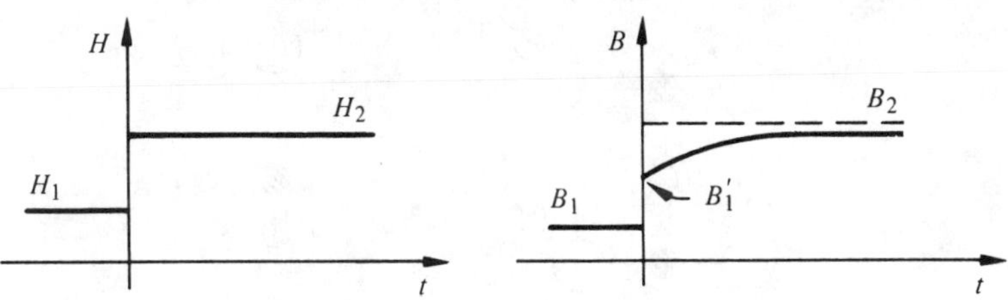

Fig. 3.73

When the field H suddenly varies from H_1 to H_2, the induction follows immediately from B_1 to B'_1 and then tends asymptotically to the value of B_2 corresponding to H_2. Two different mechanisms may be responsible for this phenomenon.

The process of *thermal fluctuation after-effect* can be described as follows. Following the initial movement of the Bloch walls due to the transition from H_1 to H_2, the forces created at certain pinning points are close enough to those necessary for unpinning that this process occurs after a certain time under the action of local fluctuations in the polarization. In fact, the polarization does not have exactly the same direction at any point in a domain, as is shown by the decrease in $I(T)$ when the temperature rises (Fig. 3.20). This direction fluctuating over time sometimes weakens the effect of the external field and at other times reinforces it, making the most stressed pinning points give way until they completely disappear and B is stabilized.

Diffusion after-effect is found in a typical manner in α iron containing carbon (Fig. 3.74). The carbon atoms are situated in an interstitial position on the edges and at the center of the faces of the cube. Three families of sites—x, y, and z—can be occupied in this way.

Let us suppose that only x sites are occupied, then the crystal will be expanded along x. If the polarization is oriented along x, the magnetostriction energy will be lowered provided that the substance has a positive magnetostriction. There is, therefore, an interaction between the structure of the magnetic domains and the positions of the carbon atoms. In particular, a change in polarization produces a diffusion of atoms lowering the internal energy of the system, which in turn modifies the polarization, *et cetera*. Because this phenomenon cannot be instantaneous, it produces after-effect by diffusion.

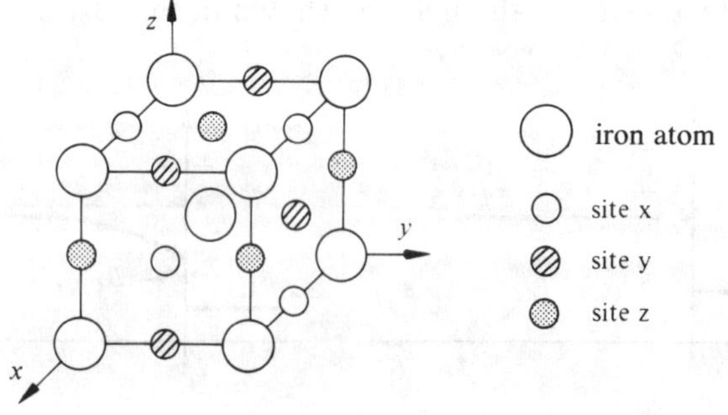

Fig. 3.74

3.8.8 Model of the Hysteresis Loop at Low Fields

Except in certain cases of little practical interest, it is not possible to establish the $B(H)$ function of a given material in a purely theoretical manner. It therefore remains to find the equations representing the $B(H)$ functions found experimentally as best as possible.

In low fields, all magnetic materials exhibit similar behavior. For this reason, general-purpose models are used in which only the characteristic parameters change from one material to another. The model used most often because of its great simplicity and the satisfactory quality of the results that it provides is the *Rayleigh model.*

This model gives an expression of the permeability as a function of the field as well as an equation describing the hysteresis loop.

As a function of the amplitude $\hat{H}$ of the field, the permeability always gives a variation of the type shown in Fig. 3.75.

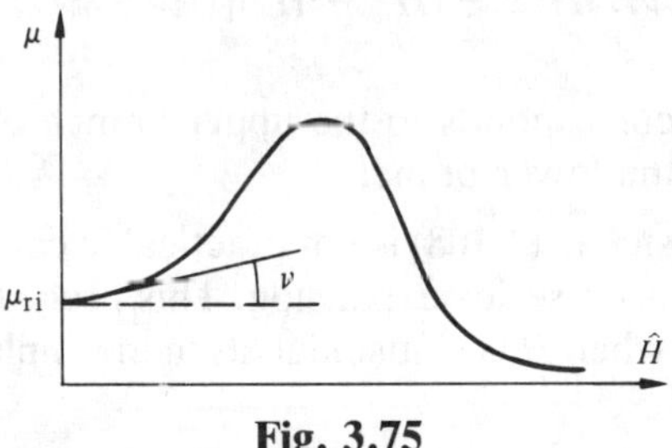

Fig. 3.75

For small values of $\hat{H}$, the permeability can therefore be represented by a linear function:

$$\mu_{rR} = \mu_{ri} + \nu\hat{H} \tag{3.105}$$

μ_{rR} is the Rayleigh relative permeability and ν the Rayleigh coefficient.

Expression (3.105) does not take into account the variation of the permeability as a function of frequency, examined in Section 3.8.12.

By multiplying both sides of (3.105) by $\mu_0\hat{H}$, we obtain the equation for the initial magnetization curve:

$$\hat{B} = \mu_0\mu_{ri}\hat{H} + \nu\mu_0\hat{H}^2, \tag{3.106}$$

which contains one endpoint of the hysteresis loop. For a given value of $\hat{H}$, the Rayleigh hysteresis loop is obtained in the following manner (Fig. 3.76).

- Starting from the point $\hat{B}$, $\hat{H}$ on the initial magnetization curve, a straight line is plotted passing through the origin. Its equation can be written

$$B = \mu_0(\mu_{ri} + \nu\hat{H})H, \tag{3.107}$$

where B and H represent the instantaneous values of fields with amplitudes $\hat{B}$ and $\hat{H}$.

- We then draw two equal segments on both sides of this straight line, parallel to the induction axis with the loop passing through their ends. The length of these segments must be zero at $H = \hat{H}$ and a maximum at $H = 0$. Rayleigh observed that this length could also be expressed as a function of ν and put the equation of the loop in the form:

$$B = \mu_0\left[(\mu_{ri} + \nu\hat{H})H \pm \frac{\nu}{2}(\hat{H}^2 - H^2)\right] \tag{3.108}$$

The plus sign corresponds to the upper branch of the loop and the minus sign to the lower branch.

The change of sign in (3.108) is not practical in calculations but it can be removed by a Fourier series expansion. This procedure is particularly suitable for the case when H is sinusoidal, as in the application example of Section 3.8.18.

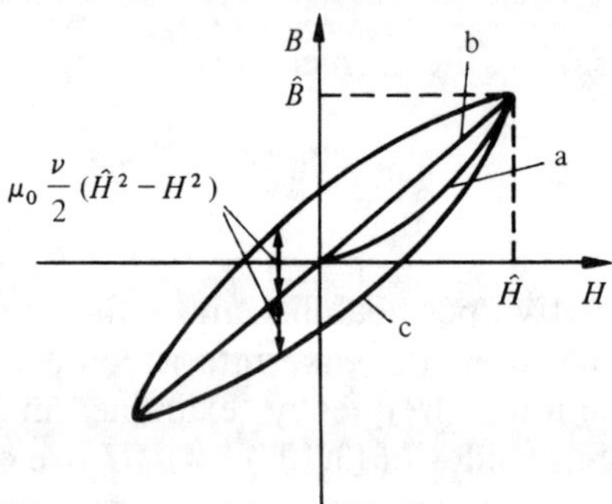

Fig. 3.76 Establishment of the Rayleigh Loop

3.8.9 Hysteresis Loss at Low Fields

The energy W_h corresponding to hysteresis loss dissipated per unit volume when we go round the loop once, is equal to

$$W_h = \int_{\text{cycle}} \boldsymbol{B} \cdot \mathrm{d}\boldsymbol{H} \tag{3.109}$$

This integral represents the area of the loop. By assuming the Rayleigh representation for it, we have, from (3.108):

$$W_h = \int_{-\hat{H}}^{+\hat{H}} \mu_0 \nu(\hat{H}^2 - H^2)\mathrm{d}H = \frac{4\mu_0 \nu \hat{H}^3}{3} \quad \text{J/m}^3 \tag{3.110}$$

In a core characterized by a cross section S of magnetic flux and an average length l of the B lines, the power P_h lost by hysteresis at the frequency f is equal to

$$P_h = W_h S l f \tag{3.111}$$

An inductance that would only present loss by hysteresis can be represented by the following equivalent circuit (Fig. 3.77):

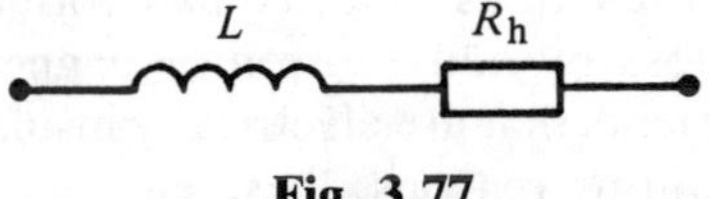

Fig. 3.77

The resistance R_h is chosen so that the power dissipated there by the Joule effect is equal to P_h. This resistance as well as the loss angle due to hysteresis defined by

$$\tan \delta_h = \frac{R_h}{\omega L} \tag{3.112}$$

can easily be deduced from the above. Let N be the number of turns of the inductance, and i be the effective value of the sinusoidal current with angular velocity ω which flows through it. We have

$$P_h = R_h i^2 \tag{3.113}$$

and

$$i = \frac{\hat{H}l}{N\sqrt{2}} \tag{3.114}$$

By substituting (3.110) into (3.111) and by identifying the result with (3.113), we have

$$\tan \delta_h = \frac{4\nu\hat{H}}{3\pi\mu_{rR}} = \frac{4\nu\hat{B}}{3\pi\mu_0\mu_{rR}^2} \tag{3.115}$$

3.8.10 Eddy Current Loss

Loss by eddy currents P_F in a core with resistivity ρ and volume V are calculated from the relationship:

$$P_F = \int_V \rho i_c^2 \, dV, \tag{3.116}$$

where i_c represents the effective value of the density of the eddy currents. The integration of (3.116) always results in tedious treatment but, except for rare cases, it is possible to use approximate formulas. In fact, with the aim of reducing the loss by eddy currents, magnetic circuits are most often divided into elements that are electrically insulated from each other. For obvious manufacturing reasons, these elements are always made in simple shapes (Fig. 3.78): sheets, strips, circular wires, fine particles in pressed cores (Section 3.9.8).

It is sufficient to calculate (3.116) for each shape that can be used for making a magnetic core.

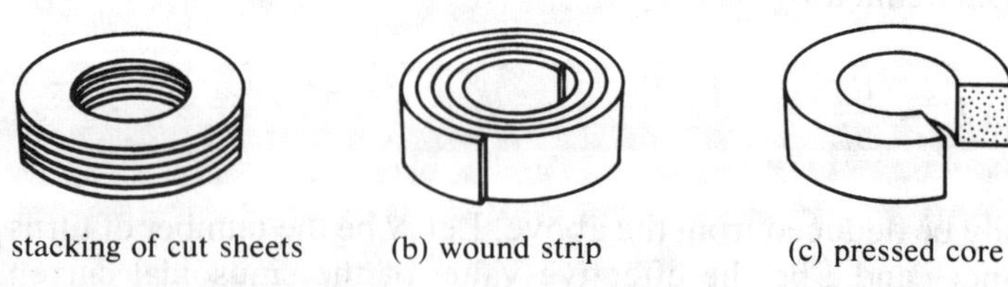

Fig. 3.78 Three Ways of Reducing Eddy Current Loss in a Core

Consider the case of a magnetic sheet subjected to a field oriented along the z-axis: $\boldsymbol{H} = \mathrm{Re}\{\mathbf{k}\hat{H}_0 \exp \mathrm{j}\omega t\}$ (Fig. 3.79).

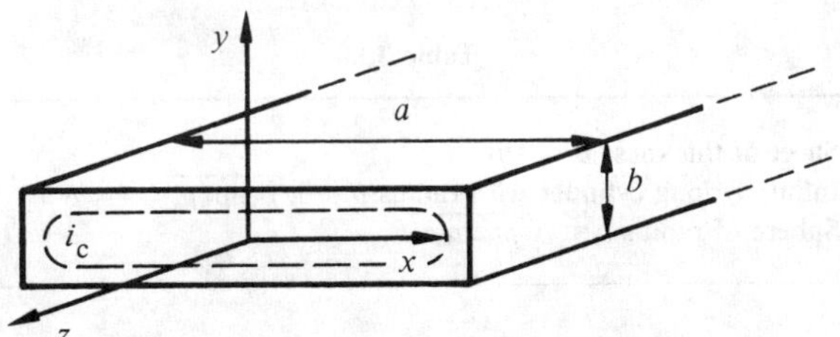

Fig. 3.79

If $b \ll a$, the solution of Maxwell's equations for the magnetic field phasor in the sheet, gives

$$H_z(y) = H_0 \frac{\cosh[(1 + \mathrm{j})\beta y]}{\cosh[(1 + \mathrm{j})\beta b/2]} \tag{3.117}$$

where

$$\beta = \sqrt{\frac{\omega \mu_0 \mu_r}{2\rho}} \tag{3.118}$$

In these expressions, ω denotes the angular velocity of H, ρ the resistivity of the sheet. Equation $\nabla \times \boldsymbol{H} = \boldsymbol{E}/\rho$ reduces to

$$\frac{\partial H_z}{\partial y} = \frac{E_x}{\rho} = i_c \tag{3.119}$$

By removing i_c from (3.119) and (3.117) we obtain from (3.116) the loss per unit volume p_F due to the eddy currents:

$$p_F \cong \frac{1}{8} \frac{(\hat{B}\omega b)^2}{\alpha \rho} \quad \mathrm{W/m^3} \tag{3.120}$$

This formula is valid as long as the reduction in eddy currents associated with the attenuation of H_z inside the sheet (3.117) remains negligible. The equality corresponds to the theoretical case of zero attenuation. Expression (3.120) is still valid in the case of a sphere or a cylinder under the conditions summarized in Table 3.80.

Table 3.80

Sheet of thickness $b < 1/\beta$	$\alpha \cong 3$
Infinitely long cylinder with radius $b < \sqrt{\rho/\omega\mu_0\mu_r}$	$\alpha \cong 8$
Sphere of radius $b < \sqrt{\rho/\omega\mu_0\mu_r}$	$\alpha \cong 10$

In a sheet of silicon-iron, typically $\mu_r = 5000$ and $\rho = 0.6 \cdot 10^{-6}\ \Omega \cdot \mathrm{m}$ so that $1/\beta = 0.75$ mm at 50 Hz and 0.075 mm at 5 kHz.

It is often worth representing eddy current loss by a resistance R_F placed in series in an equivalent circuit. We then define an eddy current loss angle δ_F by

$$\tan \delta_F = \frac{R_F}{\omega L} \tag{3.121}$$

By proceeding as in the case of hysteresis loss, we deduce from (3.120) that

$$\tan \delta_F = \frac{\mu_0 \mu_r b^2 \omega}{\alpha \rho} \tag{3.122}$$

The eddy current loss given by (3.120) is a loss *per unit volume*. It does not depend on the shape or the size of the core but only on the shape and the size of the magnetic elements that form the core.

3.8.11 Remarks

In general, the permeability of a metal sheet varies as a function of the angle of $\boldsymbol{B}$ with the direction of rolling (Fig. 3.92). Therefore, with the solution shown in Fig. 3.78(a), it is not possible to exploit the maximum permeability of the sheet because $\boldsymbol{B}$ turns through 360°, with respect to the rolling direction, when a line of $\boldsymbol{B}$ is described. The solution shown in Fig. 3.78(b) used for a material with maximum permeability in the rolling direction does not suffer from this disadvantage.

The relative permeability of a pressed core is limited by the gap between each grain, to values on the order of 100 at most. However, the reduced size of the grains makes it possible to use it at higher frequencies than the two preceding solutions.

3.8.12 Correction to the Permeability

Eddy currents not only produce loss but also an apparent reduction of the permeability associated with the attenuation of the magnetic field inside the material. The value of H inside a metal sheet can be deduced from (3.117) by using the relationship (3.123):

$$\cosh[(1 + \mathrm{j})x] = \sqrt{\frac{1}{2}\,(\cosh 2x + \cos 2x)} \qquad (3.123)$$

The result is shown in Fig. 3.81 for a few values of the product βb.

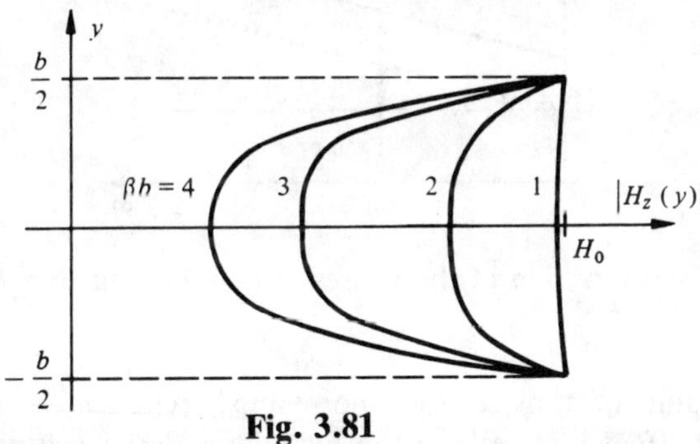

Fig. 3.81

The mean induction $\bar{B}$ in the sheet is therefore less than $\mu_0 \mu_r(H_0) H_0$. It is equal to

$$\bar{B} = \frac{\mu_0}{b} \int_{-b/2}^{+b/2} \mu_r(H_z) H_z(y)\, \mathrm{d}y \qquad (3.124)$$

This integral can only be calculated easily when $\mu_r(H_z)$ may be considered as constant for $-b/2 < y < +b/2$. This condition is not too restrictive. In fact, it corresponds to the situation suitable for using a metal sheet. We then find

$$|\bar{B}| = \mu_0 \mu_r \frac{\sqrt{2}}{\beta b} \sqrt{\frac{\cosh \beta b - \cos \beta b}{\cosh \beta b + \cos \beta b}}\, H_0 \qquad (3.125)$$

The correction for μ_r can be clearly seen in this equation, which is also used to extend the range of validity of (3.120). By replacing $\hat{B}$ by $|\hat{B}|$ calculated from (3.125) in (3.120), this expression can be used up to $b = 2/\beta$. The computed value of p_F is then 2.4%, which is too low.

3.8.13 Experimental Discrimination of the Types of Losses

Equations (3.115) and (3.122) show that $\tan \delta_h \sim \hat{B}$ and $\tan \delta_F \sim \omega$. It is therefore possible to separate these two types of losses by measuring the tangent of the loss angle corresponding to the total magnetic losses using a Wheatstone bridge. For most materials a graph is obtained called the Jordan diagram which has the following form (Fig. 3.82).

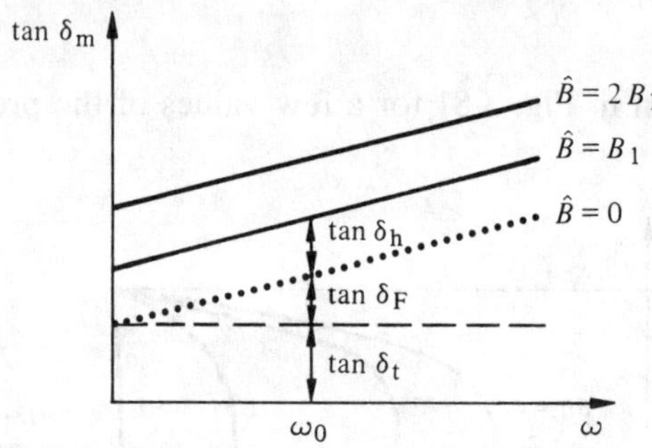

Fig. 3.82 Discrimination of the Losses at $\omega = \omega_0$ and $\hat{B} = B_1$

The value of $\tan \delta_m$ does not vanish when $\omega \to 0$ and $\hat{B} \to 0$ due to after-effect loss. When the loss depends very little on B and ω, $\tan \delta_m(\hat{B}, \omega)$ is a grid of parallel straight lines. The distance between two adjacent lines is invariant if $\hat{B}$ is increased in constant steps. However, this is an ideal case. Most often, the tangent of the after-effect loss angle $\tan \delta_t$ varies to a certain extent as a function of $\hat{B}$ and ω. Consequently, there is a more or less significant distortion of the Jordan diagram with respect to its representation in Fig. 3.82. Under these conditions, $\tan \delta_h$, $\tan \delta_F$ and $\tan \delta_t$ cannot be separated in an absolutely rigorous manner.

3.8.14 General Models for Low-Strength Field Losses

Several equivalent models in which $\tan \delta_t$ is considered as independent of $\hat{B}$ and ω, are used to evaluate the total magnetic losses. The three types of losses are treated separately from each other and then added together. By taking the results of calculation obtained in the previous sections, we can write the *loss factor* $\tan \delta / \mu$ in the form:

$$\frac{\tan \delta}{\mu_{rR}} = \frac{4\nu\hat{B}}{3\pi\mu_0\mu_{rR}^3} + \frac{\mu_0 b^2 \omega}{\alpha\rho} + \frac{\tan \delta_t}{\mu_{rR}} \tag{3.126}$$

or as

$$\frac{\tan \delta}{\mu_{rR}} = k_1\hat{B} + k_2\omega + k_3 \tag{3.127}$$

The coefficients k_1 and k_3 relate to the losses by hysteresis and aftereffect. They are characteristics of the material. The coefficient k_2 relates to the eddy current loss. It depends not only on the material but also on the shape of the core or the elements that form it.

The *Jordan model* (3.128) corresponds to equation (3.127) with both sides multiplied by $\mu_{rR} \cdot \omega/2\pi \cdot L$. It therefore provides losses in the form of a series resistance R_m placed in an equivalent circuit of the type in Fig. 3.77.

$$R_m = hHfL + Ff^2L + tfL, \tag{3.128}$$

where $h[\Omega/\mathrm{H} \cdot \mathrm{Am}^{-1} \cdot \mathrm{Hz}]$, $F[\Omega/\mathrm{H} \cdot \mathrm{Hz}^2]$, and $t[\Omega/\mathrm{H} \cdot \mathrm{Hz}]$ are the Jordan coefficients, often given in the literature in other units than those of the international system. The magnetic field H is expressed in rms value.

By multiplying both sides of (3.127) by $2\pi/\mu_0$, we arrive at the *Legg model:*

$$\frac{R_m}{\mu_0\mu_r fL} = a\hat{B} + ef + c, \tag{3.129}$$

where $a[1/\mathrm{Wb} \cdot \mathrm{m}^{-2}]$, $e[\mathrm{s}]$, and $c[1]$ are the Legg coefficients.

The magnetic losses are also expressed in terms of a complex relative permeability $\underline{\mu}_r$ defined in the case of a series equivalent circuit by

$$R_m + j\omega L = j\omega\mu_0\underline{\mu}_r N^2 \frac{S}{l} \tag{3.130}$$

By dividing $\underline{\mu}_r$ into its real and imaginary parts so that $\mu_r'' > 0$ corresponds to $R_m > 0$:

$$\underline{\mu}_r = \mu_r' - j\mu_r'' \tag{3.131}$$

We have

$$\mu_r' = \frac{L}{N^2\mu_0 \dfrac{S}{l}} \tag{3.132}$$

and

$$\mu_r'' = \frac{R_m}{\omega N^2 \mu_0 \dfrac{S}{l}} \tag{3.133}$$

The fact that a graph of the losses as a function of $\hat{B}$ and ω has the form of Fig. 3.82 demonstrates the validity of the model (3.127) for engineering calculations. However, the division of the losses into three categories independent of each other and each a function of a different variable is *artificial*. This procedure contradicts physical reality. In fact, the hysteresis loop is not only determined by $\hat{B}$. For a given $\hat{B}$, it widens along the H axis when the frequency increases and it also varies with the form of H—sinusoidal, square, triangular, *et cetera*. Cinematographic observation of the movement of the Bloch walls [42] in an alternating field gives a useful picture of these phenomena for which detailed interpretation remains complex. The velocity of displacement of the Bloch walls between two successive pinning points, and the fact that the average size of the domains reduces when the frequency increases, play an important role. In (3.127), the increase in $\tan \delta_h$ with frequency is formally transferred to $\tan \delta_F$.

3.8.15 Comments on Calculation of Losses

The calculation of $\tan \delta_F$ by (3.122) is no longer entirely correct, even in the area of application defined above for this formula. In proving it, the presence of magnetic domains has not been taken into consideration. The induction is therefore implicitly assumed to vary in a continuous manner. Now, B varies locally in an almost discontinuous manner under the effect of the abrupt movement of the Bloch walls. It is tempting to see in this phenomenon the origin of more intense eddy currents than predicted. The discrepancy between the eddy current loss determined by (3.122) and that measured from a Jordan diagram is called *anomalous* eddy current loss. These remain small at low-strength fields so that it is only necessary to correct (3.122) for medium to strong fields, which lends support to the plausibility of the interpretation just given.

In reality, losses by after-effect, hysteresis, and eddy currents depend on the dynamics of the Bloch walls. Theoretically, an adequate model of these walls should make it possible to provide an overall calculation of the losses, but the models in use in physics are still too inadequate to make it worthwhile to replace (3.127) from the engineer's point of view.

3.8.16 Magnetic Losses at Medium and Strong Fields

In medium-strength and high-strength fields, the hysteresis loop can no longer be represented by the Rayleigh model and the anomalous eddy current loss can reach several times the theoretical value given by (3.120). On the other hand, the contribution of after-effect loss becomes negligible in terms of relative value. Under these conditions, the losses are calculated from an empirical expression of the type:

$$p = \eta B^a f + eB^2 f^2 \qquad \text{W/kg} \tag{3.134}$$

The first term represents the hysteresis loss evaluated using the Steinmetz formula. In the case of silicon-iron, $a = 1.6$, and η varies approximately between $5 \cdot 10^{-4}$ and $4 \cdot 10^{-3}$ depending on the percentage of Si and the thermal treatment. The formula can be applied for $0.2\text{T} < B < 1.5\text{T}$. A large number of measurements show that its application can be extended to almost all magnetic materials including the ferrites, provided that the frequency remains low, for example less than 100 Hz. The value of a always remains close to 1.6. The range of B for which the model can be used, depends on B_{sat} and on the extent of the Rayleigh domain, in which the hysteresis loss is proportional to B^3 and not to $B^{1.6}$ (3.110). The surprising feature of the Steinmetz formula is that it does not contain the permeability.

The second term of (3.134) represents the eddy current loss. The coefficient e cannot be calculated theoretically.

When the magnetic material is in the form of a sheet, e is most often determined, then η and a if necessary, using the Epstein apparatus (Fig. 3.83).

The magnetic sheet to be examined is divided into strips that are stacked in order to form four prisms. Each provided with a primary winding and a secondary winding, these prisms are then assembled to form a closed magnetic circuit in the form of a square. Precautions have to be taken in order to avoid any gaps forming in the corners. The four primary windings and four secondary windings are connected in series and incorporated in the measurement circuit of Fig. 3.84.

The Epstein apparatus therefore operates as a transformer with an open-circuited secondary, if we disregard the consumption of the instruments. The voltage U_s is a measurement of B, when the frequency and the magnetic cross section are known. The wattmeter directly indicates the total magnetic losses. The copper losses in the primary are not taken into account by the wattmeter because its voltage circuit is connected to the secondary.

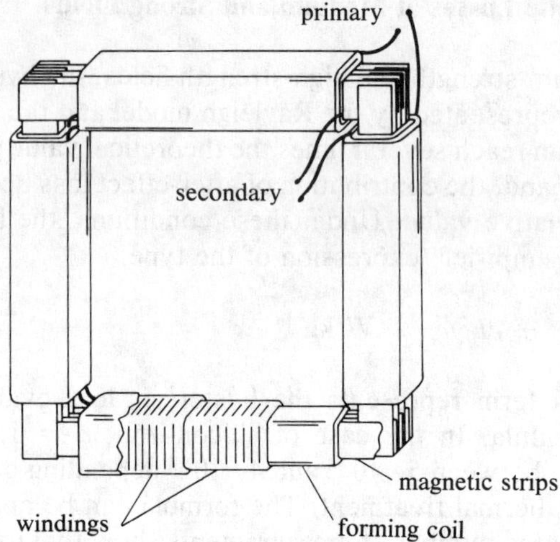

Fig. 3.83 Epstein Apparatus

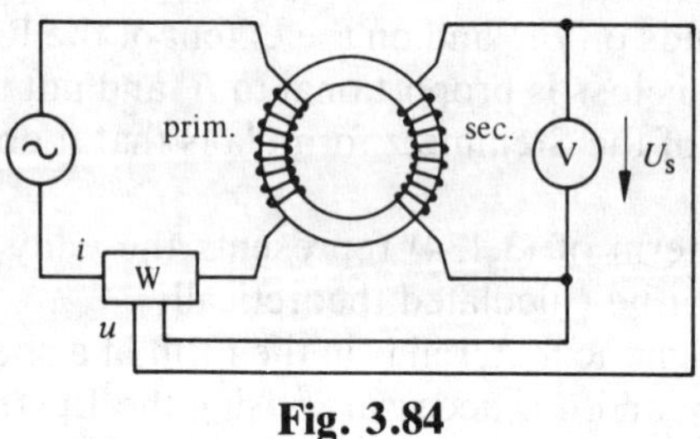

Fig. 3.84

The coefficient a being assumed to be known, hysteresis loss and eddy current loss can easily be separated from each other. By dividing (3.134) by f, we have

$$\frac{p}{f} = \eta B^a + eB^2 f \tag{3.135}$$

According to the model, the experimental points of the function p/f must therefore lie on a straight line with slope eB^2, where the intersection with the vertical axis determines η (Fig. 3.85).

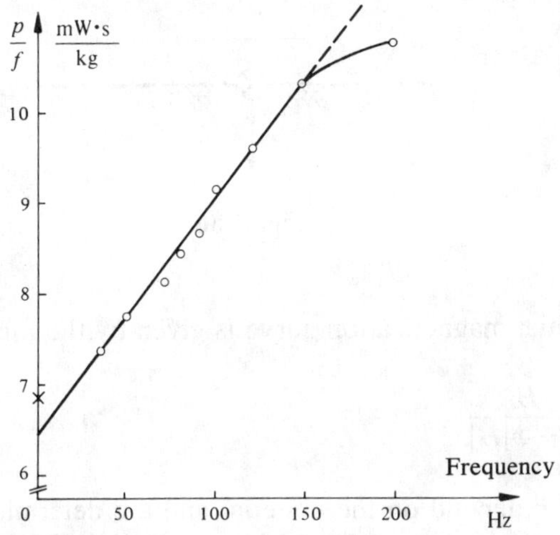

Fig. 3.85 Total Magnetic Losses per Loop in Permalloy 45; (45% Ni + 55% Fe). Sheet Thickness: 0.15 mm, μ_r = 20,000, $\hat{B}$ = 1 T [29]

Fig. 3.85 provides a confirmation of the validity of (3.134) but it also shows that the model can no longer be applied above a certain frequency. If a is not known, it is sufficient to carry out two series of measurements with different values of $\hat{B}$ in order to determine it.

Ferrites can obviously not be placed in an Epstein apparatus but the measurement principle is retained for these materials.

The calculation of medium and strong field losses that has just been presented may seem somewhat crude. However, it should not be forgotten that the requirements here are less severe than for low fields. Often the frequency is fixed as well as the amplitude of B and only the total losses have any significance. On the other hand, the possible variation of the magnetic and electrical parameters from one production batch to another may cause the greater accuracy of the results of more sophisticated calculations to be useless.

If, nevertheless, the losses have to be calculated with greater precision (maximum tolerated error on the order of a few percent, for example), no other method is available except for the numerical solution of the Maxwell equations which are made nonlinear by the presence of the function $B(H)$. A solution of this type is presented in [43] for the case of an infinite plate. The sinusoidal field H is applied along y (Fig. 3.86).

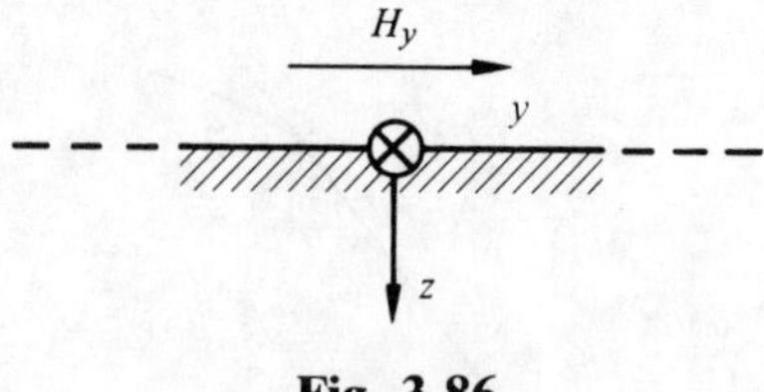

Fig. 3.86

The initial magnetization curve is given by the function:

$$B = \frac{H}{a + b|H|}, \tag{3.136}$$

where a and b depend on the material and are determined by measurement. For example, for an alloy of iron containing 36% Ni and 6% Cr, with a resistivity of $2.5 \cdot 10^{-7}\ \Omega \cdot \text{m}$, we have the following values: $a = 1.5 \cdot 10^3 \text{m/H}$ and $b = 0.47\ \text{T}^{-1}$. Outside the domain $-H_{sat} < H < H_{sat}$, (3.136) is replaced by

$$B = \mu_0 H + I_{sat} \quad \text{for} \quad H > H_{sat}$$

and (3.137)

$$B = \mu_0 H - I_{sat} \quad \text{for} \quad H < -H_{sat}$$

The hysteresis loop is divided into segments, each of which is obtained by a particular translation and deformation of sections of the initial magnetization curve.

In [44], only the description of the hysteresis loop is given. This is represented by a differential equation established in an axiomatic manner from general considerations concerning the symmetry characteristics of the loop. The model obtained in this way is aimed to be as general as possible. In principle, it is capable of representing all materials, from the transformer sheet to the ferrite with rectangular loop used in memories. All types of modulation, with and without continuous polarization can be considered.

3.8.17 Magnetic Losses in Rotating Fields

A different approach is necessary when, with the field amplitude constant, the orientation varies over time. This case is found, for example, at the junction of magnetic circuits of three-phase transformers and in the gearing of rotating machinery. Several references can be found in [45] on measurements and attempts to model losses in rotating fields.

3.8.18 Distortion of Signals

The signals that pass through elements of circuits using magnetic materials are subject to a distortion because $B(H)$ is not a linear function. This question will be treated in the light of the following problem.

Consider a transformer with a primary supplied by a current source $\boldsymbol{I} = \hat{\boldsymbol{I}} \cos \omega t$. What is the magnetic induction and the third harmonic ratio in the voltage appearing in the open-circuited secondary, knowing that low fields are involved?

The source produces a field:

$$H = \hat{H} \cos \omega t \tag{3.138}$$

By introducing this expression into the Rayleigh model (3.108), we have

$$B = \mu_0[(\mu_{ri} + \nu\hat{H})\hat{H} \cos \omega t \pm \frac{\nu}{2} \hat{H}^2 \sin^2 \omega t] \tag{3.139}$$

By noting that $B = \hat{B}$ for $\omega t = 0$ and $B = B_r$ for $\omega t = \pi/2$, we simplify the notation of (3.139) by setting

$$B = \hat{B} \cos \omega t \pm B_r \sin^2 \omega t, \tag{3.140}$$

where

$$\hat{B} = \mu_0(\mu_{ri} + \nu\hat{H})\hat{H} \qquad \text{and} \qquad B_r = \frac{\mu_0\nu\hat{H}^2}{2} \tag{3.141}$$

It remains to expand this expression, or to be more exact, the term $\pm B_r \sin^2 \omega t$ as a Fourier series in order to determine the harmonic spectrum of B. The choice of the sign $\pm$ is related to the direction through which the hysteresis loop is naturally traveled in accordance with Fig. 3.87.

The function $\pm \sin^2 \omega t$ is therefore odd and its expansion will only contain terms in sines.

$$\pm \sin^2 \omega t = \sum_{n=1}^{\infty} b_n \sin n\omega t, \tag{3.142}$$

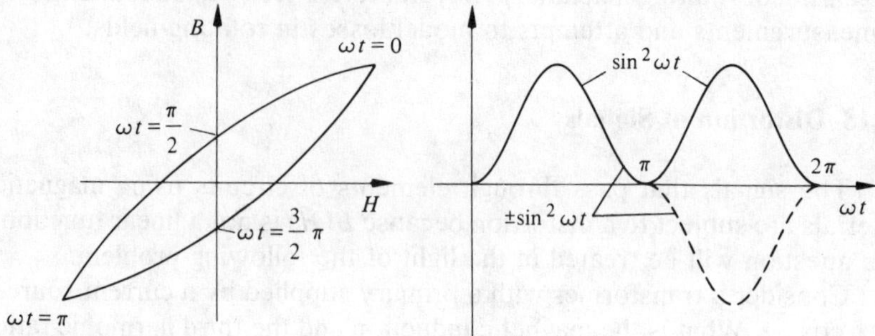

Fig. 3.87

where

$$b_n = \frac{1}{\pi}\int_0^{\pi} \sin^2 \omega t \sin n\, \omega t\, \omega\, dt - \frac{1}{\pi}\int_{\pi}^{2\pi} \sin^2 \omega t \sin n\, \omega t\, \omega\, dt \tag{3.143}$$

These integrals vanish for even n. For $n = 1$ and $n = 3$, they can be easily calculated and the results corresponding to higher values of n can be obtained by recursion. We find

$$B = \hat{B} \cos \omega t + \frac{8B_r}{3\pi} \sin \omega t - \frac{8B_r}{\pi}\left(\frac{1}{1 \cdot 3 \cdot 5} \sin 3\, \omega t + \frac{1}{3 \cdot 5 \cdot 7} \sin 5\, \omega t + \ldots\right) \tag{3.144}$$

The ratio of the amplitude of the third harmonic to that of the fundamental is obtained by rearranging the terms in sin and cos in (3.144) by the relationships

$$A \cos \omega t + B \sin \omega t = C \sin (\omega t - \varphi), \tag{3.145}$$

where

$$C = \sqrt{A^2 + B^2} \tag{3.146}$$

We have

$$\frac{B_3}{B_1} = \frac{8B_r/15\pi}{\sqrt{\hat{B}^2 + (8/3\pi)^2 B_r^2}} \tag{3.147}$$

With a few exceptions, soft magnetic materials are used in transformers so that $B_r \ll \hat{B}$. By replacing B_r and $\hat{B}$ by their values according to (3.141):

$$\left|\frac{B_3}{B_1}\right| \cong \frac{8B_r/15\pi}{\hat{B}} = \frac{4\nu\hat{H}/15\pi}{\mu_{ri} + \nu\hat{H}} \tag{3.148}$$

In the region of low fields, the increase in permeability due to the magnetic field being proportional to $\nu\hat{H}$ is always small compared to μ_{ri}. Using (3.115), (3.148) therefore reduces to

$$\left|\frac{B_3}{B_1}\right| \cong \frac{4\nu\hat{H}}{15\pi\mu_{ri}} = 0.2 \tan \delta_h \tag{3.149}$$

Finally, the third harmonic ratio in the voltage e appearing in the secondary is equal to

$$\left|\frac{e_3}{e_1}\right| = \left|\frac{3B_3}{B_1}\right| \cong \frac{4\nu\hat{H}}{5\pi\mu_{ri}} = 0.6 \tan \delta_h \tag{3.150}$$

3.9 SOFT MAGNETIC MATERIALS

3.9.1 Introduction

This section and the next one are devoted to the main categories of magnetic materials that are available on the market. We shall look at, in turn, pure iron, silicon-iron alloys, nickel-iron alloys, special alloys, the ferrites, and pressed cores. A few brief indications concerning applications will be given.

The numerical values should only be considered as a guide. In fact, it should be remembered that magnetic properties result from a large number of factors, the chemical composition, mechanical treatment, and thermal treatment being the most important. This is why two materials of different origin but identical in terms of catalogue data can have appreciable differences for certain characteristics. A complete list of the most important commercial names with the corresponding manufacturers can be found in [45].

3.9.2 Iron

Ordinary soft iron contains impurities, C, Si, P, S, N, and O, in particular. The solubility of most of these elements in iron is poor (Table 3.88).

Table 3.88
(After [29].)

Element	*Solubility in Percent*
Carbon	0.007
Oxygen	0.01
Sulfur	0.02
Nitrogen	0.001
Phosphorous	1.0

Apart from special refining, their concentration is much higher than the values in Table 3.88 and this is responsible for the appearance of various inclusions in the form of carbides, nitrides, oxides of iron, *et cetera*. (Fe_3C, Fe_4N, FeO). These inclusions hinder the movement of the Bloch walls so that the permeability remains moderate: $\mu_{ri} = 200$; $\mu_{rmax} = 5000$, and the losses are high: 10–12 W/kg at $\hat{B} = 1.5$ T and $f = 50$ Hz.

With time, certain impurities that have remained in solid solution (C and N in particular) can diffuse in the direction of the inclusions to which they become attached. The diffusion of carbon produces a slow reduction in permeability called *disaccommodation*. This process is reversible, i.e., a thermal treatment can cause the concentration of carbon in solid solution to return to its initial value. The precipitation of nitrides also produces a slow reduction in the permeability called *magnetic aging*. Here, the process is irreversible.

The purification of iron improves its magnetic properties in a spectacular manner as soon as the concentration of impurities becomes sufficiently low for them to remain in solid solution everywhere. In this form, the impurities can no longer hamper the movement of the Bloch walls whose thickness is on the order of 150 interatomic distances (Section 3.7.9). The elimination of carbon, nitrogen, and oxygen is accomplished by annealing for several hours in an atmosphere of hydrogen. Maximum relative permeabilities of $0.3 \cdot 10^6$ for a polycrystal and $1.5 \cdot 10^6$ for a monocrystal of iron have been obtained in this way in the laboratory. However, the practical use of these high permeabilities is limited because of the high cost of the techniques used to obtain them.

Often too expensive or of too mediocre quality, nonalloyed iron still has its preferred areas of application. We can give as an example the manufacture of magnetic poles with the largest specimens weighing several thousand tons. Ordinary soft iron is preferred here because of its particularly favorable permeability-cost ratio and because magnetic losses do not enter into consideration (B = constant). On a different scale, the same reasons apply to the use of soft iron in the electromechanical relay industry. Finally, the important market of consumer appliances makes considerable use of soft iron in all motors because here a low price is judged to be more important than a high yield of energy.

3.9.3 Silicon-Iron

The addition of a few percent of silicon to soft iron improves several of its qualities:

- the relative permeability increases;
- the coercive field H_c considerably decreases, which results in a reduction of the hysteresis loss;
- the electrical resistivity increases, and consequently the eddy current loss decreases;
- the stability of magnetic characteristics over time improves appreciably.

The only magnetic parameter adversely affected by the presence of silicon is the saturation induction, which slightly decreases (Fig. 3.89).

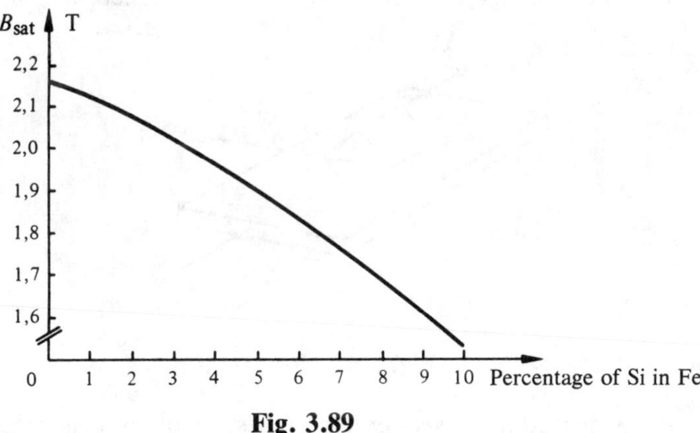

Fig. 3.89

This disadvantage is largely compensated by the above-mentioned advantages. Another effect of silicon is to make iron brittle. Above 5% silicon, it becomes so fragile that it is impossible to work with, as is shown by the reduction in elongation before fracture (Fig. 3.90).

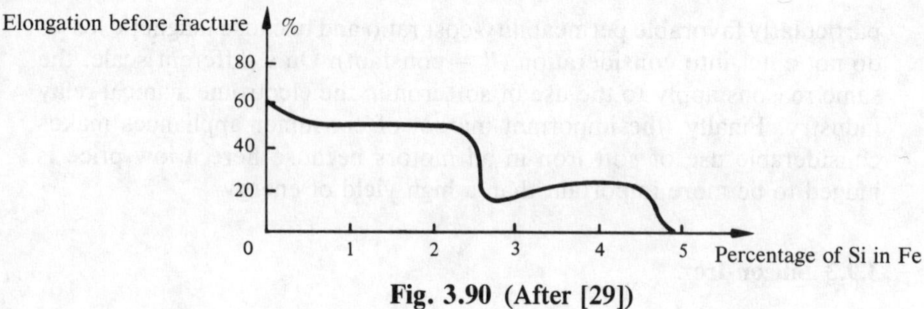

Fig. 3.90 (After [29])

Most often, alloys containing 1–2.5% silicon are used in motors, generators, and other rotating machinery, whereas alloys with 3–4.5% silicon are reserved for transformers.

The increase in the permeability and the corresponding decrease in hysteresis loss are attributed to the fact that in the presence of silicon, carbon is not precipitated in the form of iron carbide but in the form of graphite. The decrease in magnetocrystalline anisotropy and magnetostriction also contributes to the increase in μ_r. The improvement in stability over time is due to the binding of nitrogen in the form of nitrides Si_3N_4 and to graphite precipitation.

The permeability of silicon-iron can be increased by a procedure that allows the crystal axes of all the grains to be aligned (Fig. 3.91).

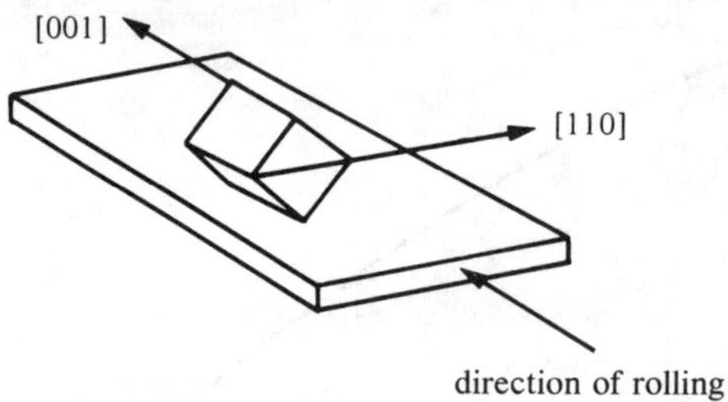

Fig. 3.91 Goss Texture

This procedure, discovered by Goss, involves a succession of cold-rolling operations and suitable thermal treatment. The easy magnetization directions of iron having the form ⟨100⟩, one of them is parallel to the direction of rolling. The variation of the magnetic properties as a function of the angle of $\boldsymbol{H}$ with the direction of rolling is illustrated in Fig. 3.92.

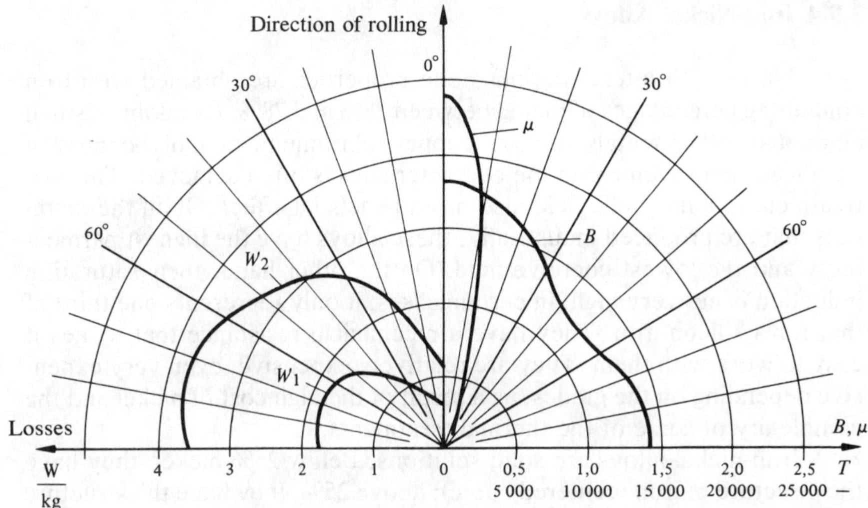

Fig. 3.92 Typical Characteristics of Grain-Oriented Fe-Si (3.5% Si), After [46]. Thickness of the Sheet: 0.35 mm; μ = Permeability for B = 1.5 T (Continuous); B = Induction for H = 800 A/m; W_1 = Total Losses at 50 Hz and B = 1 T; W_2 = Total Losses at 50 Hz and $\hat{B}$ = 1.5 T

Sheets made of grain-oriented silicon-iron are therefore only used to their fullest extent when $\boldsymbol{B}$ is parallel everywhere to the direction of rolling. This condition is perfectly satisfied in wound strip cores (Fig. 3.78(b)). In EI cores, the solution of Fig. 3.93 is usually considered to be adequate for low-power transformers. EI cores for high-power transformers are more finely divided to limit the areas where $\boldsymbol{B}$ departs from the direction of rolling.

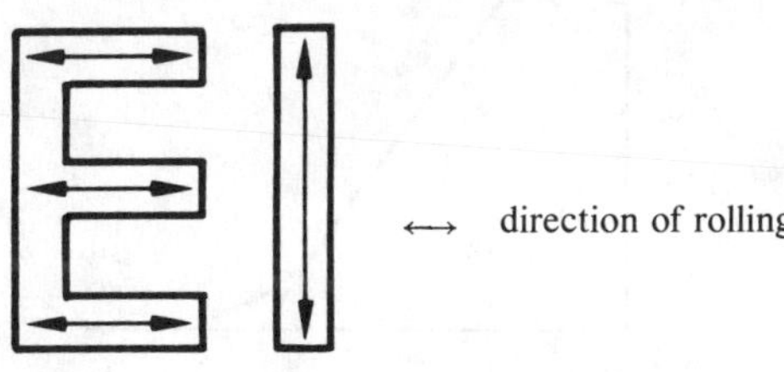

Fig. 3.93

3.9.4 Iron-Nickel Alloys

Alloys with interesting magnetic properties are obtained with iron containing percentages of nickel between 28% and 78%. By adding a small amount of other metals such as copper, chromium, or molybdenum, a significant improvement in the characteristics is often achieved. Thermal treatment in a magnetic field also has a beneficial effect. Of all the materials that are produced industrially, these alloys have the highest permeability and the lowest coercive field. On the other hand, their saturation induction is not very high; in certain cases, it only represents one third of that for a silicon-iron. They have a mechanical resistance that makes it easy to work with them. They are relatively expensive, even very expensive depending on the grades, as a result of the high cost of nickel and the complexity of some of the thermal treatments.

Iron-nickel alloys are solid solutions. Below 25% nickel, they have the structure of iron (centered cubic); above 25%, they have the structure of nickel (face-centered cubic). They are available on the market with four different concentrations of nickel.

Alloys with approximately 30% Ni have a Curie temperature close to room temperature, which can be adjusted when the alloy is manufactured. In a temperature range reaching 70°C and above, the permeability decreases almost linearly when the temperature increases, provided that the applied field H is sufficiently high, for example 10^4 A/m (Fig. 3.94).

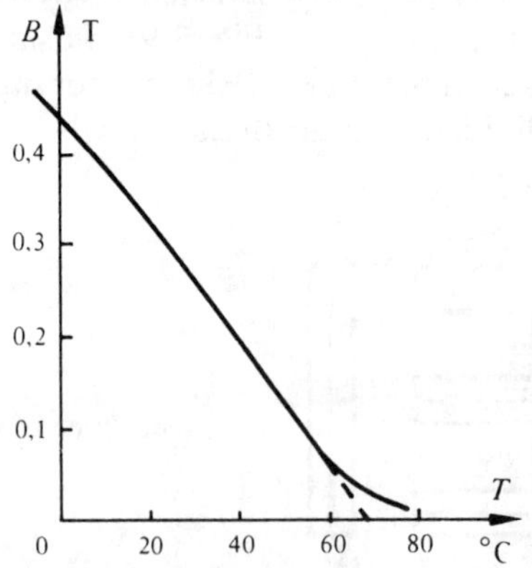

Fig. 3.94

These alloys are useful, for example, for compensating the thermal drift in the electromagnets of measuring equipment, electricity meters, *et cetera*.

Alloys with approximately 36% Ni are interesting because of their high electrical resistivity. Their permeability is relatively low, but for certain grades, it has the property of remaining very constant as a function of the field in the Rayleigh domain. Their saturation induction is rather high for an iron-nickel alloy. They are used in the manufacture of inductances and telephone isolation transformers.

Alloys with approximately 50% Ni make it possible to obtain materials with very different magnetic properties by rolling and appropriate thermal treatment.

By prolonged annealing at 1000°C in an atmosphere of hydrogen, a permeability is obtained that is close to that of alloys with 75–80% nickel. Relative to these, it then has the advantage of a higher saturation induction.

By recrystallization after strong cold-rolling, it is possible to obtain a texture of the type shown in Fig. 3.95. The material then exhibits a rectangular hysteresis loop and its area of application includes magnetic amplifiers, memories, *et cetera*.

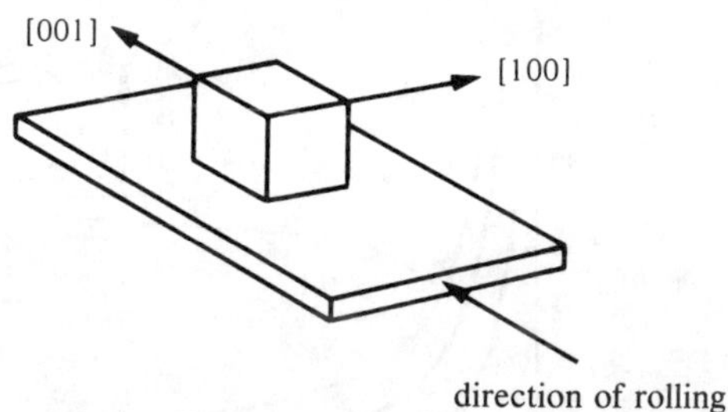

Fig. 3.95

In general, the initial permeability is higher in aluminum alloys, whereas the maximum permeability is higher in silicon-iron. The alloy with 13% aluminum has a strong magnetostriction. With 16% aluminum, the highest permeability is achieved. The use of these alloys is limited because of the costly damage that aluminum oxide formed on the surface inflicts on rolling mills.

Nickel with approximately 0.3% of manganese has strong magnetostriction and a high resistance to corrosion. It is used, for example, as a transducer for the emission of ultrasound.

Any mechanical treatment (deformation) or thermal treatment (exaggerated heating) not part of the normal manufacturing process and not allowed for in the conditions of use, may have dramatic consequences on a magnetic material. Alloys with the highest permeability are most sensitive. By way of illustration, Fig. 3.96 shows the reduction in permeability of a high-permeability alloy strip, produced by the simple transfer of this strip from a drum of diameter D_1 to a drum of diameter D_2.

By cold-rolling a strip with this texture, it is possible to obtain a material with an easy magnetization direction perpendicular to the direction of rolling. The permeability along the direction of rolling is then low, but very constant as a function of field, for example: $\mu_{ri} = 90$, $\mu_{max} = 100$.

Alloys with 75–80% Ni have the highest initial and maximum permeabilities and the lowest coercive fields. Their saturation induction is the lowest of the series. With these compositions, the magnetocrystalline anisotropy is almost zero. Consequently, there is a reduction in the energy of the Bloch walls, reducing the pinning forces exerted by the inclusions. Because the magnetostriction is also very weak, the movement of the walls is not hindered by the appearance of stresses at the scale of the domains, as discussed in Section 3.7.7. These two characteristics are responsible for the high permeability observed, which finds an application in the following areas: measurement transformers, recording heads, high-

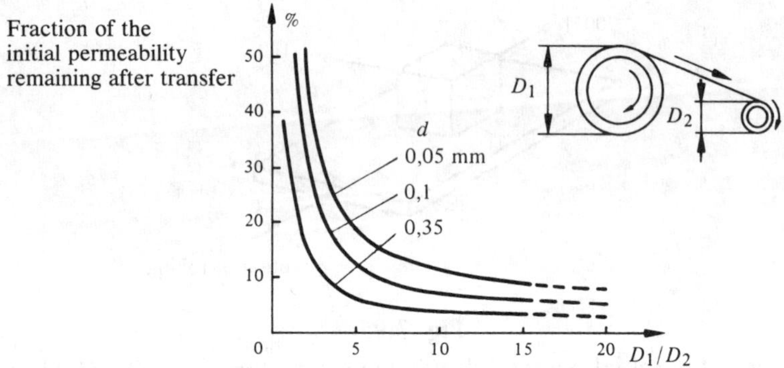

Fig. 3.96 Case of a High-Permeability Alloy (Permalloy) Strip (78% Ni, 4% Mo, 5% Cu, 13% Fe) of Thickness *d*. (After [45])

frequency transformers, pulse transformers, filters, magnetic shielding, magnetic amplifiers, *et cetera*.

3.9.5 Special Alloys

Alloys with cobalt have the highest saturation inductions and the highest Curie temperatures. Cobalt is soluble in iron up to a concentration of 75% and the structure is centered cubic.

In the neighborhood of 30% cobalt, the alloy can be cast, a forming method that is particularly suited for the manufacture of magnet poles.

With approximately 50% cobalt, the permeability is higher and the saturation induction remains close to the maximum (2.43 T). Satisfactory mechanical properties are only obtained by adding 1.5–2% vanadium. Hot- and cold-rolling thus become possible. This type of alloy is used in airplane generators where their value of B_{sat} makes it possible to save weight and space. They are also useful in the manufacture of telephone receiver membranes because they preserve a high differential relative permeability in the presence of large continuous fields.

The alloy with 7% Co, 70% Ni, and 23% Fe has a permeability that is very constant as a function of the magnetic field. On the order of 850, it does not vary by more than 1% between 0 and 0.06 T.

A concentration of aluminum in iron between 0 and 16% produces alloys with magnetic properties comparable to those of silicon-iron. Compared to the latter, they have the advantage of a better ductility. The resistivity increases quite linearly with the concentration of aluminum, and with 16% aluminum, is approximately two times that of silicon-iron.

3.9.6 Ferrites

Ferrites are approximately twice as light as iron. They contain four nonmagnetic ions (oxygen) for every three magnetic ions in the spinel structure. These characteristics, as well as the mechanism of aligning the magnetic moments peculiar to the ferrites (Section 3.6.3), limit in an absolute manner the saturation induction of these materials. It is approximately five times lower in ferrites than in magnetic alloys.

Still comparing with metals, the initial permeability and the maximum permeability at low frequency are also lower, sometimes by several orders of magnitude. The coercive field is higher. In a frequency range that may be quite extensive, the magnetic losses of the inferior ferrites may be higher than those of the thinnest magnetic strips and especially than those of pressed cores. On the other hand, the permeability of ferrites very often far exceeds that of pressed cores.

An electrical resistivity higher than that of the magnetic alloys, by a ratio that can reach 14 orders of magnitude and even more, is the hallmark of the ferrites compensating for the disadvantages just mentioned.

The ferrites and the alloys are sufficiently dissimilar for each of them to have their own specific field of application. In very broad terms, the area of large circuits is reserved for the alloys whereas that of very high frequencies is reserved for the ferrites. There is hardly any competition except for electromechanical and electronic components when the frequencies are not too high.

The medium performances of ferrites at low frequency in terms of μ_r and H_c are not attributable to fundamental limitations, as in the case of B_{sat}. They reflect the difficulties in controlling, during the manufacturing process, the concentration of impurities and the growing of crystal structures, which is much easier in the case of the metals.

In terms of their most characteristic properties, the ferrites can be divided into three classes:

- high-permeability ferrites;
- ferrites with low losses at high frequency;
- ferrites with resonance losses (Section 3.9.7).

The family of *manganese ferrites* $Mn_{1-x}Zn_xFe_2O_4$ is the most representative of the high permeability ferrites. The initial permeability is on the order of 500–10,000 depending on the grade. The resistivity is relatively low because of the presence of ferrous ions at the B sites. These ions liberate electrons according to the reaction:

$$Fe^{++} \rightleftharpoons Fe^{+++} + e^{-} \tag{3.151}$$

characterized by a low activation energy. A certain concentration of Fe^{++} ions is also necessary in order to reduce the magnetocrystalline anisotropy. It is therefore not possible to obtain a high permeability and small eddy current loss *at the same time*. By accepting that the maximum useful frequency is that for which the permeability is reduced by eddy currents to half its direct current value, manganese ferrites can be used up to a few hundred kHz.

The family of *nickel ferrites* $Ni_{1-x}Zn_xFe_2O_4$ is the most representative of the ferrites with small losses at high frequency. The presence of ferrous ions may be avoided here by producing a slight iron deficit with respect to the theoretical value of two ions per molecule. The result is a resistivity greater by several orders of magnitude than that of manganese ferrites, making the eddy current loss completely negligible in most cases. Losses related to the presence of imperfections inside the grains are also small because the grains are so small that often they only form a single magnetic domain. Polarization is therefore only established by rotation and any dissipative mechanism related to the movement of the Bloch walls is avoided. Unfortunately, the initial permeability is reduced by this. The manufacturing process for ferrites is similar to that of ceramics (Section 4.10.2). Figures 3.97 and 3.98 show samples of ferrite cores.

Fig. 3.97 Series of Ferrite Pot Cores for the Manufacture of Inductances. (After [47])

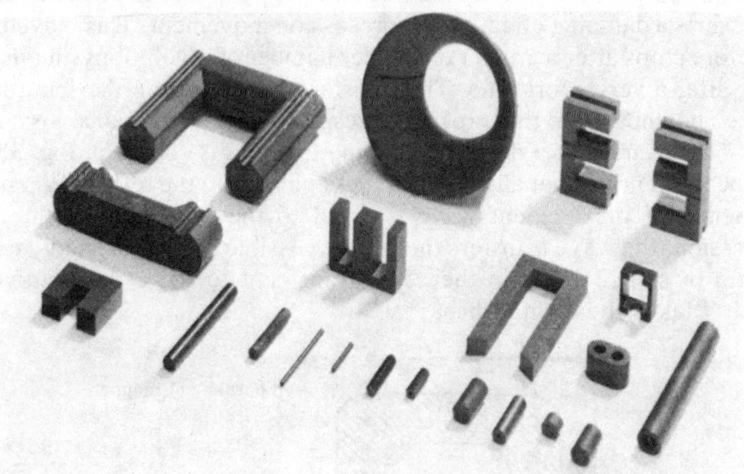

Fig. 3.98 Types of Cores Used in Television Sets. (After [47])

3.9.7 Ferrites With Resonance Loss

In addition to the losses studied thus far, which vary regularly with the frequency, magnetic materials exhibit resonance loss mainly produced by a resonance of the spins themselves.

A free magnetic moment $\boldsymbol{m}$, placed in a continuous field $\boldsymbol{H}$, is subject to a precession movement. For the spin of a free electron, the frequency f_p of precession can be deduced from the following equation ([39] Ch. 3):

$$f_p = \frac{1}{2\pi}\mu_0 \frac{e}{m_n} H = \nu H \qquad (3.152)$$

$$\nu = 35 \text{ kHz/Am}^{-1} \qquad (3.153)$$

In a ferrite where, for example, $B = 0.5$ T, we will have $H = 1/\mu_0 0.5 \text{ T} \approx 4 \cdot 10^5 \text{ Am}^{-1}$, from which we have $f_p = 14$ GHz. This crude calculation is sufficient to show that the resonance of the spins occurs in the microwave region.

The precession movement of an isolated spin could last indefinitely. The same does not apply if this spin is that of an electron belonging to an atom that forms part of a crystal because the spin-crystal lattice interaction exerts a damping effect on the precession movement. This movement therefore stops after a greater or smaller number of revolutions, but in any case, after a very short time. The transfer of energy from the spin to the lattice accompanying this process is responsible for resonance loss.

The diagram of a device exploiting this loss is shown in Fig. 3.99. The permanent magnet aligns the precession axes in the ferrite. When the frequency of the incident wave is equal to the frequency of the spin precession, the wave transfers the largest possible part of its energy to the system of spins, which in their turn transfer it to the crystal lattice in which it takes the form of heat.

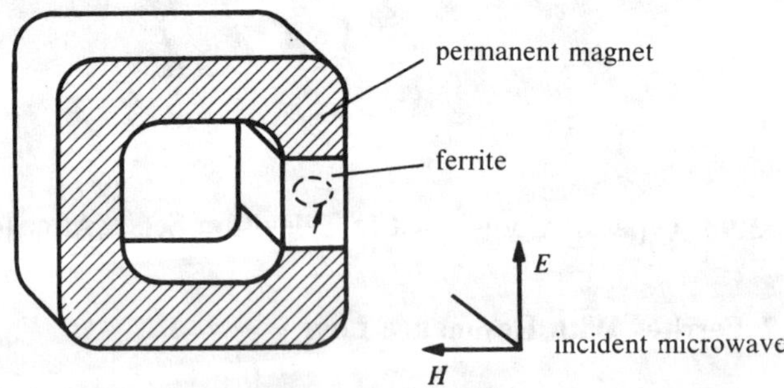

Fig. 3.99

The forced precession movement of the spins may be uniform but spin waves can also exist. These are formed by a progressive spatial variation of the orientation of the spins (Fig. 3.100).

These waves are propagated in the ferrite or, on the contrary, form a system of stationary waves. A large number of microwave components in

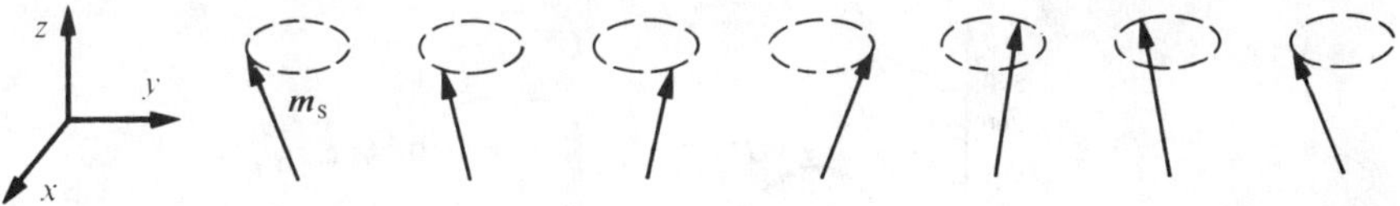

Fig. 3.100 Diagrammatic Representation of a Spin Wave. *H* is Directed Along *z*; m_s = Magnetic Moment Due to the Spin

low-power or high-power circuits use the principle shown in Fig. 3.99, to a greater or lesser degree. Examples of this are gyrators, insulators, phase-shift devices, *et cetera*.

In alloys and ferrites with relatively low resistivity, the penetration of a microwave is extremely limited and the resonance of the spins is not a property that can be used.

Ferrites with high resistivity that are most interesting for microwave applications have the garnet structure or the spinel structure.

Nickel ferrites are indicated when a large magnetic polarization is necessary. They have a broad resonance. They have the disadvantage of relatively high magnetic and dielectric losses and they are sensitive to mechanical stress.

Lithium ferrites are more stable when subjected to mechanical and thermal stress because their Curie temperature is higher. In a large number of situations, they represent the best possible choice except for cases when the effect of dielectric losses is of critical importance.

From this point of view, *ferrites with the garnet structure* are appreciably superior because they do not enclose ferrous ions if the ratio of the chemical elements is exact. They have a narrow resonance band. The most common material is the yttrium iron garnet (YIG).

The width of the resonance region not only depends on the composition, but also on other factors such as the shape of the sample, the condition of its surface, the applied field, the microstructure (size and orientation of the grains), thermal treatment, *et cetera*. The effect of the applied field is illustrated in Fig. 3.101 for the case of a ferrite $NiZnFe_2O_4$. The losses are expressed in terms of the imaginary part of the permeability μ_r'' (3.133).

The variation of the resonance frequency as a function *H* is linear, as is shown by Fig. 3.102. The same behavior was predicted for free spins by (3.152). Furthermore, the slope of the straight line is very close to the value of ν. A large bibliography on the properties and applications of ferrites in microwave technology is given at the end of [37] which provides a review of the subject.

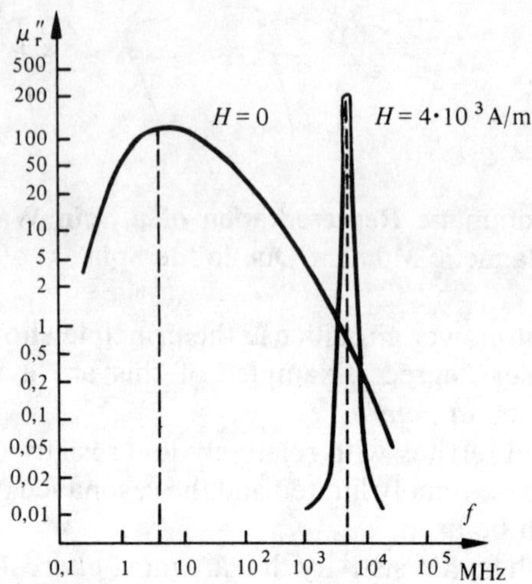

Fig. 3.101 (After [45])

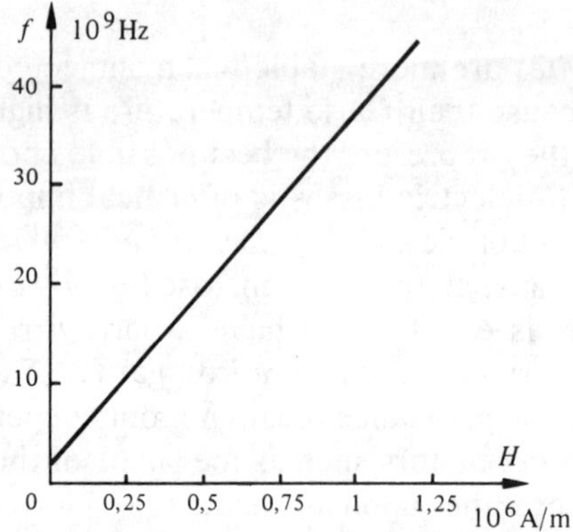

Fig. 3.102 $f_p(H)$ in a Spherical Polycrystalline Specimen of the Ferrite $NiFe_2O_4$. (After [34])

3.9.8 Pressed Cores

Pressed cores are formed from iron powder or iron-nickel alloy where the diameter of the grains is on the order of 50 μm to 100 μm. These grains coated with an insulator are pressed, in the presence of a suitable binder, to obtain the core, usually in the form of a torus. The fine division of the magnetic material strongly reduces the eddy current loss which means that pressed cores can be used up to several hundred kHz. However, it should be noted that this result is only obtained at a considerable cost in terms of the permeability the value of which usually lies between 10 and 100. The insulation of each grain produces a gap. The cumulative effect of these gaps explains why the permeability of the core is several orders of magnitude less than that of the grains themselves. Pressed cores are in the process of disappearing and they are being replaced by ferrites.

3.10 HARD MAGNETIC MATERIALS

3.10.1 Introduction

The only application for hard magnetic materials is the manufacture of permanent magnets. The use of these magnets is very widespread and, for example, affects the following areas: motors, generators and other rotating machinery, telephone receivers and loudspeakers, measurement instruments, microwave components, and recording media: magnetic bands and disks. The consumer goods industry also absorbs a considerable quantity. As a whole, the market for hard magnetic materials equals, or even slightly surpasses in financial terms, the market for soft magnetic materials.

The only function of a permanent magnet is to create an external magnetic field. This implies that the magnet has poles that are regions in which its polarization $\boldsymbol{I}$ has a component normal to the surface. The intensity of the poles is measured by the surface density of the magnetic masses σ_m (3.85) that are accumulated there. These masses produce a demagnetizing field $\boldsymbol{H}_d$ (Section 3.7.5).

A specimen of hard magnetic material only becomes a permanent magnet after being subjected to the action of a magnetic field $\boldsymbol{H}_{mag}$ aimed at producing irreversible modifications of the polarization leading to the appearance of the poles at the domain scale. On the hysteresis loop (Fig. 3.103), this operation corresponds to the O-S-P path.

The demagnetizing field, which does not exist at the beginning, is created progressively and stabilizes at the value H_d and H_{mag} returns to zero. The presence of the poles therefore reduces, by means of H_d, the induction in the magnet to a value $B_a < B_r$. We recall that B_r, a specific value of the material, is determined for a specimen that does not have any poles (torus shape).

The straight line OP is called the load line (Vol. IX, Section 3.2.5) and the part of the main loop for which $B > 0$ and $H < 0$ is called the demagnetization characteristic. Because the slope of the load line varies

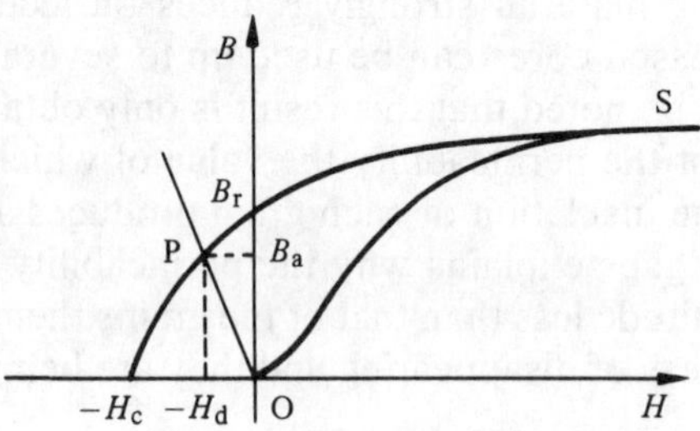

Fig. 3.103

as a function of the shape of the magnet and the occupation of its air gap, a complete knowledge of the demagnetization characteristic is necessary for computing the magnets (Vol. IX, Ch. 3).

The maximum of the product BH over the demagnetization characteristic is an important parameter. In fact, the best use of the magnetic material is obtained under static conditions when the magnet works at maximum BH.

3.10.2 Alnico and Ticonal

Alloys that are most widely used as hard magnetic materials contain in addition to iron, cobalt and nickel and then in small proportions, aluminum and copper, and sometimes other elements as well. Known under the general names of alnico or ticonal, these materials are also given a large number of different trade names. In what follows, they will simply be designated by alnico.

The alnicos are very hard and brittle so that they are formed when the alloy itself is manufactured by casting or sintering. Casting gives very large grains on the order of one mm in diameter and sintering gives much smaller grains.

The magnetic hardness of the alnicos is obtained by thermal treatment, which gives them a particular structure. The alloy is first heated to 1250°C until a homogeneous solid solution is obtained called phase α. During a cooling operation, carried out at a rate of approximately 1°C per second down to a temperature below 500°C, a second very magnetic phase α' appears, enriched in iron and cobalt. During an annealing phase at 600°C, α' is further enriched in iron and cobalt to the detriment of α to such a point that α can even become paramagnetic. Finally, α' takes the form of small elongated cylinders with a diameter of 30 nm and a height of 120 nm approximately aligned in the direction ⟨100⟩ of α. The reduced dimension of the cylinders means that they can only contain a single magnetic domain. Their elongated form gives them a very high magnetostatic energy (Section 3.7.5) which strongly binds the direction of polarization to the axis of the cylinders. The hardness of the alnicos therefore results from the nature of the monocrystalline-domain particles from which they are formed.

Once a polycrystalline specimen has undergone the treatment just described it is magnetically isotropic on the macroscopic scale. This is rarely an advantage, but by producing an anisotropy it would be possible to increase B_r and H_c. With this in mind, it is possible to apply a magnetic field during the formation of the precipitate. The development of cylinders with an axis that coincides with the direction ⟨100⟩ closest to the field is therefore preferred, because this arrangement minimizes the magnetostatic energy (in the presence of the field). The magnetic anisotropy is further accentuated if a microstructure has been produced by controlling the solidification of the alloy, in which the crystal axes of the grains are oriented preferentially in the direction of polarization. The improvements obtained are shown in Fig. 3.104.

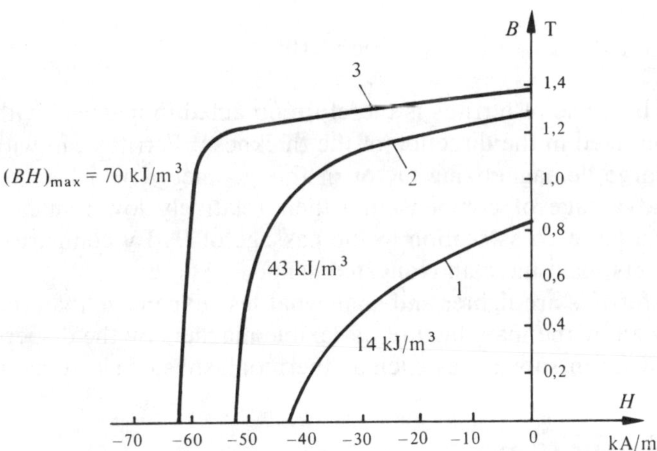

Fig. 3.104 After [48]: 1. Alnico with Isotropic Microstructure Annealed in the Absence of a Magnetic Field; 2. Alnico with Isotropic Microstructure Annealed in a Magnetic Field; 3. Alnico with Oriented Microstructure Annealed in a Magnetic Field

3.10.3 Ferrites

By virtue of their hexagonal structure, *barium and strontium ferrites* (Section 3.6.7) have a strong magnetocrystalline anisotropy energy which makes them hard magnetic materials with interesting properties. Their demagnetization characteristic is markedly different from that of the Alnicos in terms of a higher coercive field and a lower remanent induction (Fig. 3.105).

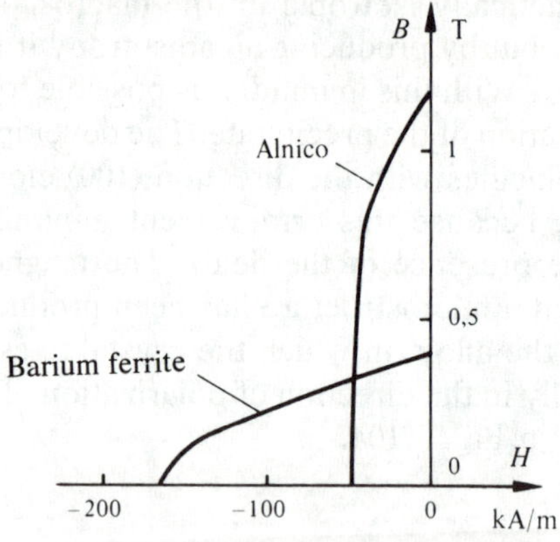

Fig. 3.105

The best use of ferrites is therefore obtained in magnets with a flat shape, polarized in the direction of the thickness. Ferrites can withstand well the large demagnetizing factor of this geometry, which in addition has the advantage of compensating their relatively low remanence by providing a large cross section to the passage of $\boldsymbol{B}$. By comparison, Alnico magnets must have an elongated shape.

The ferrites are lighter and somewhat less expensive than the Alnicos. They allow the manufacture of flexible magnets by the dispersion of ferrite powder in substances such as thermoplastics or elastomers.

3.10.4 Other Materials

In terms of the quantities used, the Alnicos and the ferrites are clearly top of the list. Nevertheless, there is a large number of other hard magnetic materials which are sometimes of key importance for certain applications.

Compounds of the lanthanide elements corresponding to one of the general formulae RM, RM_2, RM_5 and R_2M_{17} are the object of intensive research because of their extraordinarily high magnetocrystalline anisotropy energy, on the order of 20 times higher than that of barium ferrite, for example. R denotes a lanthanide or yttrium and M denotes a transition metal. The most interesting compound from the practical point of view is, at the present time, $SmCo_5$ (Table 3.109). Its structure is hexagonal, as is that of all the RM_5 compounds. Its price remains relatively high in proportion to the cost of refining samarium, because the great chemical similarity of the lanthanides makes it difficult to separate them. With the formula R_2M_{17}, it is possible to use a natural mixture of lanthanides for R_2 with a predominance of cerium, which could make it possible to lower the cost of lanthanide element magnets.

The *pressed powder* technique is also used to manufacture magnets by the following process. Elongated particles of Fe-Co are deposited electrolytically on a mercury bath. Their dimensions are such that they contain only a single magnetic domain. Lead is added which forms an amalgam. The mercury is then evaporated. The fine layer of lead remaining on the particles is used as a binder when the powder is formed by pressing. It is not necessary that the grains be insulated electrically from each other here. Depending on whether an anisotropic or isotropic material is required, a magnetic field is applied or not applied during the forming. The structure obtained, formed from elongated magnetic grains arranged in a nonmagnetic matrix, is similar to that of the alnicos with the advantage that the dimensions and proportions of the grains may be varied much more easily in order to obtain special characteristics. The powder technique is very suitable for the manufacture of magnets with complicated shapes and requiring high mechanical precision because the forming of the powder is the final operation. There is no annealing, especially sintering, shrinking is more or less well controlled. These magnets are used, for example, in micromotors, measuring instruments, *et cetera*.

Quenched steels with a high carbon content, which have been used to make compass needles for centuries, are hardly used any more. Their price is low but their performances are judged as inadequate today except in the case of steel containing 3.5% chromium. The magnetic hardness of these materials is due to the presence of inclusions and microstresses obstructing the movement of the Bloch walls.

In conclusion, we point out that there are alloys forming an exception to the rule that materials that are magnetically hard are also mechanically hard. The most well known of these is the alloy CoPt, the most expensive magnetic material ever produced on the industrial scale because of its high H_c and today replaced by lanthanide element compounds.

3.10.5 Materials for Magnetic Recording

The medium for magnetic recordings is formed by a fine layer of magnetic oxide arranged in a suitable plastic binder and deposited on a tape or a disc.

An iron oxide, *maghemite*, is very commonly used. Its coercive field is well suited to the requirements of recording, i.e., it is relatively low (~20–25 kA/m). Its chemical stability is excellent and its cost is low. It only contains trivalent ions and has the spinel structure of Fe_3O_4 in which a sixth of the B sites remains vacant. It occurs in the form of elongated particles of 0.1 microns in diameter and approximately 0.6 microns long, containing only a single magnetic domain. The magnetic hardness therefore derives from the magnetostatic energy developed by the form of the domains. In the manufacture of a magnetic tape, the particles are aligned by means of a field so that their major axis is parallel to the direction of the movement of the tape. In this way, the maximum remanent induction is achieved because the recording head polarizes the tape in the direction of its movement. Chromium dioxide CrO_2 crystallizes in the rutile structure (TiO_2) and also occurs in the form of elongated grains. This is a competitor of maghemite with a slightly higher coercive field.

3.11 TABLES OF MAGNETIC PROPERTIES

The most important properties of common magnetic materials are indicated in Tables 3.106 to 3.109.

Table 3.106
Soft Magnetic Alloys. (After [49].)

Approximate Composition Percentage by Weight				*Permeability*						
Fe	*Ni*	*Other Elements*	*Thickness (1) mm*	*Initial (2)*	*Maximum*	*Coercive Field A/m*	*Saturation Induction (3) T*	*Total Magnetic Losses at 50 Hz W/kg (4)*	*Curie Temperature °C*	*Resistivity* $10^{-6}\ \Omega \cdot m$
100			bulk	1 500	30 000	12	2.15		770	0.1
97		3 Si (5)	0.3	2 000	35 000	10	2.03	1.0 (1.5)	750	0.4
64	36		0.3	4 000	20 000	16	1.30	0.55 (1.0)	250	0.75
55–50	45–50		0.2	12 000	80 000	4	1.55	0.25 (1.0)	440	0.45
46–32	54–68		0.1	60 000	125 000	1.2	1.50	0.1 (1.0)	540	0.45
28–17	72–83	(6)	0.2	50 000	120 000	1.2	0.80	0.025 (0.5)	400	0.55
28–17	72–83	(6)	0.1	150 000	300 000	0.4	0.78	0.01 (0.5)	400	0.6
50		50 Co	0.3	1 000	12 000	110	2.35	5.5 (2.0)	950	0.35
84		16 Al	0.2	8 000	40 000	4	0.9	0.05 (0.5)	350	1.45

(1) Thickness of the specimen for which the characteristics of this table have been determined.
(2) Measured with $\hat{H} = 0.4$ A/m.
(3) Theoretical value corresponding to the saturation polarization (3.100).
(4) The values between parentheses designate the peak values of the corresponding induction in tesla.
(5) Grain-oriented Fe-Si: Goss texture (Fig. 3.91); induction in the direction of rolling.
(6) Concentration of additive elements such as Cu, Cr, and Mo, and the thermal treatments distinguish these two alloys.

Table 3.107
Soft Manganese-Zinc Ferrites. (After [50].)

Grade	*Initial Permeability (1)*	*Coercive Field A/m*	*Remanent Induction T*	*Saturation Induction (2) T*	*After-Effect Loss Factor* $tg\ \delta_r/\mu_r$ *(3)* 10^{-6}	*Legg Hysteresis Coefficient (4)* $10^9\ T^{-1}$	*Total Magnetic Losses (5)* W/dm^3	*Resistivity* $\Omega \cdot m$
I	800– 2 500	10–30	0.08–0.14	0.35–0.5	2–10	0.3–1.3	45–130	0.5–7
II	500– 1 000	40–100	0.15–0.20	0.4	5–15	0.48–1.9	250	1–20
III	1 500–10 000	2.8–24	0.07–0.14	0.3–0.5	4–60	0.1–1.3	50–150	0.02–0.5
IV	1 000– 3 000	10–30	0.1–0.25	0.35–0.52			50–120	0.2–1.0

(1) Measured at $f < 10$ kHz and $\hat{B} < 0.1$ mT.
(2) Practical value at $H = 1$ kA/m.
(3) Measured at $\hat{B} < 0.1$ mT, $f = 100$ kHz.
(4) Measured at $\hat{B} = 1$ mT, $f = 10$ kHz.
(5) Measured at $\hat{B} = 200$ mT, $f = 10$ kHz.

Application examples:
Grade I — Low-frequency inductances (<200 kHz)
Grade II — Medium-frequency inductances, antennas (100 kHz–2 MHz)
Grade III — Broad-band transformers, pulse transformers
Grade IV — Power transformers, applications requiring high B_{sat} (<100 kHz)

Note: The majority of ferrite manufacturers supply materials corresponding to each of these grades.

Table 3.108
Soft Nickel-Zinc Ferrites. (After [50].)

Grade	Initial Permeability (1)	Coercive Field A/m	Remanent Induction T	Saturation Induction and Corresponding Field T	kA/m	After-Effect Loss Factor $tg\ \delta r/\mu_r$ 10^{-6}	Corresponding Frequency MHz	Legg Hysteresis Coefficient (2) $10^9\ T^{-1}$	Resistivity $\Omega \cdot m$
I	2 000	20	0.85	0.26	1	20	0.1	6.4	10
II	500–1 000	16–50	0.15–0.19	0.28–0.34	1	150–300	1	4.8–14	$10–10^7$
III	160–490	80–160	0.12–0.16	0.3–0.36	2	50–200	3	11–16	$>10^3$
IV	70–500	160–500	0.24–0.34	0.25–0.42	4	60–120	10	1.6–48	$>10^3$
V	35–65	300–500	0.15–0.20	0.24–0.28	4	200–1 000	30	64–95	$>10^3$
VI	12–30	500–1 600	0.08–0.15	0.15–0.26	8	200–500			
VII	10	800–1 600	0.05–0.1	0.1–0.2	8				

(1) Measured at $f < 10$ kHz and $\hat{B} < 0.1$ mT.
(2) Measured at 10 kHz and $\hat{B} < 0.1$ mT.
(3) Measured at $\hat{B} < 0.1$ mT.

Application examples:

Grade I Broad-band transformers (1–300 MHz) and pulse transformers (<0.1 μs).
Grade II Broad-band transformers (5–300 MHz) and power transformers (0.1–1 MHz).
Grade III Antennas for long and medium wavelengths, power transformers (0.5–5 MHz).
Grade IV Short-wave antennaes, power transformers and inductances (2–30 MHz).
Grade V Inductances for f = 10–40 MHz.
Grade VI Inductances for f = 20–60 MHz.
Grade VII Inductances for $f > 30$ MHz.

Table 3.109
Hard Magnetic Materials. (After [45, 51].)

Material	*Composition: Weight Percent*	*Remanent Induction T*	*Coercive Field kA/m*	*B H max J/dm³*
Chrome steel	96.5 Fe 3.5 Cr	0.95	5.3	2.3
Cast Alnicos	Al Ni Cu Co Fe			
Alnico (1)	10 18 13 6 63	0.7	52	13.5
Alnico (2)	8 15 24 3 50	1.2	57	40
Alnico (3)	8 15 24 3 50	1.3	56	52
Oriented ferrites	$Ba\,O \cdot 6\,Fe_2\,O_3$	0.39	190	28
	$Sr\,O \cdot 6\,F2_2\,O_3$	0.34	260	23
Lanthanide element compound	$SmCo_5$	0.9	720	160
Anisotropic pressed powder	72 Pb 10 Co 18 Fe	0.68	76	24

(1) Isotropic microstructure annealed in the absence of a magnetic field.
(2) Isotropic microstructure annealed in a magnetic field.
(3) Anisotropic microstructure annealed in a magnetic field.

3.12 PROBLEMS

3.12.1 The relative susceptibility of nickel has been measured *versus* temperature (Table 3.110).

Show that these results obey the Curie-Weiss law.

Determine the Curie temperature. Calculate the average number of Bohr magnetons per atom.

3.12.2 How would the molecular magnetic moments of a ferrite with spinel structure corresponding to the formula MFe_2O_4 vary as a function of the relative amounts of material present in the normal structure and in the inverse structure? Consider the cases when M represents manganese, and then nickel.

Table 3.110

$T°\,C$	500	600	700	800	900
χ_r	$4.35 \cdot 10^{-4}$	$2.2 \cdot 10^{-4}$	$1.7 \cdot 10^{-4}$	$1.2 \cdot 10^{-4}$	$1.1 \cdot 10^{-4}$

3.12.3 Calculate the width l of the domains at equilibrium in the specimen of Fig. 3.56. Compare the magnetostatic energy of this structure with that of a parallelepiped of the same dimensions with only a single domain. Consider the case of iron. In this metal, $W_{SB} = 3.5 \cdot 10^{-3}$ J/m², $I_s = 2.15$ Wb/m². Numerical application for $a = 1$ cm.

3.12.4 The torus of Fig. 3.111 consists of a magnetic material characterized by the parameters μ_{ri} and ν of the Rayleigh model. Calculate the average values of these parameters that can be applied to the torus, taking into account the variation of magnetic field as a function of the radius. Express these parameters as a function of the ratio $X = r_e/r_i$.

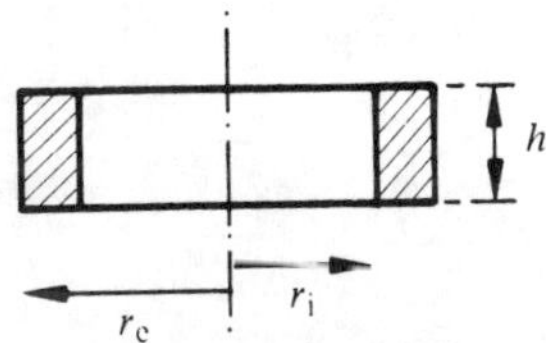

Fig. 3.111

3.12.5 Prove the relationship (3.117) by referring to Fig. 3.79, where a is made to tend to infinity.

3.12.6 Calculate the number of turns and the resistance as a function of frequency at 800, 1600, 2400, and 3200 Hz of a coil of 160 mH with a current of 1 mA flowing through it. The magnetic material has the following Jordan coefficients: $t = 2.5 \cdot 10^{-3}$ Ω/ H · Hz, $F = 4 \cdot 10^{-6}$ Ω/H · Hz², $h = 4 \cdot 10^{-2}$ Ω/H · Am⁻¹ · Hz, and its relative permeability is 80. The average length of an induction line is 12.6 cm and the magnetic cross section is 2 cm². The resistance of the winding is 5 Ω.

Chapter 4
Dielectric Properties

4.1 INTRODUCTION AND GENERAL CONSIDERATIONS

According to the energy band model (Section 2.6), matter becomes dielectric, i.e., a bad conductor of electricity, when the conduction band and the valence band are separated by an energy gap higher than 5 eV. Then, at normal temperatures only a very small number of electrons receive the thermal energy necessary to make a transition to the conduction band. When the temperature rises, the transition probability increases and the conductivity σ can be written as

$$\sigma = \sigma_0 \exp\left(-\frac{W}{k_B T}\right) \tag{4.1}$$

where W is interpreted as an activation energy of the conduction process. Relationship (4.1) is valid in crystalline and amorphous dielectrics as well as partially crystalline dielectrics such as the polymers.

Roughly speaking, dielectrics can be considered as substances in which all the electrons are so strongly bound to their atom(s) that they cannot be responsible for an electric current. Under these conditions, a current of ionic origin is also excluded.

The fact that the charges in a dielectric are not free does not mean at all that they are bound absolutely rigidly to each other. In particular, an applied electric field E slightly displaces the positive and negative charges with respect to each other causing the appearance of electric dipoles. Several different mechanisms contribute to the appearance of these dipoles (Section 4.3.3). If E varies over time, sinusoidally, for example, a phase difference is observed above certain frequencies between this field and the appearance of the dipoles (Section 4.5). This phase difference causes the dissipation of energy responsible for dielectric loss. Finally, if the electric field or the temperature exceed certain values, electrical conduction can occur abruptly, destroying the dielectric locally or in its entirety (Section 4.7).

The conductivity σ, the permittivity ε, the loss tangent tan δ, and the maximum admissible electric field E_c are the four variables that the electrical engineer most often uses to characterize a dielectric. The permittivity and one part of the dielectric loss are directly related to the polarization process (appearance of dipoles). The conductivity and the maximum admissible field are strongly dependent on the purity and the structure of the materials. These two variables are mainly studied experimentally (Section 4.7). Some special properties such as piezoelectricity and ferroelectricity are examined in Section 4.9.

In principle, the terms dielectric and insulator are synonymous, although the term insulator is more often used when the values of σ and E_c are more relevant, and the term dielectric is more often used when the values of ε and tan δ are more relevant. For example, we speak of the insulator of a cable and the dielectric of a capacitor.

Certain insulators developed and used since the last century have remarkable properties and are still used today, for example paper, mineral oil, glass, *et cetera*. The polymer industry and recent developments in the ceramics sector dominate the insulator market today. The most important industrial materials are described in Sections 4.10 to 4.14.

4.2 MACROSCOPIC LAWS AND DEFINITIONS

4.2.1 Electrostatics in Vacuum: Reminder

Two electric point charges q_1 and q_2 separated by a distance r_{12} (Fig. 4.1) exert forces $\boldsymbol{F}_1$ and $\boldsymbol{F}_2$ on each other given by Coulomb's law:

$$\boldsymbol{F}_2 = \frac{1}{4\pi\varepsilon_0} \frac{q_1 q_2}{r_{12}^2} \frac{\boldsymbol{r}_{12}}{r_{12}} = -\boldsymbol{F}_1, \tag{4.2}$$

where ε_0 is the permittivity of vacuum equal to $8.85 \cdot 10^{-12}$F/m.

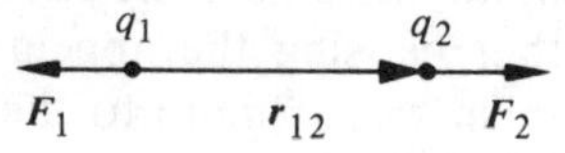

q_1 q_2

F_1 r_{12} F_2

Fig. 4.1 Case Where q_1 and q_2 Have the Same Sign

A system of charges q_i creates a displacement field $\boldsymbol{D}$ in space defined by

$$\int_A \boldsymbol{D} \mathrm{d}\boldsymbol{A} = \sum q_i \tag{4.3}$$

In this expression $\mathrm{d}\boldsymbol{A} = \boldsymbol{n}\mathrm{d}A$, $\mathrm{d}A$ being an element of the closed surface A containing the charges q_i (Fig. 4.2); $\boldsymbol{n}$ is a unit vector perpendicular to A and directed outward. The integral applies to the entire surface. The expression for the Gauss theorem can be recognized from (4.3).

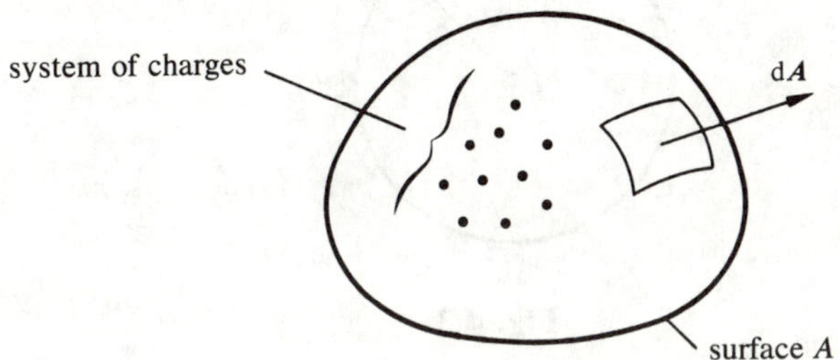

Fig. 4.2

If A is a sphere of radius r and if the system of charges is reduced to a single point charge q placed at the center of A, (4.3) gives for the value of $\boldsymbol{D}$ over the sphere:

$$\boldsymbol{D} = \frac{q}{4\pi r^2} \frac{\boldsymbol{r}}{r} \tag{4.4}$$

In the case where A contains a density of charges ρ, (4.3) can be written as

$$\int_A \boldsymbol{D} \mathrm{d}\boldsymbol{A} = \int_V \rho \, \mathrm{d}V, \tag{4.5}$$

where V is the volume bounded by A. The divergence theorem allows (4.5) to be written in the form:

$$\nabla \cdot \boldsymbol{D} = \rho \tag{4.6}$$

The effect of a system of charges can be measured not only by $\boldsymbol{D}$, but also by the electric field $\boldsymbol{E}$. At a point P in space, there exists a field $\boldsymbol{E}$ when a charge q placed at P is subject to a force $\boldsymbol{F}$ given by

$$\boldsymbol{F} = q\boldsymbol{E} \tag{4.7}$$

The fields $\boldsymbol{E}$ and $\boldsymbol{D}$ are *two different measurements of the same property of space* created by the presence of electrical charges. There must therefore be an equation relating these two variables. It is easily established by studying the system of Fig. 4.3.

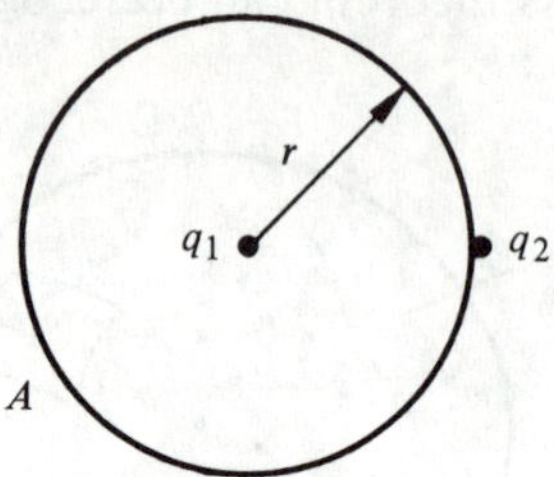

Fig. 4.3

The charge q_2 is infinitely close to the sphere but outside it and q_1 is at the center. Equations (4.2), (4.4), and (4.7) allow us to write

$$\boldsymbol{F}_2 = \frac{1}{4\pi\varepsilon_0}\frac{q_1 q_2}{r^2}\frac{\boldsymbol{r}}{r} = \frac{1}{\varepsilon_0}\boldsymbol{D}q_2 = q_2\boldsymbol{E} \tag{4.8}$$

In vacuum, the displacement field and the electric field are therefore related by

$$\boldsymbol{D} = \varepsilon_0 \boldsymbol{E} \tag{4.9}$$

These two variables play completely comparable roles; a single one of them would suffice to describe the phenomena. This will no longer be the case inside matter (Section 4.2.6).

4.2.2 The Dipole Moment

An *electric dipole* (Vol. III, Section 3.3.3) is a system formed from two point charges with the same value but opposite signs separated by a distance d (Fig. 4.4).

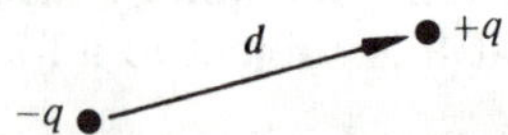

Fig. 4.4 Electric Dipole (Model)

The magnitude of this dipole is measured by its *dipole moment* $\boldsymbol{p}$ defined by

$$\boldsymbol{p} = q\boldsymbol{d} \tag{4.10}$$

If $\boldsymbol{d} \rightarrow 0$ while $\boldsymbol{p}$ remains constant, we obtain a *point dipole* at the limit. Dielectrics have different types of dipoles described in Section 4.3.3. None of them is exactly like the one in Fig. 4.4, but this can serve as a model for each of the types that are found.

4.2.3 The Polarization Vector *P*

Consider a volume V bounded by the closed surface A and containing N dipoles. The total charge of V is 0. If the orientation of the dipoles is perfectly random, V does not have any special property on the macroscopic scale. If, on the other hand, the orientation of the dipoles is not random, V will have a macroscopic dipole moment $\boldsymbol{P}_V$.

$$\boldsymbol{P}_V = \sum_{i=1}^{N} \boldsymbol{p}_i \tag{4.11}$$

In a volume $\mathrm{d}V$, in principle infinitesimal, but assumed to be large enough to contain a large number of dipoles, it is possible to define a *dipole moment density* $\boldsymbol{P}$, thus making the *transition from a system of discrete dipoles to a continuous distribution of dipole moment.* If $\mathrm{d}\boldsymbol{p}$ is the vector sum of the dipole moments contained in $\mathrm{d}V$,

$$\boldsymbol{P} = \frac{\mathrm{d}\boldsymbol{p}}{\mathrm{d}V} \tag{4.12}$$

$\boldsymbol{P}$ is also called the *polarization vector* or simply the *polarization*. On the scale of the volume V in the continuous model, (4.11) can be written as

$$\boldsymbol{P}_V = \int_V \boldsymbol{P}\,\mathrm{d}V \tag{4.13}$$

4.2.4 Representation of a Polarization State

It can be shown [52] that the potential created by the dipole moments contained in V is identical to that created by a density ρ' of charges distributed in V plus a surface density σ' of charges distributed over A, such that

$$\sigma' = \boldsymbol{P} \cdot \boldsymbol{n} \tag{4.14}$$

$$\rho' = -\nabla \cdot \boldsymbol{P}, \tag{4.15}$$

where $\boldsymbol{n}$ is still the unit vector normal to A and directed outward. The primes serve to indicate that these charge distributions are imaginary. Only the dipoles contained in V have a physical reality. The value of (4.14) and (4.15) lies in the fact that it is always possible to *replace* the effect of the dipole moments by that of σ' and ρ' in a system containing an electrically polarized medium. This results in a simplification of the calculations and a better understanding of the phenomena.

Because the total charge of V is zero,

$$\int_A \sigma' \, \mathrm{d}A + \int_V \rho' \, \mathrm{d}V = 0 \tag{4.16}$$

In the most common case where $\nabla \cdot \boldsymbol{P} = 0$, only the surface charge remains (Fig. 4.5).

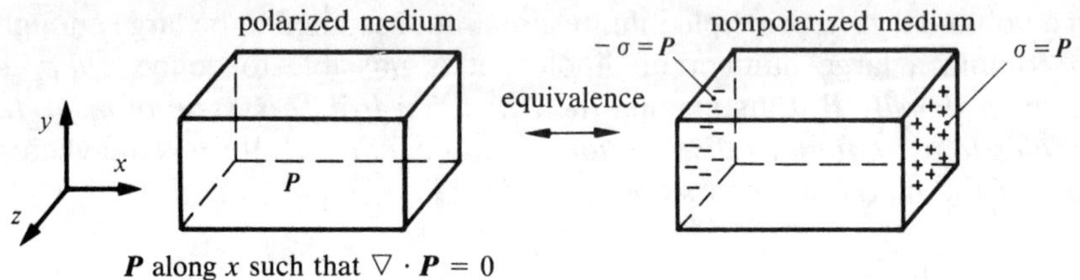

Fig. 4.5

4.2.5 Realized Charges: Definition

In the general case, a dielectric may contain surface and volume charge densities, σ and ρ, corresponding to the presence of actual charges in addition to the dipoles, the effect of which can be described by σ' and ρ'. The surface and volume charge densities are called *realized charges* [53] in order to distinguish them from the imaginary charges corresponding to σ' and ρ'.

4.2.6 Electric Fields and Displacement Field Inside a Dielectric

$\boldsymbol{D}$ and $\boldsymbol{E}$ can be defined inside a dielectric in the same way as in vacuum because a dielectric can be considered as an assembly of dipoles and point charges placed in vacuum. Let $\boldsymbol{D}_{\mathrm{m}}$ and $\boldsymbol{E}_{\mathrm{m}}$ be the displacement and field that result from this procedure. Equation (4.6) will take the form:

$$\nabla \cdot \boldsymbol{D}_{\mathrm{m}} = \rho + \rho' \tag{4.17}$$

The equation relating $\boldsymbol{E}_{\mathrm{m}}$ to $\boldsymbol{D}_{\mathrm{m}}$ will have the same form as (4.9):

$$\boldsymbol{D}_{\mathrm{m}} = \varepsilon_0 \boldsymbol{E}_{\mathrm{m}} \tag{4.18}$$

Inside matter, as in vacuum, the roles of the displacement field and the electric field are totally comparable. Just one of these variables is therefore sufficient.

However, *by convention,* the displacement field in a dielectric is the result of only the realized charges. The field $\boldsymbol{D}_{\mathrm{m}}$ as defined by (4.17) is not used in electricity. Therefore, equation (4.9) does *not* take the form of (4.18) in matter. We have

$$\nabla \cdot \boldsymbol{D} = \rho = \nabla \cdot \boldsymbol{D}_{\mathrm{m}} - \rho' \tag{4.19}$$

Using (4.15) and excluding one constant equal to zero, we have

$$\boldsymbol{D} = \boldsymbol{D}_{\mathrm{m}} + \boldsymbol{P} \tag{4.20}$$

Still, *by convention,* the electric field inside a dielectric results from the presence of polarization *and* the presence of the realized charges. Consequently,

$$\boldsymbol{E} = \boldsymbol{E}_{\mathrm{m}} \tag{4.21}$$

which, from (4.18) and (4.20), we have

$$\boldsymbol{D} = \varepsilon_0 \boldsymbol{E} + \boldsymbol{P} \tag{4.22}$$

4.2.7 Dielectric Susceptibility: Definition

In dielectrics (except ferroelectrics), the polarization is produced by the electric field and disappears with it. It is therefore natural to set

$$P = \chi_r \varepsilon_0 E \tag{4.23}$$

The *relative dielectric susceptibility* is written as χ_r. It is a dimensionless quantity. When χ_r is independent of E we are dealing with a linear dielectric.

4.2.8 Permittivity: Definitions

The form of (4.23) suggests combining the two terms on the right-hand side of (4.22). We write

$$D = \varepsilon_0 \varepsilon_r E \tag{4.24}$$

where ε_r is the *relative permittivity* of the dielectric. The product of ε_0 and ε_r, often denoted simply ε, is called the *absolute permittivity* of the dielectric.

By eliminating D from (4.22) and (4.24), we have

$$P = \varepsilon_0(\varepsilon_r - 1)E \tag{4.25}$$

so that

$$\chi_r = \varepsilon_r - 1 \tag{4.26}$$

4.3 POLARIZATION ON THE MICROSCOPIC SCALE

4.3.1 Local Field E_L

In a polarized medium, the electric field varies over the distance separating one dipole from the next one. This variation on the *microscopic scale* is not described by the Maxwell equations, which only relate the quantities representative on the *macroscopic scale*. These quantities are the *average* measurements of phenomena taking place on the atomic scale.

In order to calculate the contribution of a dipole to the polarization P as a function of microscopic data, it is essential to know the local electric field E_L directly acting on this dipole. This field may be calculated as a function of E and P.

Without restricting the generality of the result, we establish the expression for E_L in the case of an homogeneous electric field such as it exists in a planar capacitor (Fig. 4.6(a)).

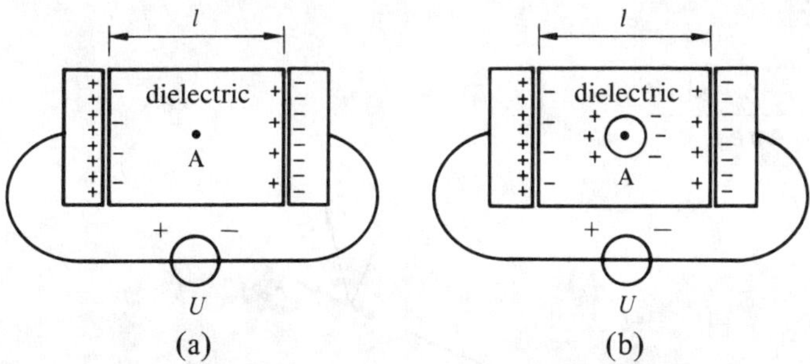

Fig. 4.6 The Effect of *P* is Replaced by the Charges Represented Diagrammatically in the Dielectric

The local field at the point A consists of two components:

$$\boldsymbol{E}_{\mathrm{L}} = \boldsymbol{E} + \boldsymbol{E}_{\mathrm{dip}}, \tag{4.27}$$

where $\boldsymbol{E}$ is the field imposed by the voltage source U. This field does not depend on the presence of the dielectric between the plates of the capacitor and is equal to U/l; $\boldsymbol{E}_{\mathrm{dip}}$ is the field resulting from the presence of dipoles over the entire volume of the dielectric. $\boldsymbol{E}_{\mathrm{dip}}$ can easily be calculated by the following method due to Lorentz. The dielectric is formally divided into two regions by a sphere S with center at A (Fig. 6(b)). Inside S, the effect of each dipole is individually taken into account by first considering the nearest neighbors of A and then the next nearest neighbors and so on until the neighbors situated on S are reached. The radius of the sphere is large enough for the distance of the dipoles outside S to the point A to be very large with respect to the distance separating two neighboring dipoles. The effect at A of dipoles outside S can therefore be replaced by that of a continuous dipole moment distribution. We therefore set

$$\boldsymbol{E}_{\mathrm{dip}} = \boldsymbol{E}_{\mathrm{i}} + \boldsymbol{E}_{\mathrm{e}}, \tag{4.28}$$

where $\boldsymbol{E}_{\mathrm{i}}$ and $\boldsymbol{E}_{\mathrm{e}}$ are the fields created by the dipoles situated inside and outside S respectively. The field $\boldsymbol{E}$ (Fig. 4.7) produced by a point dipole $\boldsymbol{p}$ is expressed in polar coordinates (Section 4.16.1) by

$$E_{\mathrm{r}} = \frac{2p}{4\pi\varepsilon_0}\,\frac{\cos\theta}{r^3} \tag{4.29}$$

and

$$E_\theta = \frac{p}{4\pi\varepsilon_0}\frac{\sin\theta}{r^3} \tag{4.30}$$

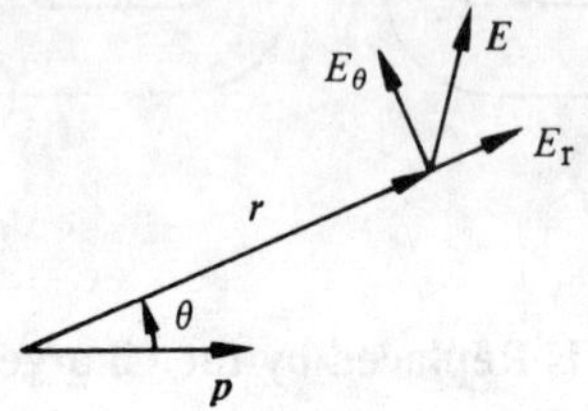

Fig. 4.7

In a crystal where the neighbors of A are arranged according to Fig. 4.8 (cubic lattice), we deduce from (4.29) and (4.30) that $\boldsymbol{E}_i = 0$.

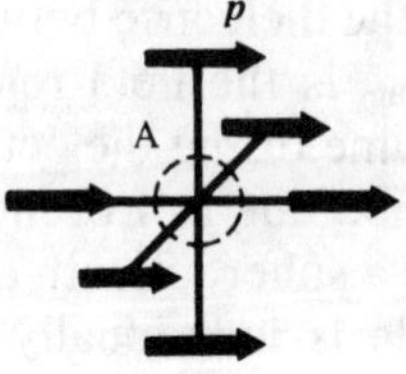

Fig. 4.8

The effect of the dipoles outside S can be replaced by a surface distribution of charges $\sigma'(\theta)$ situated on S (Fig. 4.9).

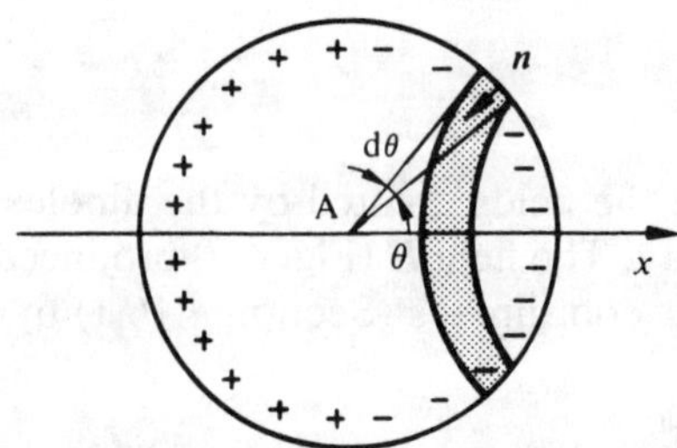

Fig. 4.9 ***E*** **and** ***P*** **are Carried by** ***x***

According to (4.14), the charge dq' carried by the elementary segment of Fig. 4.9 is equal to

$$dq' = -P \cos \theta 2\pi r^2 \sin \theta \, d\theta \tag{4.31}$$

The elementary force dF due to dq' acting on a positive charge q placed at A is, by reason of symmetry, carried by x. By considering dF positive in the direction of x, we have, by Coulomb's law,

$$dF = \frac{dq'q}{4\pi\varepsilon_0 r^2} \cos \theta, \tag{4.32}$$

from which we have the total force, on q:

$$F = \frac{Pq}{2\varepsilon_0} \int_0^\pi \cos^2\theta \sin \theta \, d\theta = -\frac{Pq}{2\varepsilon_0} \frac{\cos^3\theta}{3} \bigg|_0^\pi \tag{4.33}$$

The field created at A by the dipoles outside S is therefore equal to

$$E_e = \frac{P}{3\varepsilon_0} \tag{4.34}$$

Finally, the local field acting on a dipole is given by

$$E_L = E + \frac{P}{3\varepsilon_0} \tag{4.35}$$

Other approximations to the local field (Section 4.16.2) are not based on the assumption of a cubic lattice.

4.3.2 Polarization Factor α: Definition

Let $\boldsymbol{p}$ be the dipole moment of an atom or a molecule projected onto the local field $\boldsymbol{E}_L$ to which this atom or this molecule is subjected. The polarization factor α is defined by the ratio:

$$\alpha = \frac{p}{E_L} \qquad F \cdot m^2 \tag{4.36}$$

If N is the number of carriers of $\boldsymbol{p}$ per unit volume, the polarization due to the number of carriers is equal to

$$P = N\alpha E_L \tag{4.37}$$

4.3.3 Polarization Mechanisms

The most important mechanisms contributing to the appearance of ***p*** are represented diagrammatically in Table 4.10.

Table 4.10

Type of polarization	$E = 0$	$\overrightarrow{E}$
Electronic		
Ionic		
By orientation		
Interfacial		

atomic nucleus; anion; cation; polar molecule; grain

The displacements of the charges are very strongly exaggerated.

- *Electronic polarization* is due to a relative displacement of the nucleus of the atom with respect to all the electrons that surround it. All atoms present this type of polarization to different extents. It is established in a very short time and remains appreciable until the frequency exceeding that of visible light (10^{15} Hz). This is why electronic polarization is often called optical polarization.
- *Ionic polarization* is only found in ionic crystals. It results from the displacement of ions of opposite signs in opposite directions. This

polarization is established more slowly than the preceding polarization. It is manifested up to frequencies lying between microwave and infrared frequencies.

- Most often, a molecule formed from different atoms has a *spontaneous* dipole moment, i.e., independent of the existence of an external field. In the absence of such a field, these moments are oriented in a random manner so that there is no observable microscopic polarization. Under the effect of a field, on the other hand, the moments have a tendency to align. The result is a polarization called *orientation polarization* which occurs up to frequencies lying between 1 kHz and 1 MHz. In addition to this orientation phenomenon, the field can make the moment of the molecule vary by deforming it and its orbits.
- The charge carriers, never totally absent in a dielectric, migrate under the effect of the fields and have a tendency to concentrate around imperfections such as impurities, vacancies, grain boundaries, *et cetera*. They are known collectively as *interfacial polarization,* the polarization resulting from local accumulations of charges due to all the migration phenomena. This polarization takes the longest to build up: it may take several minutes or even more.

Electronic, ionic, and orientation polarizations are covered by general-purpose models, the simplest of which are studied in the following section. The diversity of the mechanisms responsible for interfacial polarization does not allow a general model to be obtained for it. We shall therefore not look for an explicit form for the interfacial polarization factor α_{if}.

The subject of electrical polarization is treated in a fundamental manner in [54]. The end of this section is devoted to polarization in a constant electric field. The next section will examine the polarization in a sinusoidal field.

4.3.4 Electronic Polarization

Electronic polarization can be described by the following classical model. The atom is considered as formed from a point nucleus carrying a charge Ze. This nucleus is surrounded by electrons confined to a sphere of radius R, inside which they produce a uniform charge density ρ. In the absence of a field, the nucleus is situated at the center O of the sphere (Fig. 4.11).

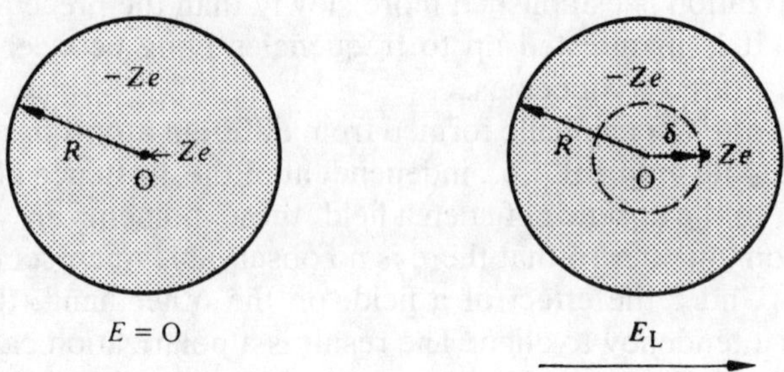

Fig. 4.11

A local field $\boldsymbol{E}_L$ creates forces F_r tending to remove the nucleus from the charged sphere. However, a displacement δ of the nucleus with respect to O creates, in its turn, Coulomb attraction forces F_a tending to return the nucleus to O. The displacement of the nucleus is therefore determined by the equilibrium condition $F_a = F_r$. We have

$$\boldsymbol{F}_r = Ze\boldsymbol{E}_L \tag{4.38}$$

Let us suppose that the sphere containing the electrons is not deformed by the action of $\boldsymbol{E}_L$. According to the theorem of Gauss, only the electric charge contained in the sphere of radius δ produces an attractive force on the nucleus. From (4.5), the displacement field due to this charge at the surface of the sphere is equal to

$$\boldsymbol{D}(\delta) = \frac{-Ze\boldsymbol{\delta}}{4\pi R^3} \tag{4.39}$$

from which we have

$$\boldsymbol{F}_a = -\frac{-(Ze)^2\boldsymbol{\delta}}{4\pi\varepsilon_0 R^3} \tag{4.40}$$

The equilibrium condition becomes

$$Ze\boldsymbol{\delta} = 4\pi\varepsilon_0 R^3 \boldsymbol{E}_L = \boldsymbol{p} \tag{4.41}$$

where $\boldsymbol{p}$ represents the dipole moment of the atom. From (4.41), it immediately follows that the electronic polarization factor α_{el} is given by

$$\alpha_{el} = 4\pi\varepsilon_0 R^3 \tag{4.42}$$

It is interesting to note that α_{el} does not depend on the atomic number but only on R, which in fact represents the radius of the atom. Despite the extreme simplicity of the model, (4.42) gives correct orders of magnitude for α_{el}, in particular in the case of the noble gases, the electronic structure of which is the closest to the starting assumptions.

Since the electronic structure of atoms varies little with temperature, we have to expect that the same applies to the electronic polarization. This fact is confirmed by experience.

4.3.5 Ionic Polarization

Let us consider the case of an ionic monocrystal with a simple cubic structure, the direction ⟨100⟩ of which is parallel to the applied field and the resulting local field.

Under the action of $\boldsymbol{E}_L$, the ions are displaced (Fig. 4.12) by a quantity Δx with respect to their positions at rest.

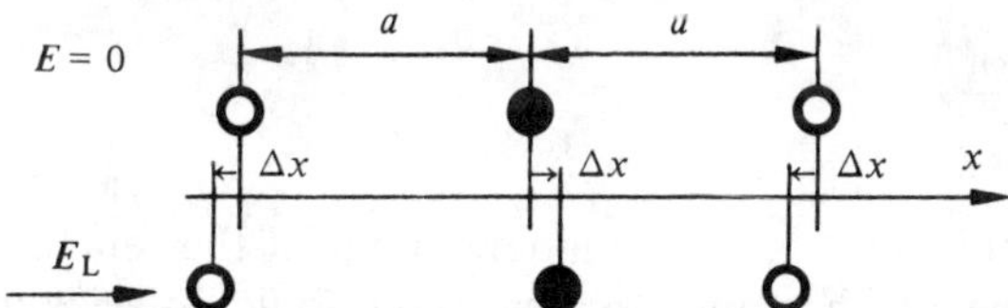

Fig. 4.12 •, Cation (Reference Ion); ○, Anion; *a*, Distance Between the Ions in the Resting Position ($E = 0$)

The equilibrium between the restoring forces (valence) and the force due to the local field is reached because of displacements Δx. For small displacements that are involved here, the restoring forces can be assumed to be proportional to Δx. By considering only one of the two neighbors of the reference ion, the restoring force is equal to

$$F_r = 2\gamma\Delta x \tag{4.43}$$

The restoring constant γ can be deduced from the bonding energy $W(x)$ of these two ions. By means of (1.36):

$$\gamma \cong \frac{1}{2} \frac{\partial^2 W}{\partial x^2}\bigg|_{x=a} \tag{4.44}$$

By considering the two neighbors of the reference ion, the equilibrium of the forces can be written as

$$4\gamma\Delta x = neE_L, \tag{4.45}$$

where n is the number of electronic charges carried by the ions.

The dipole moment resulting from the displacements Δx of the ions of Fig. 4.12 is equal to

$$p = ne(a + 2\,\Delta x) - ne(a - 2\,\Delta x) = 4ne\,\Delta x \tag{4.46}$$

from which we have

$$p = \frac{(ne)^2 E_L}{\gamma}. \tag{4.47}$$

The ionic polarization factor α_{io} is therefore equal to

$$\alpha_{io} = \frac{(ne)^2}{\gamma} \tag{4.48}$$

Since the energy $W(x)$ is nearly independent of the temperature (except for phase changes), the same applies to the ionic polarization factor.

4.3.6 Orientation Polarization

For two different atoms A and B to form a molecule, it is necessary that one of them has a more marked tendency to share one or more of its electrons with the other. Supposing that it is A that shows this tendency. One says that A is electropositive and B is electronegative. The charge transfer from A to B does not necessarily involve a whole number of electronic charges. It can happen that the electron supplied by A only spends some of its time in the vicinity of B and that the remaining time, it stays in the vicinity of A (covalency). In all cases, because there is no charge transfer at the level of the nuclei, the molecule has a spontaneous dipole moment $\boldsymbol{p}_m$.

In a molecule containing more than two atoms, several bonds may have a spontaneous dipole moment. The dipole moment of the molecule is then equal to the vector sum of these. Figure 4.13 shows two possible configurations for a hypothetical molecule A_2B. Fig. 4.13(a) corresponds to the molecule CO_2, which, being perfectly symmetrical, does not have a spontaneous dipole moment. Figure 4.13(b) corresponds to the

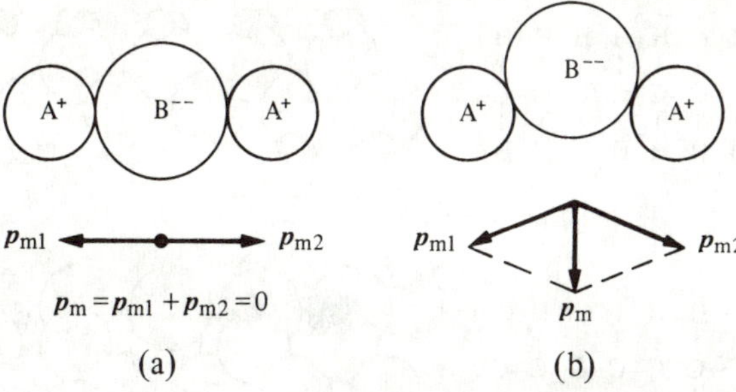

Fig. 4.13

molecule H_2O, which, with an angle A-B-A of 104°, has a dipole moment $p_m = 6.1 \cdot 10^{-30}$ C · m.

A molecule with a symmetrical chemical formula *cannot be polar*. This is the only information of an electric nature provided by the formula itself (Fig. 4.14).

Methane	Methyl chloride	Methylene chloride	Chloroform	Carbon tetrachloride
H	H	H	H	Cl
H–C–H	H–C–Cl	H–C–Cl	Cl–C–Cl	Cl–C–Cl
H	H	Cl	Cl	Cl
$p_m = 0$	$p_m = 6{,}2 \cdot 10^{-30}$ C·m	$p_m = 5{,}2 \cdot 10^{-30}$ C·m	$p_m = 3{,}8 \cdot 10^{-30}$ C·m	$p_m = 0$

Fig. 4.14

It is obvious that only methane and carbon tetrachloride cannot be polar and in reality they are not. The chlorine atom being larger than that of hydrogen, it would be expected that the spontaneous dipole moment of the other molecules would be higher for CH_3Cl and lower for $CHCl_3$, which is confirmed by experience.

The two polymers, the linear molecules of which are shown in Fig. 4.15, are dielectrics that are very widely used. The shape of the polyethylene molecule is not as simple as would be assumed from its chemical formula. However, the symmetry is perfect and polyethylene is nonpolar.

The formula for polyvinyl chloride (PVC) resembles that of polyethylene except that one atom of carbon out of two has one of its H atoms replaced by a Cl atom. The substitution occurs in one or the other of the valence bonds involved, which gives a certain random character to the

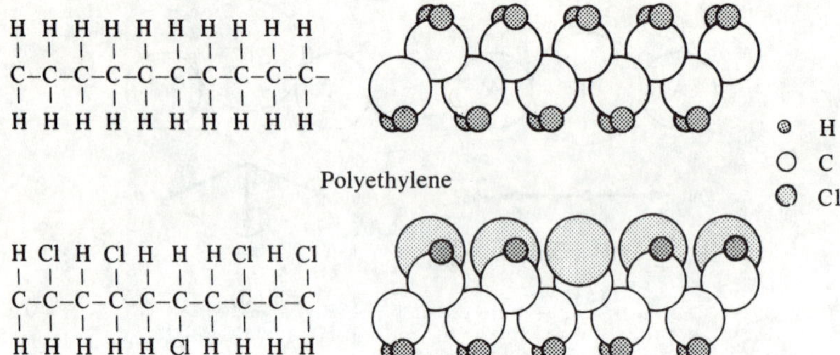

Polyvinyl chloride

Fig. 4.15

PVC molecule. This does not prevent it from remaining completely asymmetric so that it has a quite considerable dipole moment.

Subjected to a local field $\boldsymbol{E}_{\mathrm{L}}$ with which it makes an angle γ, a dipole moment $\boldsymbol{p}_{\mathrm{m}}$ is subject to a torque $\boldsymbol{C}$:

$$\boldsymbol{C} = \boldsymbol{p}_{\mathrm{m}} \times \boldsymbol{E}_{\mathrm{L}}, \tag{4.49}$$

tending to align this dipole with $\boldsymbol{E}_{\mathrm{L}}$. Thermal motion tends, on the contrary, to randomly redistribute the orientation of $\boldsymbol{p}_{\mathrm{m}}$. Formally, this situation can be compared to that found in the study of paramagnetism. By transferring the results of the Langevin theory (Section 3.4.2), we obtain

$$P = N_{\mathrm{or}} p_{\mathrm{m}} \mathrm{L}\left(\frac{p_{\mathrm{m}} E_{\mathrm{L}}}{k_{\mathrm{B}} T}\right), \tag{4.50}$$

where N_{or} is the number of molecules carrying $\boldsymbol{p}_{\mathrm{m}}$ per unit volume. The argument of the Langevin function always remains small, in practice, compared to unity. In fact, for a field $E_{\mathrm{L}} = 10^8$ V/m which is a very high value, at 20°C and for a typical dipole moment of $5 \cdot 10^{-30}$ C · m, the ratio $p_{\mathrm{m}} E_{\mathrm{L}} / k_{\mathrm{B}} T$ is only equal to 0.12. We can therefore represent the Langevin function by its expansion (3.30) restricted to the first term. We have

$$P = N_{\mathrm{or}} \frac{p_{\mathrm{m}}^2 E_{\mathrm{L}}}{3 k_{\mathrm{B}} T} \tag{4.51}$$

The orientation polarization factor α_{or} is therefore equal to

$$\alpha_{\text{or}} = \frac{p_{\text{m}}^2}{3k_{\text{B}}T} \tag{4.52}$$

Substances that contain molecules with a permanent dipole moment are called *polar*.

4.4 RELATIVE PERMITTIVITY: STEADY-STATE CONDITIONS

4.4.1 Summation of the Types of Polarization

To the extent that several different polarization mechanisms exist at the same time, their effects are added together. Characterizing by a subscript i the density of dipole moment carriers, the polarization factor, and the local field for each of these mechanisms, the total polarization is obtained by a generalization of (4.37) in the form:

$$\boldsymbol{P} = \sum N_i \alpha_i E_{\text{L}i} \tag{4.53}$$

If the dielectric can be considered isotropic, or if it has a cubic symmetry such as that resulting from Fig. 4.8, the calculation of the local field given in Section 4.3.1 can be applied. It follows that E_{L} has the same value for all types of dipoles so that

$$\boldsymbol{P} = \sum N_i \alpha_i \boldsymbol{E}_{\text{L}} \tag{4.54}$$

It is now possible to express the relative permittivity as a function of the polarization factors. By substituting (4.35) in (4.54), we have

$$\boldsymbol{P} = \left(\boldsymbol{E} + \frac{1}{3\varepsilon_0}\boldsymbol{P}\right) \sum N_i \alpha_i \tag{4.55}$$

By identifying this expression for the polarization with that given by (4.25), we can deduce for ε_{r} the expression:

$$\frac{\varepsilon_{\text{r}} - 1}{\varepsilon_{\text{r}} + 2} = \frac{1}{3\varepsilon_0} \sum N_i \alpha_i, \tag{4.56}$$

known under the name of the Clausius-Mosotti equation. This equation relates the microscopic properties α_i to a macroscopic quantity, the permittivity.

4.4.2 Remark

One could expect that several polarization mechanisms would interact with each other. In most dielectrics, this is not the case. A linear variation of the polarization as a function of the applied field is evidence for this. On the other hand, such an interaction exists in ferroelectric materials.

4.4.3 Permittivity of a Homogeneous Mixture

The dielectrics used in practice are often in the form of mixtures. It is thus possible to vary certain parameters and to obtain an optimum combination of properties corresponding to a given application.

For example, in liquids, the viscosity, the melting point, the permittivity, *et cetera* are modified in this way. In thermoplastics, the addition of various substances allows us to improve the chemical stability, lower the price per unit volume, increase the mechanical resistance, *et cetera*.

The usefulness of a formula giving the permittivity of a mixture is therefore obvious. If the mixture is perfectly homogeneous, relationship (4.56) can be used in the form:

$$\frac{\varepsilon_m - 1}{\varepsilon_m + 2} = \sum_{i=1}^{n} Y_i \frac{\varepsilon_i - 1}{\varepsilon_i + 2}, \tag{4.57}$$

where ε_m is the relative permittivity of the mixture and ε_i is the relative permittivity of component i. There are n components each occupying a fraction Y_i of the total volume.

In a large number of cases, the mixture *cannot* be considered as homogeneous on the microscopic scale. Then, there is no absolutely rigorous theory making it possible to obtain a valid formula in a general manner. Several approximate formulae have been established on a theoretical or an empirical basis, or both at the same time.

4.4.4 Permittivity of a Mixture in the Form of a Dispersion

An approximate formula may be obtained theoretically in this case. In order to establish it, we start from a first dielectric (index 1), called the host dielectric, initially pure with a relative permittivity ε_1. The effect of a spherical inclusion of the additional dielectric (index 2), with relative permittivity ε_2, into this dielectric is first of all evaluated in terms of the perturbation of the potential (Fig. 4.16). The mixture of dielectrics being then considered as a dispersion of small spheres of the additional dielectric in the host dielectric, the permittivity of the mixture is obtained by summing the effect of all the spheres.

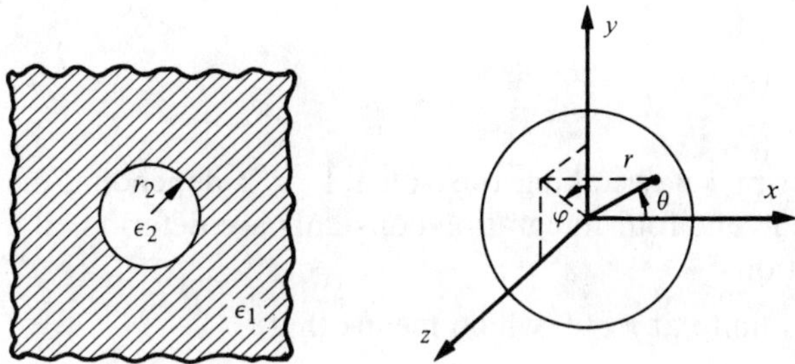

Fig. 4.16 Inclusion in the Host and Frame of Reference Used to Calculate the Potential

The potential V in the media 1 and 2 is governed by the Poisson equation:

$$\nabla^2 V + \frac{\rho_q}{\varepsilon} = 0 \tag{4.58}$$

Without further ado, it can be assumed that the density of free charges ρ_q in the dielectric is small enough to be neglected. Let $\boldsymbol{E}_0$ be the homogeneous electric field directed along x existing in the pure host dielectric. After introducing the sphere, the potential has a symmetry of revolution about the x-axis. Equation (4.58) therefore takes the form:

$$\frac{\partial}{\partial r}\left(r^2 \frac{\partial V}{\partial r}\right) + \frac{1}{\sin\theta}\frac{\partial}{\partial\theta}\left(\sin\theta\,\frac{\partial V}{\partial\theta}\right) = 0, \tag{4.59}$$

which is valid in both media. The general solution of (4.59) can be written as

$$V = \sum_{n=0}^{\infty}\left(A_n r^n + \frac{B_n}{r^{n+1}}\right) \mathrm{P}_n(\cos\theta), \tag{4.60}$$

where the A_n and B_n are sets of constants determined by the coupling conditions imposed at the interface of the media 1 and 2, P_n designating the Legendre polynomial of order n. Equation (4.60) is valid for an inclusion of any shape but assuming a symmetry of revolution about x. In the case of a spherical inclusion, only the terms corresponding to $n = 0$ and $n = 1$ remain. We can therefore write (4.60) (apart from an additive constant, which is of no importance because this is a potential), in the form:

$$V_i = \left(A_i r + \frac{B_i}{r^2}\right) \cos \theta, \tag{4.61}$$

the subscript i now taking the value 1 or 2 depending on the medium considered. The four integration constants are determined by the following conditions:

- V_2 is finite at $r = 0$ which means that

$$B_2 = 0 \tag{4.62}$$

- the tangential component of $\boldsymbol{E}$ is continuous at the surface of the sphere:

$$\left.\frac{\partial V_1}{\partial \theta}\right|_{r=r_2} = \left.\frac{\partial V_2}{\partial \theta}\right|_{r=r_2} \tag{4.63}$$

- the normal component of $\boldsymbol{D}$ is continuous at the surface of the sphere:

$$\varepsilon_1 \left.\frac{\partial V_1}{\partial r}\right|_{r=r_2} = \varepsilon_2 \left.\frac{\partial V_2}{\partial r}\right|_{r=r_2} \tag{4.64}$$

- at infinity, $\boldsymbol{E}$ is no longer perturbed by the presence of the inclusion and is equal to $\boldsymbol{E}_0$:

$$E_1(r) = -\frac{\partial V_1}{\partial r} = -\left(A_1 - \frac{2B_1}{r^3}\right) \cos \theta \tag{4.65}$$

By setting $r = \infty$ and $\theta = 0$, we have

$$A_1 = -E_0 \tag{4.66}$$

By expressing (4.63) and (4.64) by means of (4.61), we finally obtain

$$V_1 = -E_0 \left(r + \frac{\varepsilon_1 - \varepsilon_2}{2\varepsilon_1 + \varepsilon_2} \frac{r_2^3}{r^2}\right) \cos \theta \tag{4.67}$$

and

$$V_2 = -E_0 \frac{3\varepsilon_1}{2\varepsilon_1 + \varepsilon_2} r \cos \theta \tag{4.68}$$

It should be noted that V_2 only depends on $r \cos \theta$, i.e., on x. Therefore, in inclusions, the equipotential surfaces are planes perpendicular to x. It follows that E_2 has the same value at every point of the inclusion (Fig. 4.17).

$$E_2 = -\frac{\partial V_2}{\partial x} = E_0 \frac{3\varepsilon_1}{2\varepsilon_1 + \varepsilon_2} \tag{4.69}$$

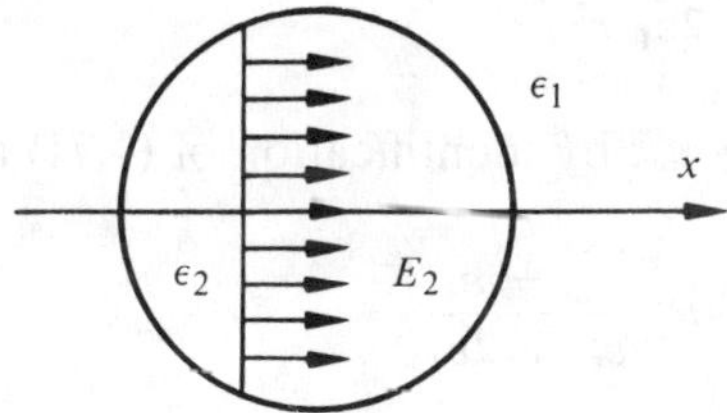

Fig. 4.17 Electric Field in the Inclusion

When $\varepsilon_1 > \varepsilon_2$, the field in the inclusion is higher than in the host dielectric. This result is important in the case of dielectrics with inhomogeneities such as bubbles.

We now consider N inclusions distributed uniformly inside a sphere S with radius $R \gg r_2$, still situated inside the host dielectric (Fig. 4.18).

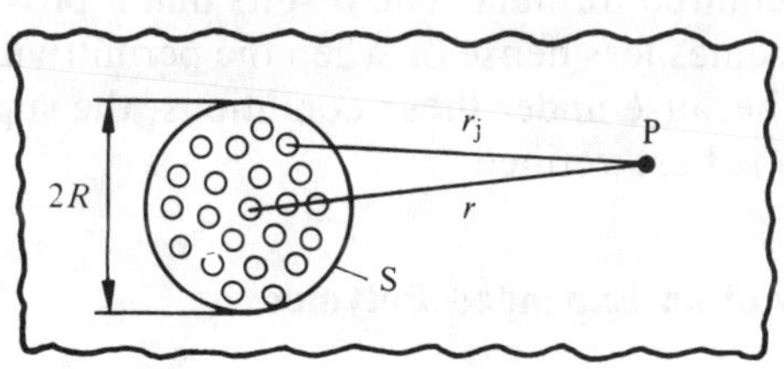

Fig. 4.18

If the density of the inclusions in S is not too large, their contribution to the potential at P, V_{NP}, can be expressed by the sum of the contributions of each inclusion taken individually. The contribution of an inclusion is given by the second term in (4.67). We therefore have

$$V_{NP} = \sum_{j=1}^{N} E_0 \frac{\varepsilon_2 - \varepsilon_1}{\varepsilon_2 + 2\varepsilon_1} \frac{r_2^3}{r_j^2} \cos\theta, \tag{4.70}$$

an expression that can be simplified provided P is placed sufficiently far from S so that R is small relative to r. Then,

$$V_{NP} = NE_0 \frac{\varepsilon_2 - \varepsilon_1}{\varepsilon_2 + 2\varepsilon_1} \frac{r_2^3}{r^2} \cos\theta \tag{4.71}$$

By considering S as made from a homogeneous dielectric with unknown relative permittivity ε_m, the contribution of this sphere to the potential at P can be written as

$$V_{NP} = E_0 \frac{\varepsilon_m - \varepsilon_1}{\varepsilon_m + 2\varepsilon_1} \frac{R^3}{r^2} \cos\theta, \tag{4.72}$$

from which we have ε_m, by identification of (4.71) and (4.72):

$$Nr_2^3 \frac{\varepsilon_2 - \varepsilon_1}{\varepsilon_2 + 2\varepsilon_1} = R^3 \frac{\varepsilon_m - \varepsilon_1}{\varepsilon_m + 2\varepsilon_1} \tag{4.73}$$

By denoting the fraction by volume of the additional dielectric by $Y = Nr_2^3/R^3$, we have

$$\frac{\varepsilon_m}{\varepsilon_1} = \frac{\dfrac{\varepsilon_2}{\varepsilon_1} + 2 + 2Y\left(\dfrac{\varepsilon_2}{\varepsilon_1} - 1\right)}{\dfrac{\varepsilon_2}{\varepsilon_1} + 2 - Y\left(\dfrac{\varepsilon_2}{\varepsilon_1} - 1\right)} \tag{4.74}$$

This is the required formula. The results that it provides improve as the dispersion becomes less dense or when the permittivities of the media 1 and 2 are close because under these conditions, the superimposition of the effects, (4.70) is best verified.

4.4.5 Permittivity of an Expanded Polymer

In expanded polymers, experiment shows that a logarithmic relationship of the type:

$$\ln \varepsilon_m = \sum_{i=1}^{n} Y_i \ln \varepsilon_i \tag{4.75}$$

gives satisfactory results. If ε_r is the permittivity of the homogeneous dielectric and d is its density, and d' is the density of the expanded material, (4.75) is simply written as

$$\ln \varepsilon_m = \frac{d'}{d} \ln \varepsilon_r \tag{4.76}$$

Fig. 4.19 shows the variation of the permittivity of expanded polystyrene.

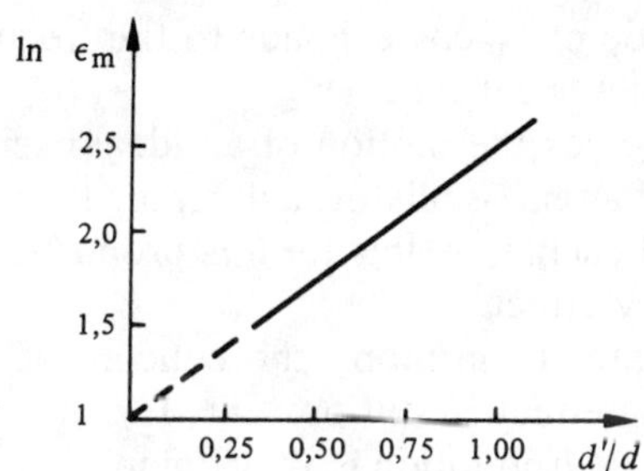

Fig. 4.19

4.5 POLARIZATION UNDER SINUSOIDAL CONDITIONS

4.5.1 Introduction

Most frequently, dielectrics are subject to variable fields, in particular sinusoidal fields. This section reexamines the polarization models presented in Section 4.3 from the dynamic point of view in order to establish expressions for the polarization factors under sinusoidal conditions.

4.5.2 Electronic Polarization in a Sinusoidal Field

According to the model developed in Section 4.3.4, the atom consists of a mechanical oscillator of which one mass is the nucleus and the other is the system of electrons. The nucleus, at least 1800 times heavier than the electrons, can be considered immobile. By still assuming that the local field displaces the orbits with respect to the nucleus without deforming them, the dynamic equation for the oscillator can be written according to (4.40) as

$$m\frac{d^2x}{dt^2} = -\frac{(Ze)^2}{4\pi\varepsilon_0 R^3}x = -kx, \tag{4.77}$$

where m is the total mass of the Z electrons surrounding the nucleus and k is the restoring force constant. Equation (4.77) can be recognized as the equation of a harmonic oscillator whose eigenfrequency is

$$\omega_0 = \sqrt{\frac{k}{m}} \tag{4.78}$$

The order of magnitude of ω_0 corresponds to the frequencies situated in the visible or ultraviolet spectrum.

Equation (4.77) gives the motion of an ideal oscillator. In order to describe the motion of a real oscillator, a damping term must be added. In the case of a classical oscillator, this term is proportional to the velocity and reflects a viscosity effect.

When the oscillator is an atom, the concept of viscosity loses its sense but the damping effect is still present. It can have more than one origin and its detailed interpretation is very complex. Here, it is sufficient for us to assume the following very simplistic mechanism. When an external field is applied, the corresponding local field transfers the necessary energy to the electrons to make the transitions resulting in orbital changes creating the dipole moment of the atom. This energy transfer takes place in a discontinuous manner because the energy levels are quantized. Furthermore, a certain time can elapse between the time when the energy necessary for a transition is available and the time when this transition actually takes place. These are the two reasons why there is a delay between cause and effect and therefore hysteresis and dissipation of energy in the form of heat. The phenomenological description of this process by a term proportional to dx/dt is an acceptable approximation.

The atom being assumed to be subject to a local sinusoidal field acting along x, $E_L(t) = \text{Re}\{\underline{E}_{L0}\exp(j\omega t)\}$, the dynamic equation for the oscillator becomes, using complex notation:

$$-\omega^2 m\underline{X}_0 + j\omega d\underline{X}_0 + k\underline{X}_0 = Ze\underline{E}_{L0} \tag{4.79}$$

with

$$X = \text{Re}\{\underline{X}_0 \exp(j\omega t)\}, \tag{4.80}$$

where d is the damping factor. From (4.79), we derive

$$\underline{X}_0 = -\frac{Z\dfrac{e}{m}\underline{E}_{L0}}{\omega^2 - \omega_0^2 - j\dfrac{d}{m}\omega} \tag{4.81}$$

This expression allows the dipole moment to be calculated

$$p(t) = \text{Re}\,\{Ze\underline{X}_0 \exp(j\omega t)\} = \text{Re}\,\{\underline{p} \exp(j\omega t)\} \tag{4.82}$$

by definition of the phasor vector $\underline{p}$.

The electronic polarization factor is now a complex variable:

$$\underline{\alpha}_{el} = \frac{\underline{p}}{\underline{E}_{L0}} = -\frac{\dfrac{Z^2e^2}{m}}{\omega^2 - \omega_0^2 - j\dfrac{d}{m}\omega} \tag{4.83}$$

By separating the real and imaginary parts, we have

$$\underline{\alpha}_{el} = \frac{Z^2\dfrac{e^2}{m}}{(\omega_0^2 - \omega^2)^2 + \dfrac{d^2}{m^2}\omega^2}\left((\omega_0^2 - \omega^2) - j\frac{d}{m}\omega\right), \tag{4.84}$$

or, by definition of α'_{el} and α''_{el}:

$$\underline{\alpha}_{el} = \alpha'_{el} - j\alpha''_{el} \tag{4.85}$$

The discussion of (4.83) as a function of frequency is simplified because

$$\frac{d}{m} \ll \omega_0 \tag{4.86}$$

Therefore, up to angular velocities close to ω_0, α'_{el} remains constant and retains the value given by (4.42) for the steady-state conditions, α''_{el} being negligible. Close to ω_0, α'_{el} rapidly varies and becomes negative for $\omega > \omega_0$. α''_{el} has a maximum for $\omega = \omega_0$ and again tends to 0 for $\omega \to \infty$ (Fig. 4.20). By setting

$$\Delta\omega = \omega_0 - \omega, \tag{4.87}$$

and provided that

$$\Delta\omega \ll \omega_0 \tag{4.88}$$

we can write

$$\alpha'_{\rm el} \cong \frac{Z^2e^2}{m}\frac{\Delta\omega/2\omega_0}{\Delta\omega^2 + \dfrac{d^2}{4m^2}} \tag{4.89}$$

and

$$\alpha''_{\rm el} \cong \frac{Z^2e^2}{m}\frac{d/4m\omega_0}{\Delta\omega^2 + \dfrac{d^2}{4m^2}} \tag{4.90}$$

The extreme values of $\alpha'_{\rm el}$ are situated at $\omega = \omega_0 \pm d/2m$. The width of the peak of $\alpha''_{\rm el}$ measured at half the height is equal to d/m.

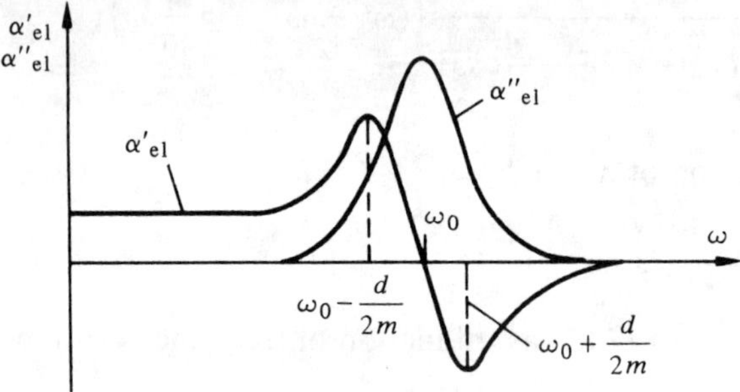

Fig. 4.20

It should also be noted that $\alpha''_{\rm el}$ is nonzero only at places where $\alpha'_{\rm el}$ varies and that, *vice versa*, there is no variation of $\alpha'_{\rm el}$ if $\alpha''_{\rm el}$ is zero.

4.5.3 Ionic Polarization in a Sinusoidal Field

The dipole moments corresponding to ionic polarization are also mechanical oscillators (Fig. 4.12). Since the masses involved are larger, the frequencies are lower than in the case of electronic polarization. They are situated in the infrared. By analogy with (4.85), we set

$$\underline{\alpha}_{io} = \alpha'_{io} - j\alpha''_{io} \qquad (4.91)$$

The frequency behavior of α'_{io} and α''_{io} is at all points similar to that described in Fig. 4.20.

4.5.4 Orientation Polarization in a Sinusoidal Field

Two factors affect the orientation of permanent dipoles:

- the local field E_L;
- thermal motion, the effect of which can be described in terms of the action of the collisions of the dipoles.

These collisions tend to produce a random orientation of the dipoles. Except in ferroelectric materials, their effect clearly predominates over the local field (Section 4.3.6) which tends to align the dipoles parallel to each other. Therefore, it can be assumed as a first approximation that the orientation of a dipole does not vary between two successive collisions.

On the other hand, since the dipoles are not coupled to each other, it is reasonable to assume that immediately after a collision, the energy of a dipole is governed by Boltzmann's distribution (Section 7.3) corresponding to the value of the local field at the time of the collision. The polarization dP, resulting from the dn dipoles having experienced their last collision between t_0 and $t_0 + dt$, is calculated under these conditions by the Langevin theory (Section 3.4.2). Since the local field has the form:

$$E_L = \text{Re}\,\{\underline{E}_{L0} \exp(j\omega t)\} \qquad (4.92)$$

rearrangement of (3.28) gives, taking (3.30) into account,

$$dP = \text{Re}\left(\frac{p_m^2\, dn}{3k_B T}\,\underline{E}_{L0} \exp(j\omega t_0)\right) \qquad (4.93)$$

The polarization at time t resulting from *all* the dipoles is obtained by integrating (4.93) as a function of t_0 over the interval $-\infty$ to t. For this, we set (Fig. 4.21):

$$t_0 = t - \theta \qquad (4.94)$$

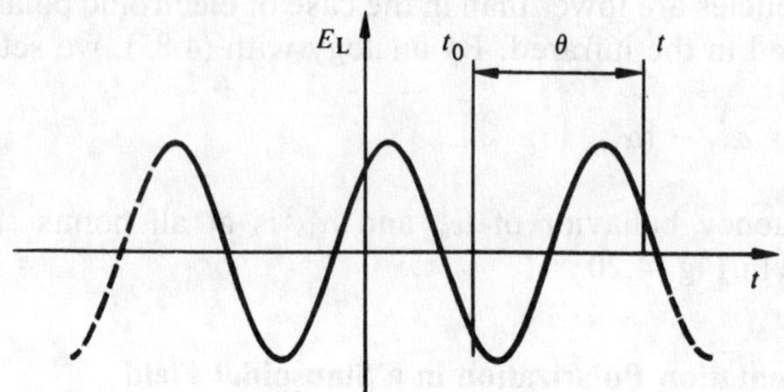

Fig. 4.21

We have

$$P = \text{Re}\left(\int_0^\infty \frac{p_m^2\, dn}{3k_B T}\, \underline{E}_{L0} \exp j\omega(t - \theta)\right) \tag{4.95}$$

The average frequency of the collisions does not depend on E_L nor consequently on the time. In (4.95), dn therefore does not depend on t but only on the time interval θ.

The number of dipoles experiencing their first collision between t and $t + dt$ (beginning of the observation at $t = 0$) is equal to the number of dipoles that, having experienced their last collision between $-t$ and $-(t + dt)$, have not experienced any more collisions up to $t = 0$. The considerations of Section 2.2.4 therefore apply to the calculation of dn. From (2.5) and (2.6), we derive

$$dn = N_{or} \frac{1}{\tau} \exp(-\theta/\tau)\, d\theta \tag{4.96}$$

N_{or} represents the number of dipoles per unit volume and τ represents the average time separating two consecutive collisions experienced by one dipole. We call τ the *relaxation time*. By substituting (4.96) in (4.95) and carrying out the integration, we find

$$P = \text{Re}\left(\frac{N_{or} p_m^2}{3k_B T} \frac{1}{1 + j\omega\tau} \underline{E}_{L0} \exp(j\omega\tau)\right) \tag{4.97}$$

This expression is known as Debye's formula. The orientation polarization factor can be deduced from it as

$$\underline{\alpha}_{\rm or} = \frac{p_{\rm m}^2}{3k_{\rm B}T}\frac{1}{1 + {\rm j}\omega\tau} = \alpha'_{\rm or} - {\rm j}\alpha''_{\rm or} \tag{4.98}$$

Equation (4.98) defines the decomposition of $\alpha_{\rm or}$ into real and imaginary parts. We have

$$\alpha'_{\rm or} = \frac{p_{\rm m}^2}{3k_{\rm B}T}\frac{1}{1 + \omega^2\tau^2} \tag{4.99}$$

$$\alpha''_{\rm or} = \frac{p_{\rm m}^2}{3k_{\rm B}T}\frac{\omega\tau}{1 + \omega^2\tau^2} \tag{4.100}$$

The variation of these parameters is shown in Fig. 4.22, to be compared with Fig. 4.20.

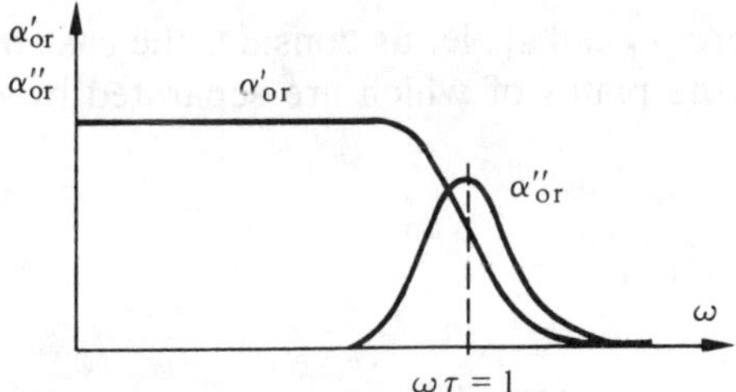

Fig. 4.22

As long as $\omega \ll 1/\tau$, $\alpha''_{\rm or}$ is negligible. There are no dielectric losses and the polarization is in phase with the electric field. For $\omega \gg 1/\tau$, the dipoles can no longer follow the variations in the field so that the orientation polarization vanishes. In the region where $\omega \approx 1/\tau$, the phase shift between the electric field and the polarization is maximum and the dielectric losses are at a maximum as well.

Debye set

$$\tau = \eta(T)/k_{\rm B}T, \tag{4.101}$$

where $\eta(T)$, a decreasing function of temperature, represents the viscosity hindering the movement of the dipole from aligning with the electric field. Today, τ is assumed to have the form:

$$\tau \sim \exp(-W_{\rm a}/k_{\rm B}T) \tag{4.102}$$

where W_a is an activation energy characterizing the dipole's orientation process.

4.6 RELATIVE PERMITTIVITY: SINUSOIDAL CONDITIONS

4.6.1 Complex Permittivity and Dielectric Losses: Definitions

A dielectric subject to a sinusoidal field must be characterized by complex polarization factors that are functions of the angular frequency. The Clausius-Mosotti equation (4.56) shows that the relative permittivity is then itself a complex variable that is a function of the angular frequency. By analogy with the permeability (3.131), we set

$$\underline{\varepsilon}_r(\omega) = \varepsilon_r' - j\varepsilon_r'' \tag{4.103}$$

In order to interpret ε_r' and ε_r'', let us consider the case of a planar capacitor with area S, the plates of which are separated by d. Its admittance $Y(\omega)$ is equal to

$$Y(\omega) = j\omega\varepsilon_0\underline{\varepsilon}_r(\omega)\,\frac{S}{d} \tag{4.104}$$

The current I flowing in this capacitor, subject to a sinusoidal voltage $\underline{U}$ with angular frequency ω, is equal to

$$\underline{I} = \underline{Y}(\omega)\underline{U} = \varepsilon_0\,\frac{S}{d}\,\omega(\varepsilon_r'' + j\varepsilon_r')\underline{U} \tag{4.105}$$

It therefore has a component in phase with $\underline{U}$, reflecting the power dissipation in the dielectric. This power dissipation, which in general cannot be recovered, is referred to as *dielectric loss*. These losses may have two origins:

- the irreversible work needed to build up the polarization (the polarization factors are complex);
- the residual ohmic conduction of the dielectric.

In the first case, we talk more exactly of *polarization dielectric loss,* and in the second case of *conduction dielectric loss*.

Instead of describing a dielectric by a complex permittivity, it is possible to use the real component of the permittivity plus a loss angle.

By definition, the *loss* δ of a dielectric is called the phase displacement between the total current and the current in quadrature with the voltage. From (4.105), we have

$$\tan \delta = \frac{\varepsilon_r''}{\varepsilon_r'} \tag{4.106}$$

The polarization dielectric losses can also be expressed in terms of the equivalent conductivity σ_p. Let ε_{rp}'' be the imaginary part of ε_r representing this loss. We obtain, by (4.105):

$$\sigma_p = \omega\varepsilon_0\varepsilon_{rp}'' \tag{4.107}$$

4.6.2 Permittivity and Losses as a Function of the Frequency

The interfacial polarization mainly occurring in continuous and quasi-continuous fields is neglected in this section.

Each polarization factor varies strongly in a frequency domain that is specific to it, whereas it can be considered constant outside this domain. It follows from the above (Section 4.5) that the ranges of variation of $\underline{\alpha}_{or}(\omega)$, $\alpha_{io}(\omega)$ and $\alpha_{el}(\omega)$ are clearly separate from each other. From a practical point of view, $\underline{\alpha}_{or}(\omega)$ is of particular importance because its range of variation may lie between 0 and several thousands of megahertz depending on the material. The variation of $\underline{\varepsilon}_r(\omega)$, as well as the dielectric loss found in most applications, therefore directly result from orientation polarization. In certain cases, the variation of $\underline{\alpha}_{io}(\omega)$ may affect the micro wave region whereas that of $\underline{\alpha}_{el}(\omega)$ is always above the frequencies used in electricity.

In order to distinguish the role of $\underline{\alpha}_{or}(\omega)$, it is useful to write the Clausius-Mosotti equation in the form:

$$\frac{\underline{\varepsilon}_r - 1}{\underline{\varepsilon}_r + 2} = \frac{1}{3\varepsilon_0}\left(\sum_i N_i\alpha_i + \frac{N_{or}}{1 + j\omega\tau}\,\frac{p_m^2}{3k_BT}\right) \tag{4.108}$$

The sum over i includes all of the polarization mechanisms other than that of orientation, the factors α_i being considered as real constants.

In order to discuss the function ε_r given by (4.108), let us set

$$\varepsilon_{r0} = \lim_{\omega\to 0} \underline{\varepsilon}_r \tag{4.109}$$

and

$$\varepsilon_{r\infty} = \lim_{\omega\to\infty} \underline{\varepsilon}_r \tag{4.110}$$

These two variables satisfy the equations:

$$\frac{\varepsilon_{r0} - 1}{\varepsilon_{r0} + 2} = \frac{1}{3\varepsilon_0}\left(\sum_i N_i\alpha_i + N_{or}\frac{p_m^2}{3k_B T}\right) \tag{4.111}$$

and

$$\frac{\varepsilon_{r\infty} - 1}{\varepsilon_{r\infty} + 2} = \frac{1}{3\varepsilon_0}\sum_i N_i\alpha_i \tag{4.112}$$

By introducing these two expressions into (4.108), we have

$$\underline{\varepsilon}_r = \varepsilon_{r\infty} + \frac{\varepsilon_{r0} - \varepsilon_{r\infty}}{1 + j\beta}, \tag{4.113}$$

where

$$\beta = \frac{\varepsilon_{r0} + 2}{\varepsilon_{r\infty} + 2}\omega\tau \tag{4.114}$$

By separating the real and imaginary parts of (4.113), we obtain

$$\varepsilon_r' = \varepsilon_{r\infty} + \frac{\varepsilon_{r0} - \varepsilon_{r\infty}}{1 + \beta^2}, \tag{4.115}$$

$$\varepsilon_r'' = \frac{\varepsilon_{r0} - \varepsilon_{r\infty}}{1 + \beta^2}\beta, \tag{4.116}$$

and

$$\tan\delta = \frac{\varepsilon_{r0} - \varepsilon_{r\infty}}{\varepsilon_{r0} + \varepsilon_{r\infty}\beta^2}\beta \tag{4.117}$$

It can be seen in the above expressions that for $\omega = 0$, $\varepsilon_r' = \varepsilon_{r0}$ and $\tan\delta = 0$. For $\omega \to \infty$, $\varepsilon_r' = \varepsilon_{r\infty}$ and similarly $\tan\delta = 0$. The polarization dielectric loss therefore vanishes in two situations:

- at sufficiently low frequencies for there to be no phase difference between $\boldsymbol{P}$ and $\boldsymbol{E}$;
- at sufficiently high frequencies so that the period of $\boldsymbol{E}$ becoming small compared to the relaxation time of the permanent dipoles, their orientation is no longer influenced by $\boldsymbol{E}$ and remains random.

Between these two extremes, the variation of ε_r' and $\tan\delta$ as a function of ω takes the form shown in Fig. 4.23.

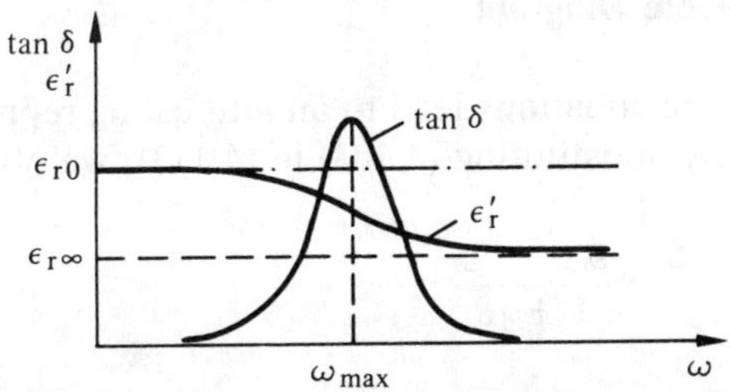

Fig. 4.23

The maximum of tan δ is obtained by the condition:

$$\frac{d}{d\beta}(\tan \delta) = 0 \tag{4.118}$$

leading to

$$\beta_{max} = \sqrt{\frac{\varepsilon_{r0}}{\varepsilon_{r\infty}}}, \tag{4.119}$$

from which, due to (4.114), we have

$$\omega_{max} = \frac{1}{\tau}\sqrt{\frac{\varepsilon_{r0}}{\varepsilon_{r\infty}}\frac{\varepsilon_{r\infty}+2}{\varepsilon_{r0}+2}} \tag{4.120}$$

Finally, from (4.117):

$$\tan \delta_{max} = \frac{1}{2}\frac{\varepsilon_{r0} - \varepsilon_{r\infty}}{\sqrt{\varepsilon_{r0}\varepsilon_{r\infty}}} \tag{4.121}$$

As long as only the variation of $\underline{\alpha}_{or}$ (ω) is involved, we note the following facts, from Fig. 4.23 in particular:

- when the frequency increases, the permittivity can only decrease;
- when the frequency increases, the loss angle can increase, or on the contrary decrease, depending on whether one is below or above ω_{max}.

4.6.3 The Cole-Cole Diagram

The preceding equations lead to an interesting representation of the function $\varepsilon_r''(\varepsilon_r')$. By substituting (4.103) in (4.113), we obtain

$$\varepsilon_r' - \varepsilon_{r\infty} - j\varepsilon_r'' = \frac{\varepsilon_{r0} - \varepsilon_{r\infty}}{1 + j\beta} \tag{4.122}$$

The equality of the moduli of both sides of this equation leads to

$$(\varepsilon_r' - \varepsilon_{r\infty})^2 + \varepsilon_r''^{\,2} = \frac{(\varepsilon_{r0} - \varepsilon_{r\infty})^2}{1 + \beta^2} \tag{4.123}$$

After elimination of β^2 by relationship (4.115), we have

$$(\varepsilon_r' - \varepsilon_{r\infty})^2 + \varepsilon_r''^{\,2} = (\varepsilon_r' - \varepsilon_{r\infty})(\varepsilon_{r0} - \varepsilon_{r\infty}) \tag{4.124}$$

or

$$\varepsilon_r'^2 + \varepsilon_r''^2 - \varepsilon_r'(\varepsilon_{r\infty} + \varepsilon_{r0}) + \varepsilon_{r0}\varepsilon_{r\infty} = 0 \tag{4.125}$$

This equation can be recognized as a circle with its center on the ε_r' axis. The top half of this circle (Fig. 4.24), which is the only part having physical meaning, is called the Cole-Cole diagram. The bottom half corresponds to an active material.

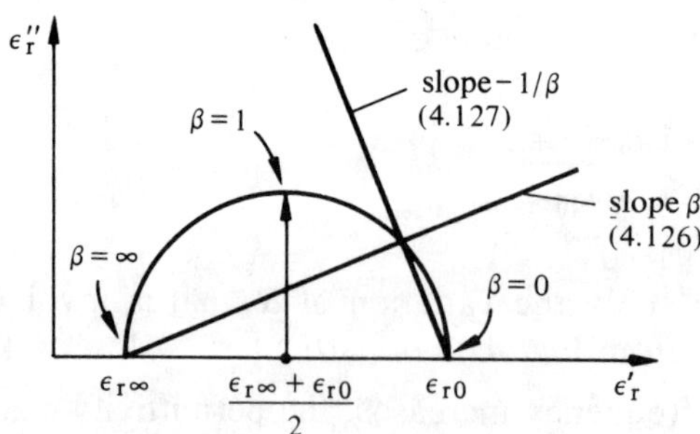

Fig. 4.24 Cole-Cole Diagram for a Single Relaxation Time

The Cole-Cole diagram gives a more accurate value of $\varepsilon_{r\infty}$ than Fig. 4.23 because the corresponding point is not discarded at infinity. It also makes it possible to determine the relaxation time τ. By writing (4.115) and (4.116) as

$$\varepsilon_r'' = \beta(\varepsilon_r' - \varepsilon_{r\infty}) \tag{4.126}$$

and

$$\varepsilon_r'' = -1/\beta(\varepsilon_r' - \varepsilon_{r0}), \tag{4.127}$$

respectively, it is found that β, and therefore τ, may be derived from the slope of one of the straight lines (4.126) or (4.127) passing through a point of the semicircle for which ω is known (Fig. 4.24).

In the case where the dielectric does not have a single relaxation time but a relaxation time spectrum, Cole and Cole [55] suggest a generalization of (4.113) in the form:

$$\varepsilon_r' - j\varepsilon_r'' = \varepsilon_{r\infty} + \frac{\varepsilon_{r0} - \varepsilon_{r\infty}}{1 + (j\beta_a)^{1-a}}, \qquad 0 \leq a < 1, \tag{4.128}$$

where a is a constant characterizing the distribution of values of β (related to the relaxation time by (4.114)), around the most probable value of β denoted β_a. For $a = 0$, (4.128) coincides with (4.113); there is only a single relaxation time; the more a increases, the more the relaxation time distribution is spread out.

Expression (4.128) also leads to a representation of $\varepsilon_r''(\varepsilon_r')$ is the form of a circle (Fig. 4.25). By separating the real and imaginary parts of (4.128), and then eliminating β_a from the expressions obtained, it can be shown that this circle always passes through the points ε_{r0} and $\varepsilon_{r\infty}$ and that its center is at a distance Δ below the axis ε_r'.

$$\Delta = \frac{\varepsilon_{r0} - \varepsilon_{r\infty}}{2} \tan\left(a\,\frac{\pi}{2}\right) \tag{4.129}$$

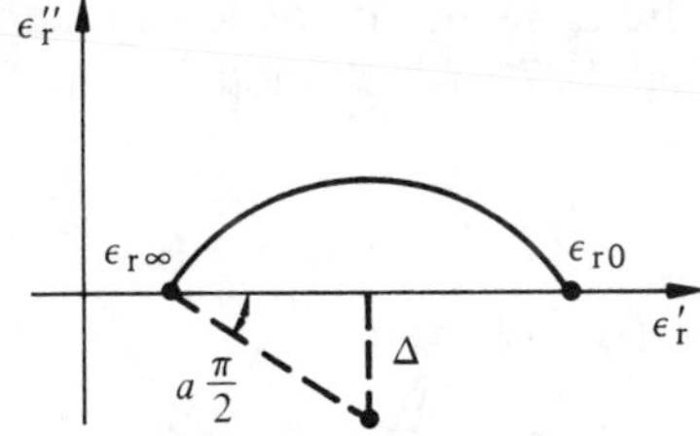

Fig. 4.25 Cole-Cole Diagram for a Relaxation Time Spectrum

4.6.4 Permittivity and Losses *versus* Temperature

According to the models studied above, it would be expected that only variations of temperature involving $\underline{\alpha}_{or}$ and the density would have an effect on $\underline{\varepsilon}_r$ and tan δ. Experience confirms this prediction. In most dielectrics it is found that after correcting the density, $\varepsilon_{r\infty}$ does not vary by more than a few 10^{-4}/°C, whereas ε_{r0} slightly decreases when the temperature increases, which is in agreement with Langevin's theory (Section 3.4.2). In general, the variations in ε_{r0} and $\varepsilon_{r\infty}$ can therefore be neglected.

In *polar* substances, the variations in $\underline{\varepsilon}_r$ and tan δ at frequencies close to $1/\tau$ are, on the contrary, likely to become very large. In order to analyze them, it is sufficient to note that ω and τ always appear in the form of the product $\omega\tau$. It is known that τ decreases as a function of the temperature (4.102). The variation of ε_r' and tan δ as a function of ω (Section 4.6.2) can therefore be immediately rearranged as a function of temperature (Fig. 4.26).

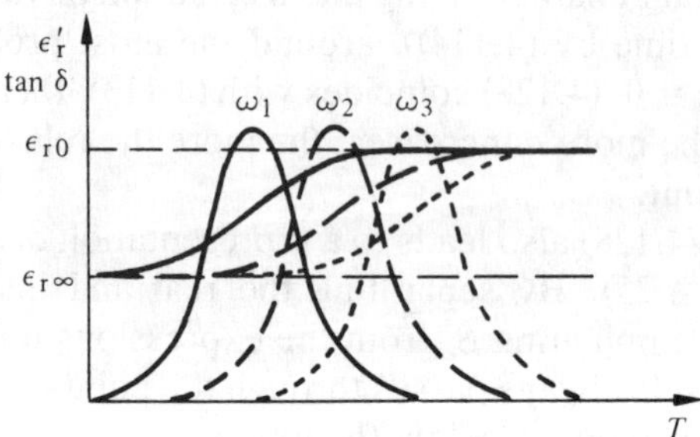

Fig. 4.26 Diagrammatic Representation of $\varepsilon_r'(T)$ and tan $\delta(T)$ for Three Angular Frequencies $\omega_1 < \omega_2 < \omega_3$. The Effect of Density is Neglected

If the rise in temperature is such that conduction in the dielectric is no longer a negligible phenomenon, it has to be taken into account in the evaluation of losses (Fig. 4.27), which modifies the form of tan δ (T).

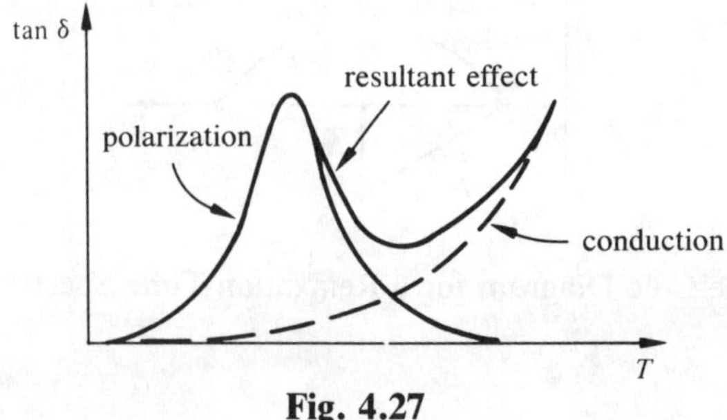

Fig. 4.27

In a Cole-Cole diagram, a rise in temperature produces a displacement of the frequencies in the direction of the point ε_{r0} to the point $\varepsilon_{r\infty}$ (Fig. 4.28).

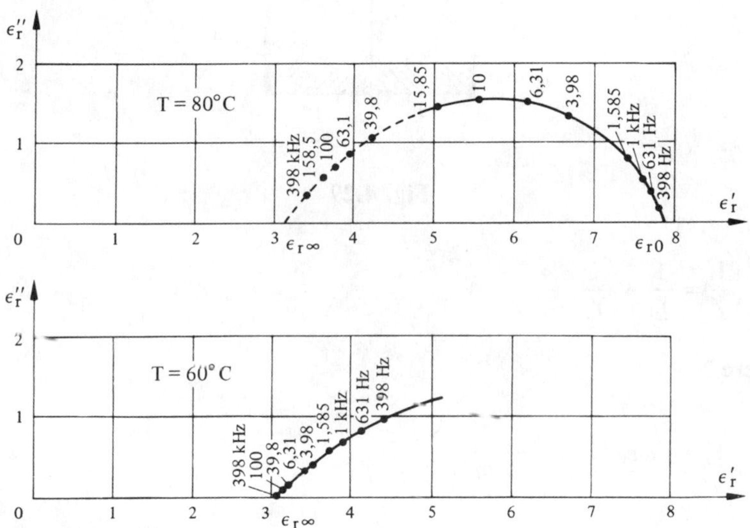

Fig. 4.28 Measurement of the Complex Permittivity of Polyvinyl Acetate as a Function of Temperature (After [56])

4.6.5 Losses in Nonpolar Dielectrics

Below the frequencies where the losses due to ionic polarization begin to appear, the only losses of nonpolar dielectrics result from residual conduction. If the dielectric medium is homogeneous, these losses do not depend on the frequency and are simply calculated from the resistivity. The loss tangent is inversely proportional to the frequency.

If the dielectric is heterogeneous, i.e., formed from several nonpolar materials, its behavior is more complex. By way of illustration, we present the simple case of an arrangement of two dielectrics with identical thickness placed in a planar capacitor (Fig. 4.29).

Let ε_k be the real part of the relative permittivity of dielectric No. k and ρ_k be its resistivity. The problem is to find the properties ε_r and $\tan\delta$ of a virtual homogeneous dielectric equivalent to the dielectrics No. 1 and No. 2 being placed next to each other. This equivalence in terms of the admittance Y is given by

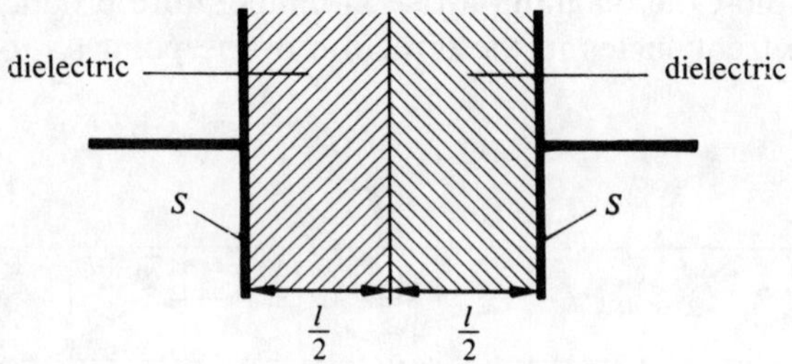

Fig. 4.29

$$\frac{1}{\underline{Y}} = \frac{1}{\underline{Y}_1} + \frac{1}{\underline{Y}_2}, \tag{4.130}$$

where

$$\underline{Y}_k = \mathrm{j}\omega\varepsilon_0 \left(\varepsilon_k - \frac{\mathrm{j}}{\omega\varepsilon_0\rho_k}\right) \frac{S}{l/2} \tag{4.131}$$

and

$$\underline{Y} = \mathrm{j}\omega\varepsilon_0\underline{\bar{\varepsilon}}_\mathrm{r} \frac{S}{l} \tag{4.132}$$

the permittivity of the equivalent dielectric being divided into its components and related to its resistivity by the equations:

$$\underline{\bar{\varepsilon}}_\mathrm{r} = \bar{\varepsilon}_\mathrm{r}' - \mathrm{j}\bar{\varepsilon}_\mathrm{r}'' = \bar{\varepsilon}_\mathrm{r}' - \mathrm{j}\frac{1}{\omega\varepsilon_0\bar{\rho}} \tag{4.133}$$

By introducing (4.131) and (4.132) in (4.130), we obtain, taking into account (4.133):

$$\bar{\varepsilon}_\mathrm{r}' = \varepsilon_{\mathrm{r}\infty}\left(1 + \frac{K}{1 + \omega^2\tau^2}\right), \tag{4.134}$$

where

$$\varepsilon_{\mathrm{r}\infty} = \frac{2\varepsilon_1\varepsilon_2}{\varepsilon_1 + \varepsilon_2}, \tag{4.135}$$

$$K = \frac{(\varepsilon_1\rho_1 - \varepsilon_2\rho_2)^2}{\varepsilon_1\varepsilon_2(\rho_1 + \rho_2)^2}, \tag{4.136}$$

$$\tau = \varepsilon_0 \frac{\rho_1\rho_2}{\rho_1 + \rho_2} (\varepsilon_1 + \varepsilon_2), \tag{4.137}$$

and

$$\bar{\varepsilon}_r'' = \frac{1}{\omega\varepsilon_0\rho_0} + \frac{\omega\tau K\varepsilon_{r\infty}}{1 + \omega^2\tau^2}, \tag{4.138}$$

where

$$\rho_0 = \frac{\rho_1 + \rho_2}{2} \tag{4.139}$$

In (4.138), the first term gives a component of $\bar{\varepsilon}_r''$ decreasing monotonically with ω (hyperbola). The second term has a maximum for $\omega = 1/\tau$.

The resistivities of the two dielectrics may easily differ from each other by several orders of magnitude (Tables 4.75 and 4.76). In this case, and if the permittivities remain relatively close to each other, we find by expanding the two terms of (4.138) that the first term is small compared to the second, in the neighborhood of $\omega = 1/\tau$. By neglecting this first term, we obtain a simplified expression for tan δ:

$$\overline{\tan\delta} = \frac{\bar{\varepsilon}_r''}{\bar{\varepsilon}_r'} \cong \frac{\omega\tau K}{1 + K + \omega^2\tau^2}, \tag{4.140}$$

which is valid as long as we do not depart too far from $\omega = 1/\tau$ in the direction of lower angular frequencies. It is interesting to note that (4.134) and (4.140) have the same dependence on ω as (4.115) and (4.117), and that we find again, as in the case of a polar substance, that tan δ has a maximum. At first sight, this result is surprising because the loss angles of dielectrics No. 1 and No. 2 decrease regularly when ω increases (Fig. 4.30).

The angular frequency ω_{max} corresponding to the maximum determined by setting the derivative of (4.140) to 0, with respect to ω, is equal to

$$\omega_{max} = \frac{\sqrt{1 + K}}{\tau}, \tag{4.141}$$

from which we obtain

$$\overline{\tan \delta_{max}} = \frac{1}{2} \frac{K}{\sqrt{1 + K}} \tag{4.142}$$

By comparing, at the angular frequency ω_{max}, the loss angles $\tan \delta_k$ of each of the dielectrics, according to (4.131) being given by

$$\tan \delta_k (\omega_{max}) = \frac{1}{\omega_{max} \varepsilon_0 \varepsilon_k \rho_k} = \frac{1}{\varepsilon_0 \varepsilon_k \rho_k} \frac{\tau}{\sqrt{1 + K}}, \tag{4.143}$$

it is found that

$$\overline{\tan \delta_{max}} \leqslant \frac{1}{2} \tan \delta_k (\omega_{max}), \tag{4.144}$$

where $\tan \delta_k(\omega_{max})$ refers to the dielectric with the highest loss angle of the two.

The function $\overline{\tan \delta(\omega)}$ always lies between the hyperbolas $\tan \delta_k(\omega)$ characterizing the two dielectrics.

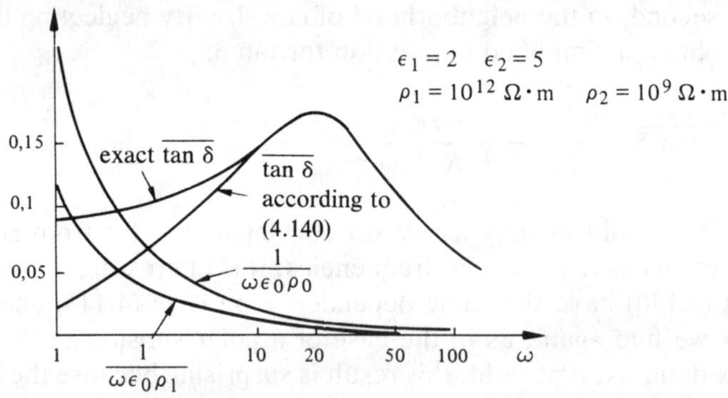

Fig. 4.30

4.6.6 Dielectric Viscosity

After a heterogeneous dielectric has been subjected to a voltage step, the electric field may still vary for a certain time, sometimes up to several minutes and even more in its various zones. This phenomenon is commonly called dielectric viscosity. It requires special precautions (standardization) when tests on dielectric strength and insulation resistance are carried out.

In order to examine dielectric viscosity, we once again consider the arrangement of two dielectrics in a planar capacitor shown in Fig. 4.29 Let U be the voltage applied to this capacitor. The sum of the polarization and conduction currents is the same in each dielectric:

$$G_1 U_1 + C_1 \frac{\mathrm{d}U_1}{\mathrm{d}t} = G_2 U_2 + C_2 \frac{\mathrm{d}U_2}{\mathrm{d}t}, \tag{4.145}$$

where

$$C_i = \varepsilon_0 \varepsilon_i \frac{2S}{l} \qquad i = 1,2 \tag{4.146}$$

and

$$G_i = \frac{1}{\rho_i} \frac{2S}{l} \qquad i = 1,2 \tag{4.147}$$

By substituting (4.146) and (4.147) in (4.145) and by replacing the voltages U_i by the circulations of the corresponding fields, we have

$$\frac{1}{\rho_1} E_1 + \varepsilon_0 \varepsilon_1 \frac{\mathrm{d}E_1}{\mathrm{d}t} = \frac{1}{\rho_2} E_2 + \varepsilon_0 \varepsilon_2 \frac{\mathrm{d}E_2}{\mathrm{d}t}, \tag{4.148}$$

from which, by eliminating the unknown E_2 by means of

$$U = E_1 \frac{l}{2} + E_2 \frac{l}{2}, \tag{4.149}$$

we obtain

$$\left(\frac{1}{\rho_1} + \frac{1}{\rho_2}\right) E_1 + \varepsilon_0(\varepsilon_1 + \varepsilon_2) \frac{\mathrm{d}E_1}{\mathrm{d}t} = \frac{2}{\rho_2 l} U \tag{4.150}$$

The general solution of this differential equation of the first order with constant coefficients is

$$E_1 = A \exp\left(-\frac{t}{\alpha}\right) + \frac{2}{\rho_2 l} U \left(\frac{1}{\rho_1} + \frac{1}{\rho_2}\right), \tag{4.151}$$

where the first term represents a general solution of (4.150) without the right-hand side and the second term represents a particular solution of (4.150) with the right-hand side. On the other hand,

$$\alpha = \varepsilon_0 \frac{\varepsilon_1 + \varepsilon_2}{(1/\rho_1) + (1/\rho_2)}, \tag{4.152}$$

where A is an integration constant to be determined by the fact that at time $t = 0 + \mathrm{d}t$, the conduction current is negligible compared to the polarization current, theoretically infinite. The voltages in the dielectrics No. 1 and No. 2 are therefore distributed in the ratio:

$$\frac{U_1}{U_1 + U_2} = \frac{\varepsilon_2}{\varepsilon_1 + \varepsilon_2} \tag{4.153}$$

so that

$$E_1 \; (t = 0) = \frac{\varepsilon_2}{\varepsilon_1 + \varepsilon_2} \frac{2U}{l} \tag{4.154}$$

A is calculated by imposing this initial condition on (4.151) which finally takes the form:

$$E_1(t) = E_{1\infty} \left[1 + K_1 \exp\left(-\frac{t}{\alpha} \right) \right] \tag{4.155}$$

with

$$E_{1\infty} = \frac{2U}{l} \frac{\rho_1}{\rho_1 + \rho_2} \tag{4.156}$$

and

$$K_1 = \frac{\rho_2 \varepsilon_2 - \rho_1 \varepsilon_1}{\rho_1(\varepsilon_1 + \varepsilon_2)} \tag{4.157}$$

$E_2(t)$ is obtained by permuting the subscripts 1 and 2 in the three expressions above. Fig. 4.31 shows the variation of E_1 and E_2 in the two cases corresponding to different combinations of permittivitie and resistivitie.

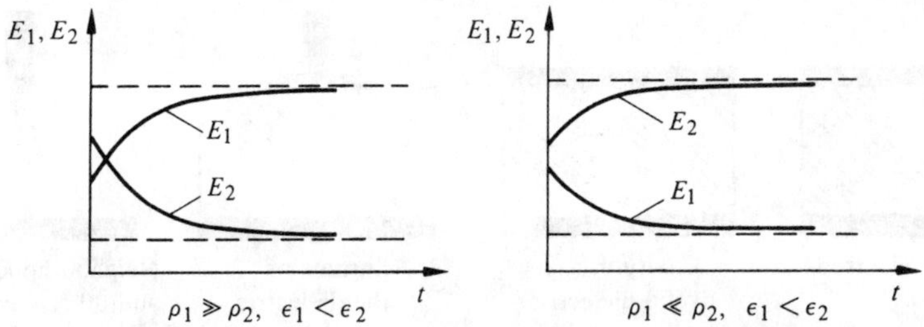

Fig. 4.31

4.7 DIELECTRIC STRENGTH AND BREAKDOWN MECHANISMS

4.7.1 Introduction and Definitions

A dielectric no longer has linear behavior when the applied field exceeds a certain value. The conductivity is no longer given by (4.1) and becomes strongly dependent on the electric field. If the field is sufficiently high, a very high conductivity may suddenly appear, causing an often irreversible modification of the dielectric, if not its complete destruction by the Joule effect.

Breakdown is the name given to the sudden loss of the insulating property of a dielectric subjected to a field.

The *dielectric strength* E_c is the maximum value of the field to which a dielectric may be subjected without breakdown occurring.

An *electric discharge,* or simply a *discharge,* is the name given to the current flowing in a dielectric at the time of a breakdown. In a discharge, there is at least one high-conductivity path connecting the electrodes.

Less spectacular, but no less important for certain materials, are the *partial discharges*. These are discharges that do not directly connect the electrodes. They are produced in a localized manner inside the dielectric itself in the vicinity of the electrodes and in the vicinity of discontinuities where the field is very inhomogeneous (Fig. 4.32).

The current carried by partial discharges usually only represents a small fraction of the displacement current. It is sometimes difficult to detect it and special circuits are needed in the case of solid and liquid insulators subjected to an alternating field [57].

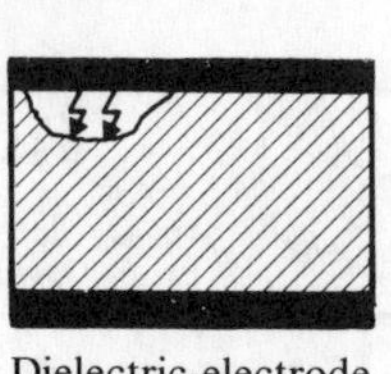

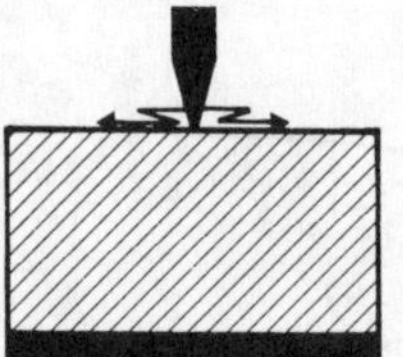

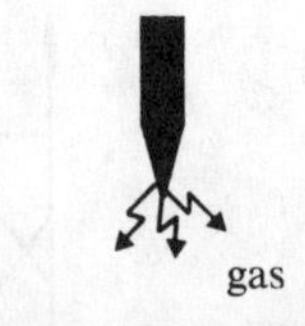

Fig. 4.32 Some Examples of Places where Partial Discharges can Take Place

The dielectric strength of a material may vary by several orders of magnitude depending on how it is used. Under identical operating conditions, it is not rare to see E_c vary by a factor of two. The purity, the implementation, the shape of the dielectric, and its environment are only a few examples of factors that may have a strong effect on E_c. This explains why, in general, it is not possible to determine E_c purely theoretically. Instead, *practical tests* are carried out using extremely precisely defined procedures so as to guarantee a certain standardization of the results obtained. These tests can be divided into two classes:

- *Tests on a sample* in which an insulator is tested for itself. The form of the sample and electrodes are specified as well as the environment and the function of time for which the field is applied: continuous field varied by steps, continuously increased sinusoidal field, pulses of the same polarity or alternating polarity, *et cetera*. These tests take place in the laboratory and serve as a control of the raw material and contribute to their technological development.
- *Tests on the finished installation.* Here it is simply a matter of demonstrating that the dielectric does not have any imperfection or that it is not stressed in any way that a breakdown can take place under normal operating conditions at any point of the installation. These tests taken as a whole involve subjecting the installation for a certain time to a voltage equal to the nominal voltage U_0 multiplied by a factor larger than one. Factors of 1.25 or 2.5 are currently used. If the insulator is sensitive to partial discharge, the absence of such discharges will be checked, for example, at 1.25 U_0.

The mechanisms responsible for breakdown can be classified into two categories: thermal breakdown and intrinsic breakdown.

4.7.2 Thermal Breakdown

The polarization and conduction dielectric losses produce a release of heat in insulators. As long as the amount of heat produced in this way is higher than that which the insulator can remove, the temperature increases. The conductivity increases with it, causing an increase in the heat produced by the Joule effect. The heat released by the polarization dielectric loss also increases, or it may decrease, depending on whether we observe from the left or the right of the maximum of tan δ (T) (Fig. 4.27).

If the improvement in cooling conditions, resulting from the rise in temperature of the insulator with respect to the environment, does not stop the increase in temperature, or if the temperature for this is too high, a breakdown takes place which is called *thermal breakdown*.

The fact that the dielectric rigidity decreases when the temperature increases favors the appearance of thermal breakdown. Sometimes, this is preceded by chemical decomposition or melting of the insulator.

By way of illustration, we shall calculate the relative dielectric strength to thermal breakdown of a cylindrical sample (4.33).

The following simplifying assumptions are made:

- there is no heat exchange through the bases of the cylinder;
- the temperature is uniform in the cylinder;
- the heat transmission coefficient λ (W/m^2°C) at the surface of the cylinder is independent of temperature;
- the polarization loss is negligible compared to the conduction losses.

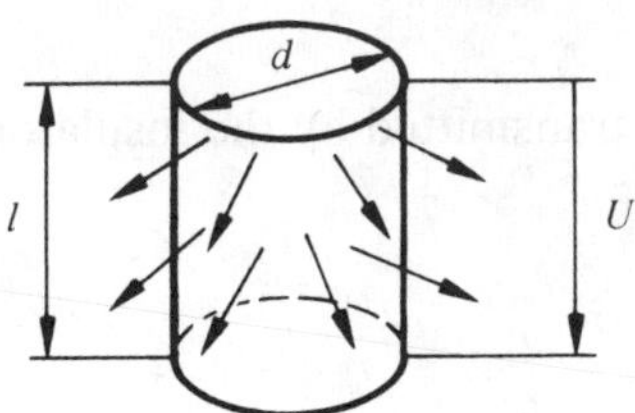

Fig. 4.33 The Field is Applied Parallel to the Axis. It Produces a Potential Difference *U* Between the Bases of the Cylinder

The power P_J dissipated in the cylinder is equal to

$$P_J \quad U^2/R, \tag{4.158}$$

where

$$1/R = \sigma \frac{\pi d^2}{4l} \tag{4.159}$$

Let T_0 be the temperature of the environment, assumed to be constant, and T be the temperature of the cylinder. We set

$$\Delta T = T - T_0 \tag{4.160}$$

From (4.1), we have

$$\sigma = \sigma_0 \exp\left(-\frac{W_a}{k_B(T_0 + \Delta T)}\right)$$

or

$$\sigma = \sigma_0 \exp\left(-\frac{W_a\, T_0}{k_B(T_0^2 - \Delta T^2)}\right) \exp\left(\frac{W_a\, \Delta T}{k_B(T_0^2 - \Delta T^2)}\right) \tag{4.161}$$

By neglecting the relative variation of $(T_0^2 - \Delta T^2)$ compared to ΔT in this expression, (4.161) can be written, with an obvious notation:

$$\sigma = \sigma_0' \exp(\gamma\, \Delta T) \tag{4.162}$$

from which we have

$$P_J = \frac{U^2 \sigma_0' \pi d^2}{4l} \exp(\gamma\, \Delta T) \tag{4.163}$$

The thermal power P_R transmitted by the insulator to the environment is equal to

$$P_R = \lambda \pi\, dl\, \Delta T \tag{4.164}$$

At equilibrium,

$$P_J = P_R \tag{4.165}$$

from which we have

$$\Delta T = \frac{U^2 \sigma_0' d}{4\lambda l^2} \exp(\gamma\, \Delta T) \tag{4.166}$$

Let us assume that the solution of this equation gives, for example, $\Delta T = \Delta T_1$ (Fig. 4.34). The cooling conditions of the specimen therefore make it possible to apply a higher voltage and therefore a larger ΔT. Let ΔT_c correspond to a voltage U_c, the maximum value of ΔT, above which the temperature of the specimen will, in theory, increase indefinitely. For $\Delta T = \Delta T_c$, we have in addition to (4.165) (see Fig. 4.34), the relationship:

$$\frac{\partial P_J}{\partial \Delta T} = \frac{\partial P_R}{\partial \Delta T} \tag{4.167}$$

From (4.165) and (4.167) we deduce that

$$\Delta T_c = \frac{1}{\gamma} \tag{4.168}$$

from which, due to (4.166), the required dielectric strength is

$$E_c = \sqrt{\frac{4\lambda}{\gamma \sigma_0' de}} = 1.21 \sqrt{\frac{\lambda}{\gamma \sigma_0' d}} \tag{4.169}$$

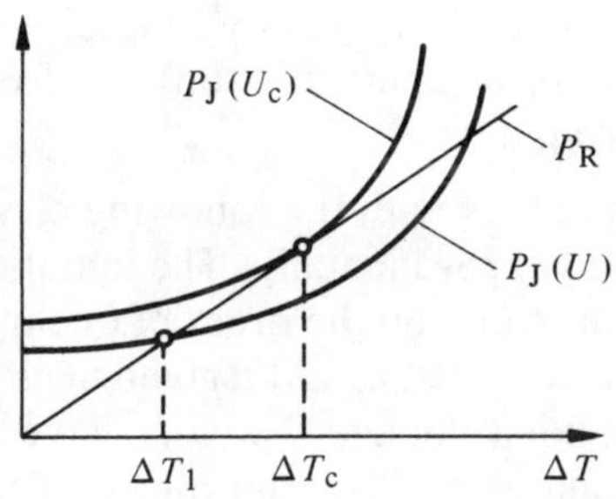

Fig. 4.34

4.7.3 Intrinsic Breakdown

Intrinsic breakdown is defined as a breakdown the triggering of which does not involve the Joule effect, associated with a current preceding the discharge itself.

In very general terms, it can be said that an avalanche mechanism is responsible for this breakdown. The conduction electrons and the electrons from isolated levels close to the conduction band are involved in the process. The isolated levels result from imperfections such as point imperfections, dislocations, and grain boundaries in the case of crystalline dielectrics. The analysis of the situation in partially crystalline and partially amorphous substances such as polymers is not so simple.

Several models [58] have been developed to account for intrinsic breakdown. Unfortunately, their value for the electrical engineer remains very limited. Furthermore, the processes themselves producing this breakdown still remain poorly understood today.

According to O'Dwyer [58], the initiation of an intrinsic breakdown in a dielectric subjected to an initially uniform field E passes through the following stages:

- for $E > E_c$, a significant proportion of ionizing collisions appears in the dielectric;
- these collisions produce relatively mobile electrons and holes with less mobility. In moving towards the cathode, the holes create a space charge that deforms the field, reinforcing it in the vicinity of the cathode and reducing it near the anode;
- the proportion of ionizing collisions therefore increases next to the cathode, the electron emission current of which is increased by the local reinforcement of the field;
- the conditions for divergent evolution of the process thus being met, the dielectric is rapidly destroyed by a massive emission of electrons from the cathode itself (field emission) and from the dielectric that immediately surrounds it.

This explanation agrees with the following characteristics of intrinsic breakdown observed experimentally. The initiation of the discharge is produced in a very short time, on the order of one microsecond and even less, after the field has exceeded E_c. At a given peak value, the frequency and the form of the applied voltage therefore have a small effect on the appearance of the breakdown. The importance of the dielectric and the electrodes is appreciably less than in the case of thermal breakdown, which justifies the use of the adjective intrinsic.

4.7.4 Progressive Degradation of the Dielectric Strength

A large number of phenomena are capable of altering a dielectric over time (aging), to produce, in particular, a reduction in the dielectric strength. They are responsible for the majority of breakdowns occurring months, even years, after voltage has been applied.

Such reductions in E_c occur, for example, when the insulator has homogeneity imperfections, cavities, inclusions of foreign particles, *et cetera*. Partial discharges arising in the vicinity of these imperfections as soon as the field is sufficient, can, by erosion, localized melting, induced chemical transformations, or other processes, create channel networks in the insulator which are more or less conducting. These networks are called arborescences because of their resemblance to the branches of a tree. The arborescences grow over time, producing a breakdown as soon as they are large enough.

The presence of humidity in certain polymers seems to favor the onset and growth of arborescences.

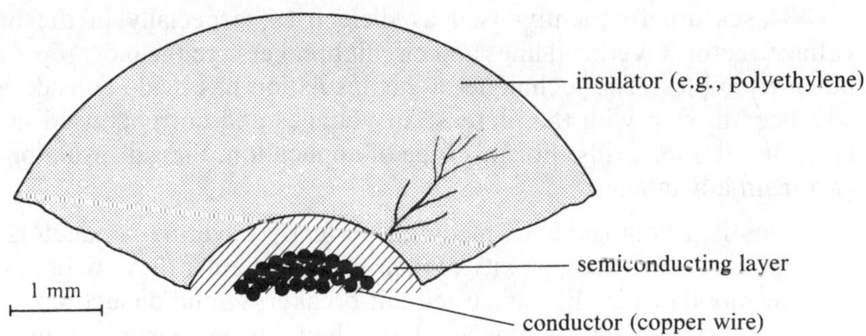

Fig. 4.35 Arborescence in the Insulator of a Cable

Mechanical stresses causing microscopic cracks to appear in certain insulators, also lower their strength.

Ionizing radiation always has an adverse effect although certain dielectrics are capable of withstanding high doses without any damage.

The shape of the electrodes also plays an important part, as well as their surface condition. Small irregularities may be enough to cause partial discharges with catastrophic effects. This is why the copper wire forming the conductor in high-voltage extruded cables is separated from the dielectric by a layer of extruded semiconductor polymer. The semiconductor polymer provides a smoother surface for the dielectric than the wire (Fig. 4.35).

Although theoretically, intrinsic breakdown only depends on the peak value of the applied field, it is found in practice that E_c is lower for an alternating or pulsed voltage than it is for a steady-state voltage. The explanation is to be found in the lower efficiency of the degradation mechanisms of E_c in a direct field.

These few considerations should suffice to show the extraordinary complexity of the problems associated with dielectric strength and the almost total impossibility of using models based on fundamental theories.

This explains the importance in this area of working methods such as statistical analysis for studying the distribution of the breakdown voltage, accelerated aging for the degradation mechanism approach, and the use of standardized tests for comparing dielectrics. The *IEEE Transactions on Electrical Insulation* [59] provide useful information in this area.

4.8 DIELECTRIC BEHAVIOR OF GASES

4.8.1 Introduction

Gases are frequently used as dielectrics, especially in the high-voltage sector. Overhead lines and circuit breakers are the most representative examples. The technique of gas insulation has made considerable advances in step with the increase in voltages and currents used in the large distribution grids. For this type of application, the gas insulator has two main advantages.

- Gas is a homogeneous medium, perfectly covering conductors regardless of the complexity of their shape, whether they are immobile or mobile as in the case of circuit-breakers. Solid dielectrics under such circumstances always allow chinks to appear which may be responsible for destructive discharges.
- After an arc has been passed and broken, a gas is the dielectric that most quickly recovers its insulating properties. Most solid dielectrics are definitely degraded by carbonization after the passage of an arc.

Although conductivity in the gas is not usually desirable in the high-voltage sector, it is exploited in other sectors such as lighting by fluorescent tubes, fire detection or radioactive particle detection by ionization probes, certain lasers, various electronic components, *et cetera*.

4.8.2 Electrical Properties of Gases: General Considerations

The low density of gases explains their special electrical properties listed below:

- relative permittivity very close to unity;
- negligible polarization dielectric losses;
- extremely nonlinear current-voltage characteristic;
- variable dielectric strength and resistivity with pressure.

A perfect gas subjected to a low field would also be a perfect insulator. However, cosmic rays and natural radioactivity produce a certain level of ionization. Every gas, therefore, has a certain conductivity that is modified through several stages when the electric field increases. Let us consider the case of helium as representative for most gases (Fig. 4.36).

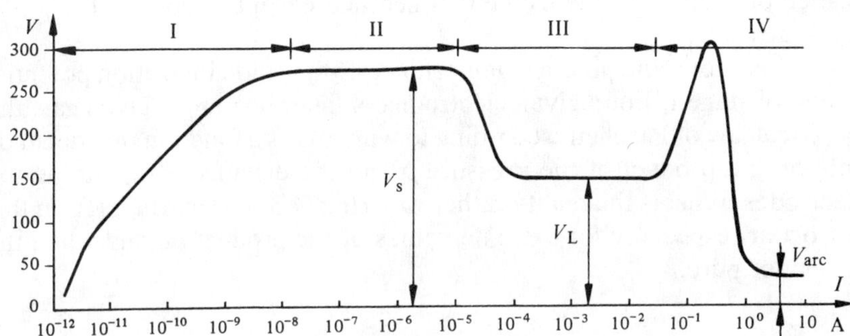

Fig. 4.36 Current-Voltage Characteristic of Helium. Planar Silver Electrodes Separated by 0.8 cm. Gas Pressure: 5.2 · 10^3 Pa. (After [60])

As long as the applied voltage remains less than V_S (stage I), a current varying in a random manner is established between the electrodes. It depends directly on the externally caused ionizing events that are produced in the gas. When these events stop, the current disappears.

In stage II, the current is maintained by ionization that it creates itself. This is the first type of autonomous discharge, called *dark discharge,* because there is almost no emission of light during discharge. The voltage V_S remains constant over several current decades and is called the *static ionization potential.*

Above a certain current, the dark discharge is transformed to the *luminous discharge* which is also autonomous. The transition from one type of discharge to another is accompanied by a considerable variation in the current and voltage after which the luminous discharge is again characterized by a constant voltage V_L over a certain current interval (stage III).

When the current increases further there is a rapid increase in voltage immediately followed by a collapse of the latter to a few tens of volts. This phenomenon corresponds to the establishment of a *very luminous arc* in which the ionized gas reaches temperatures of 10^3 to 10^5 K. This is the third type of autonomous discharge.

A characteristic such as that of Fig. 4.36 can only be recorded using a suitable circuit provided especially with an adjustable current limiter. This is a laboratory measurement. In practice, such a limiter is not always present in the circuit. In a high-voltage system in particular, the idea is only to limit short-circuit currents! Therefore if the static ionization potential is reached in an element of such a system, stages II and III are almost instantaneously reached resulting in an arc that has a strong chance of destroying the element concerned except of course if this is a spark arrester.

This shows the practical importance of the static ionization potential V_s and of stage I. For a given electrode configuration and a given gas, the empirical law of Paschen, according to which V_s is a function that depends only on the product of the pressure p and the distance d separating the electrodes, reflects the reality rather well (Fig. 4.37). Departures from this law occur, especially for extreme values of the product pd and when the gas is not pure.

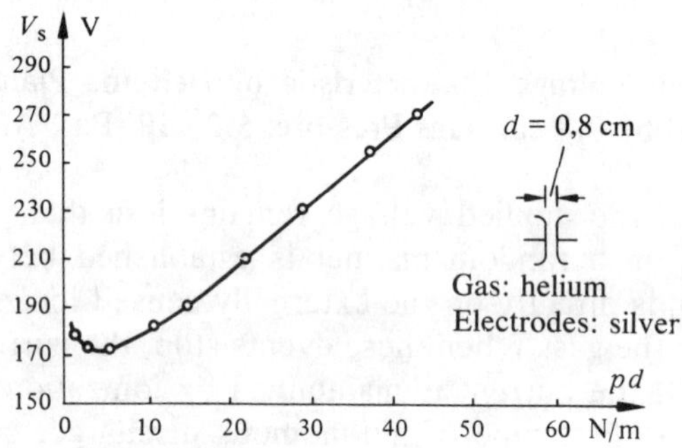

Fig. 4.37 Paschen's Law for Helium (After [60])

The currents observed in the four stages are the result of avalanche mechanisms caused by accelerated electrons in the field. The mean free path of the electrons is in the gas long enough for them to acquire enough energy to produce ionization between two collisions.

4.8.3 Avalanche Mechanism

The free electrons accelerated in a gas by an electric field are capable of producing three types of reactions by their collision on the molecules of this gas. By denoting the electron by e^-, and the molecule of the gas by Z with the asterisk indicating an excited (unstable) state, these reactions can be written as

$$e^- + Z \rightarrow (Z^-)^* \rightarrow Z^- \tag{4.170}$$

$$e^- + Z \rightarrow Z^* + e^- \rightarrow Z^+ + 2e^- \quad (4.171)$$

$$e^- + Z \rightarrow Z^* + e^- \rightarrow Z_1^+ + Z_2^- + e^- \quad (4.172)$$

The first reaction is the capture of an electron with the formation of a negative ion. It is typical of gases composed of halogens such as sulphur hexafluoride SF_6 (Section 4.10.4). The second reaction is an ionization with the production of a secondary electron. The third reaction is a dissociation of the gas molecule into two ionized molecules Z_1^+ and Z_2^- with the production of a secondary electron.

Because of their mass, the ions do not acquire sufficient velocity in order to themselves produce ionization by collision. The probability of the occurrence of each of these three reactions is characterized by a parameter called the effective cross section whose unit is m^{-1}. The effective cross section is equal to the mean number of times that an electron produces the reaction per unit of path traversed in the direction of the electric field.

By way of illustration, we shall study the development of an avalanche in stage I (Fig. 4.36) in a planar capacitor (Fig. 4.38).

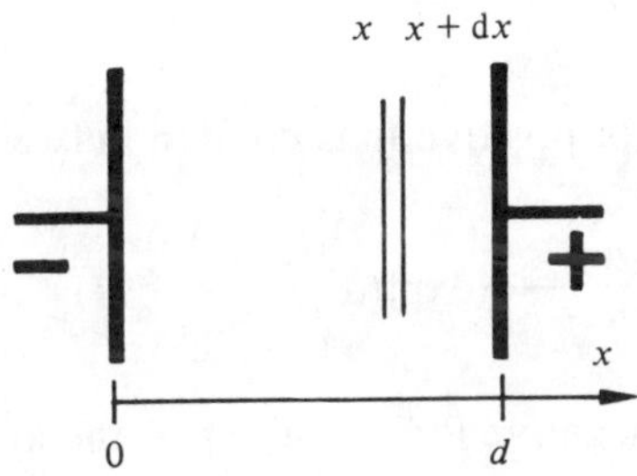

Fig. 4.38

In stage I, the charge density is low enough that it does not perturb the applied electric field locally. The billiard balls model (Section 2.2) can therefore be applied. Let η, α, and ξ be the effective cross sections of the reactions (4.170), (4.171), and (4.172), respectively. We assume that a positive ion arriving at the cathode detaches from it γ secondary electrons which in their turn will be accelerated by the electric field, *et cetera*.

We shall calculate

- the amplification coefficient k of the gas defined as the number of negative charges reaching the anode for an initial electron emitted by the cathode;
- the condition for which an autonomous discharge is produced.

The number of electrons created in the section dx is equal to

$$\mathrm{d}n_e = n_e(\alpha - \eta)\,\mathrm{d}x \tag{4.173}$$

where n_e is the number of electrons arriving at x. For one electron initially leaving the cathode, we have

$$n_e(x) = \exp[(\alpha - \eta)x] \tag{4.174}$$

The number of negative ions created in dx is equal to

$$\mathrm{d}n_- = n_e(\eta + \xi)\,\mathrm{d}x \tag{4.175}$$

The total number of negative charges created between 0 and x is given by

$$n_-(x) = \int_0^x \mathrm{d}n_- = \frac{(\eta + \xi)}{(\alpha - \eta)}(\exp[(\alpha - \eta)x] - 1) \tag{4.176}$$

The number of positive ions created in dx is equal to

$$\mathrm{d}p = n_e(\alpha + \xi)\,\mathrm{d}x \tag{4.177}$$

and the total number of positive ions created between 0 and x is equal to

$$p(x) = \int_0^x \mathrm{d}p = \frac{\alpha + \xi}{\alpha - \eta}(\exp[(\alpha - \eta)x] - 1) \tag{4.178}$$

The total number of negative ions resulting at the anode at the end of the first generation is equal to

$$n_1 = n_e(d) + n_-(d) \tag{4.179}$$

Similarly, the total number of positive ions reaching the cathode is equal to

$$p_1 = p(d) \tag{4.180}$$

There are therefore γp_1 new electrons which will create a second generation and so on. After an infinite number of generations, the total number of negative charges having reached the anode amounts to

$$N_- = n_1 \sum_{i=0}^{\infty} (\gamma p_1)^i = \frac{n_1}{1 - \gamma p_1} \tag{4.181}$$

Since we have assumed that a single electron initially left the cathode, N_- also represents the amplification coefficient k of the gas. By expanding (4.181), we obtain

$$N_- = \frac{\dfrac{\alpha + \xi}{\alpha - \eta} \exp[(\alpha - \eta)d] - \dfrac{\eta + \xi}{\alpha - \eta}}{1 - \gamma \dfrac{\alpha + \xi}{\alpha - \eta} (\exp[(\mathrm{d} - \eta)d] - 1)} = k \tag{4.182}$$

The condition for an autonomous discharge is $\gamma p_1 \geqslant 1$, that is,

$$\gamma \frac{\alpha + \xi}{\alpha - \eta} (\exp[(\alpha - \eta)d] - 1) \geqslant 1 \tag{4.183}$$

4.9 FERROELECTRICITY, PIEZOELECTRICITY, AND OTHER PROPERTIES

4.9.1 Ferroelectricity: Introduction and Definitions

A *ferroelectric material* is a dielectric material in which the dipole moments are coupled.

This coupling phenomenon is known as *ferroelectricity*.

The prefix ferro has been chosen because of the very large number of analogies between these materials and ferromagnetic materials.

Ferroelectric materials are characterized by an extremely high relative permittivity on the order of 10^3 and even 10^4. All their electric properties are very sensitive to temperature. Ferroelectricity disappears above a temperature called the *ferroelectric Curie temperature*.

4.9.2 Ferroelectric Domains and Hysteresis Loop

Locally, ferroelectric coupling produces an alignment of dipole moments with respect to each other.

A *ferroelectric domain* is the name given to each region in which all the dipole moments are aligned parallel to each other.

In a homogeneous sample at a uniform temperature, each domain has a spontaneous polarization $\boldsymbol{P}_s(\text{C} \cdot \text{m}^{-2})$ with the same magnitude. However, the orientation of $\boldsymbol{P}_s$ varies from one domain to the other so that the total dipole moment of the sample may be zero.

Although the mechanisms involved have a different nature, formally, it is possible to compare the appearance of ferroelectric domain structures and ferromagnetic domain structures (Section 3.7). The electrostatic and electrostriction energies (Section 4.9.6) correspond to the magnetostatic and magnetostriction energies. The dipole moments also have exchange and anisotropy energies.

The application of an electric field increases the energy of the domains, the more the direction of their spontaneous polarization deviates from that of the applied field. As the field increases, the fraction by volume of the domains aligned parallel to E therefore increases until the complete sample becomes a single domain for $|E| \geqslant E_s$ (Fig. 4.39).

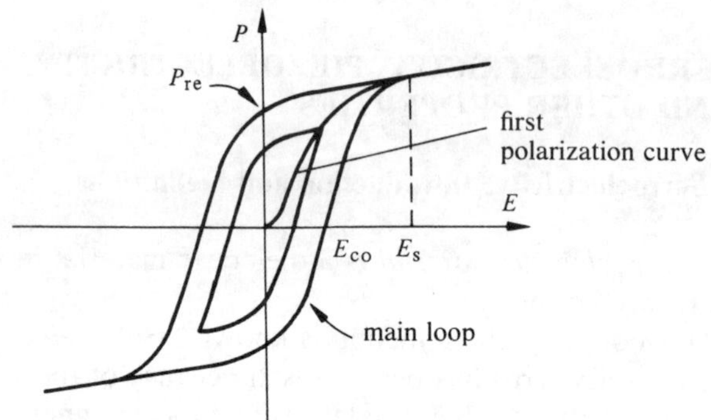

Fig. 4.39 Ferroelectric Hysteresis Loop

Because the variation in the domain structure is an irreversible phenomenon, the function $P(E)$ is strongly nonlinear in the region $0 \leqslant |E| \leqslant E_s$. In the case of an alternating field, $P(E)$ takes the form of a *ferroelectric hysteresis loop*.

Various characteristic parameters are defined over this loop like over the B-H loop for ferromagnetic materials. These are, in particular:

- the initial permittivity defined over the first polarization by the relationship:

$$\varepsilon_{ri} = \frac{1}{\varepsilon_0} \lim_{E \to 0} \frac{P}{E}; \tag{4.184}$$

- the *main hysteresis loop,* which is the loop obtained in an alternating field with an amplitude at least equal to E_s;
- the *remanent polarization* P_{re} and the *coercive field* E_{co} defined by the intersection of the main loop with the axes.

4.9.3 Spontaneous Polarization *versus* Temperature

The potential existence of ferroelectricity is contained in the expressions leading to the Clausius-Mosotti equation (4.56). By taking P from (4.55), we have

$$\boldsymbol{P} = \frac{\sum_i N_i \alpha_i}{1 - \frac{1}{3\varepsilon_0} \sum_i N_i \alpha_i} \boldsymbol{E} \tag{4.185}$$

In an ordinary (isotropic) dielectric, $\boldsymbol{P}$ and $\boldsymbol{E}$ are parallel so that the denominator of (4.185) is positive. If this denominator decreases until it vanishes,

$$\frac{1}{3\varepsilon_0} \sum_i N_i \alpha_i = 1 \tag{4.186}$$

a spontaneous polarization becomes possible. However if $P \neq 0$ whereas $E = 0$, the expression for the local field (4.35) reduces to

$$E_L = \frac{\boldsymbol{P}}{3\varepsilon_0} \tag{4.187}$$

and (4.185) is no longer valid, (4.55) reducing to exactly (4.186) which does not contain $\boldsymbol{P}$. The condition for the existence of spontaneous polarization must therefore be written as

$$\frac{1}{3\varepsilon_0} \sum N_i \alpha_i \geqslant 1 \tag{4.188}$$

In a given dielectric, the factors α_i are nearly constant, except for the orientation polarization factor α_{or}, which varies as a function of temperature. Let us separate this factor in (4.188). By virtue of (4.51), we have

$$\frac{1}{3\varepsilon_0}\left(N_{\text{or}}\frac{p_{\text{m}}^2}{3k_{\text{B}}T}+\sum_i{}' N_i\alpha_i\right) \geqslant 1 \tag{4.189}$$

In this expression Σ' includes all the polarization processes present except for the orientation polarization. The maximum temperature Θ, which still satisfies (4.189), is the ferroelectric Curie temperature.

$$\Theta = \frac{N_{\text{or}}p_{\text{m}}^2}{9\varepsilon_0 k_{\text{B}}\left(1-\dfrac{1}{3\varepsilon_0}\sum_i{}' N_i\alpha_i\right)} \tag{4.190}$$

The interpretation of this result is very simple in principle. The coupling that produces the alignment of dipole moments can only occur for $T < \Theta$. Above Θ, it is completely masked by thermal motion. Here also, the analogy with ferromagnetism is worth noting.

In the case where the polarization by orientation predominates, the dielectric susceptibility χ_{r} is easily calculated above Θ. Equation (4.190) reduces to

$$\Theta = N_{\text{or}}\frac{p_{\text{m}}^2}{9\varepsilon_0 k_{\text{B}}} \tag{4.191}$$

By substituting this expression into (4.185), we have, taking (4.23) into consideration,

$$\chi_{\text{r}} = \frac{1}{\varepsilon_0}\frac{N_{\text{or}}p_{\text{m}}^2}{3k_{\text{B}}(T-\Theta)} \tag{4.192}$$

This is the Curie-Weiss law (3.53) already encountered for ferromagnetic materials.

4.9.4 Origin of Ferroelectricity

Ferroelectricity is only found in a few particular crystal structures. The mechanism responsible for the phenomenon varies from one structure to another. The case of barium titanate $BaTiO_3$ will be chosen as an example because of the practical importance of this material.

Above the Curie temperature (120°C), $BaTiO_3$ has the Perovskite structure (Fig. 4.40).

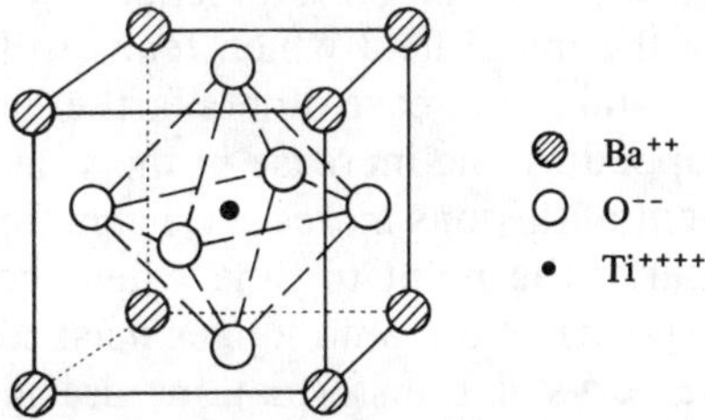

Fig. 4.40 Perovskite Structure

The eight ions of barium are situated at the vertices of a cube with the oxygen ions at the centers of the faces. Since the titanium ion occupies the center of the cube, the dipole moment of the structure is equal to zero.

Below the Curie temperature, this structure is slightly deformed, passing from the cubic structure to the tetragonal structure (Fig. 4.41). The displacements δ' and δ'' are on the order of 10^{-11} m, the side of the cube being approximately $4 \cdot 10^{-10}$ m. They give the deformed structure a dipole moment which, due to symmetry, may have six different orientations. The interaction between neighboring dipoles produces an alignment responsible for the appearance of ferroelectric domains.

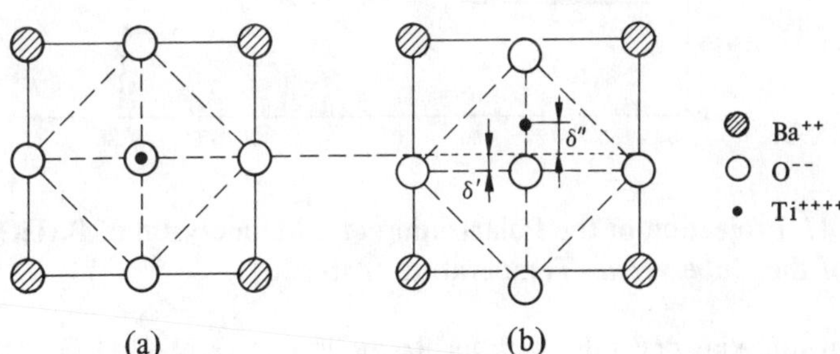

Fig. 4.41 Perovskite Structure: (a) Original ($T > \Theta$); (b) Deformed ($T < \Theta$). Projection on a Plane {1.0.0}

It is equivalent to saying that the crystal structure of $BaTiO_3$ is simply more deformable than another. A chance local deformation producing a dipole moment, it also creates an initial local field that produces the appearance of other dipole moments in the vicinity. In their turn, these reinforce the initial field which tends to further increase the local dipole moments and to increase others farther away. This development is only finally stopped by the increase in the valence forces when the relative displacement of the ions increases. From this point of view, the Curie temperature marks the point in time when the effect of the local field created by the polarization is no longer masked by thermal motion.

Two other crystal transformations are also produced in $BaTiO_3$. Close to 0°C, the structure passes from tetragonal to monoclinic, the polarization being oriented along the diagonal of a face. Around −80°C, it becomes rhombohedral and the polarization is aligned along the major diagonal of the rhombohedron.

By taking into account the angle between the edge of the cube and the polarization, it is found that the modulus of the spontaneous polarization is nearly the same in the three structures (Fig. 4.42). Explanation of ferroelectricity by ion displacement is not only valid for $BaTiO_3$ but also for other materials with Perovskite or a different structure.

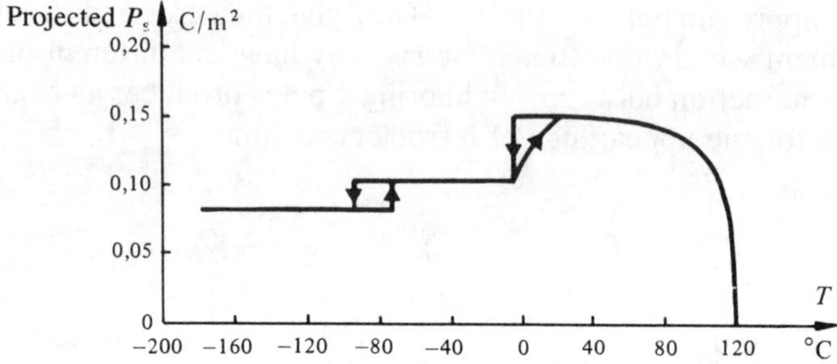

Fig. 4.42 Projection of the Polarization of a Monocrystal of $BaTiO_3$ on One Edge of the Cube *versus* Temperature (After [61])

In uniaxial crystals such as Rochelle salt ($KNaC_4H_4O_6$), triglycine sulphate (TGS) corresponding to the formula $(CH_2NH_2COOH)_3 \cdot H_2SO_4$, and the dihydrogen phosphates of potassium and other elements (KDP) (KH_2PO_4), ferroelectricity is due to a property of the hydrogen bond.

4.9.5 Piezoelectricity

Certain crystalline dielectrics have the following reciprocal properties:

- they are polarized by the action of a mechanical stress;
- in the absence of mechanical stress, their dimensions are modified when they are polarized by the action of an external electric field.

These properties constitute *piezoelectricity* and the materials in which they are manifested are called piezoelectrics. Quartz SiO_2, barium titanate $BaTiO_3$, and aluminum phosphate $AlPO_4$ are piezoelectric materials that are very commonly used.

Over the entire range of useful stresses, piezoelectricity is characterized by a linear relationship between cause and effect justifying the use of the theory of elasticity in order to describe the mechanical aspect of the problem. As is ferroelectricity, piezoelectricity is a property closely linked to the crystal structure. The study of the symmetries of the crystal structure makes it possible, in particular, to find materials that are likely to be piezoelectric. Let us consider, for example, the two two-dimensional structures shown in Figs. 4.43 and 4.44.

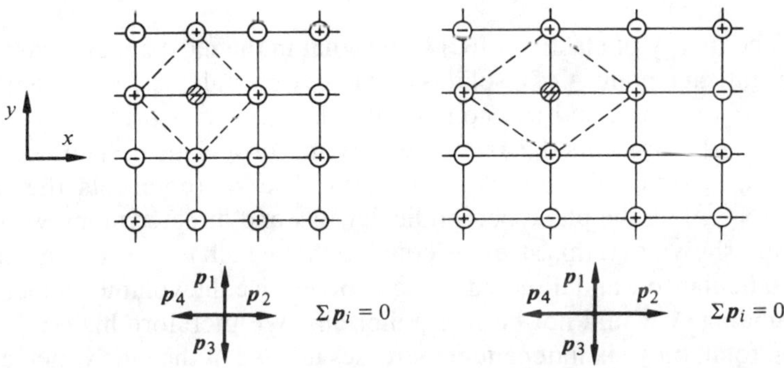

Fig. 4.43 Structure 1: ⊖ (Cross-Hatched), Reference Ion

In order to study their polarization, it is sufficient to examine the dipole moments associated with the unit cells shown in dotted lines in the figures. It can be seen that $\boldsymbol{P} = 0$ in the two nondeformed structures. The same applies to structure 1 deformed along any direction in the xy plane. This is due to the fact that the reference ion remains the center of symmetry in the deformed cell.

On the other hand, by deforming structure 2, the reference ion is no longer a center of symmetry and the cell has a resulting dipole moment whereby $\boldsymbol{P} \neq 0$. A traction along x (in the case of the figure) makes a polarization directed along $-y$ appear and, conversely, a compression along x causes a polarization along $+y$.

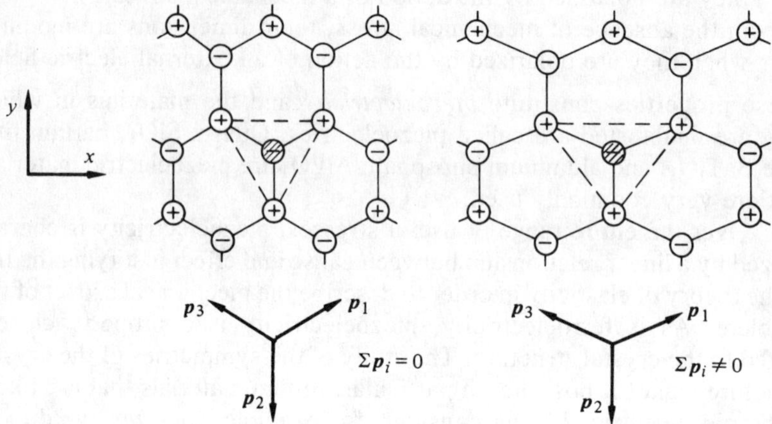

Fig. 4.44 Structure 2: ⊖ (Cross-Hatched), Reference Ion

The theory of elasticity [62] shows that in the most general case, the stress state at a point A of a solid is defined by six independent parameters σ_x, σ_y, σ_z, τ_{yz}, τ_{zx}, and τ_{xy}, constituting the stress tensor.

The σ_x represents the axial stress [N/m^2] on a plane perpendicular to the x-axis passing through A (Fig. 4.45). The τ_{xy} represents the shear stress [N/m^2] on the plane perpendicular to x and directed along y. It can be easily shown that this stress is equal to that which is exerted on a plane perpendicular to y and directed along x, otherwise the volume element dV surrounding A would not be in equilibrium. We therefore have $\tau_{xy} = \tau_{yx}$ and, in total, only six independent stresses instead of the nine which could have been expected. The other stresses have corresponding meanings for the y- and z-axes.

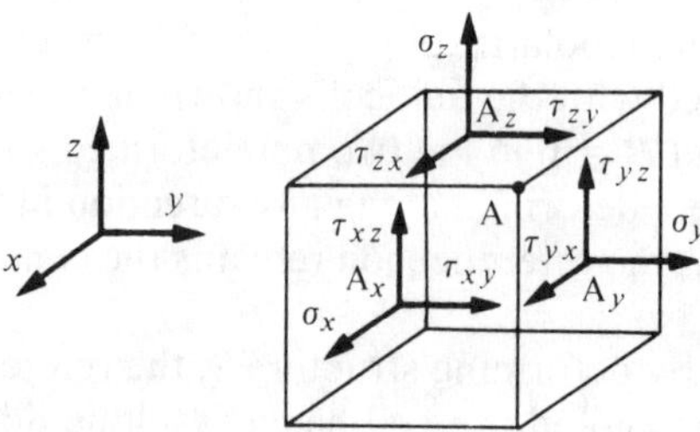

Fig. 4.45 Stress in a Solid. The Points A_x, A_y, and A_z in Fact Coincide with the Point A. They are Only Represented Separately for the Sake of Clarity. In Crystals, the Reference System *xyz* is Oriented Along the Crystal Axes

In a general form, the components P_x, P_y, and P_z of the polarization $\boldsymbol{P}$ are therefore expressed by the following linear relationship:

$$\begin{aligned} P_x &= d^x_{xx}\sigma_x + d^x_{yy}\sigma_y + d^x_{zz}\sigma_z + d^x_{yz}\tau_{yz} + d^x_{zx}\tau_{zx} + d^x_{xy}\tau_{xy} \\ P_y &= d^y_{xx}\sigma_x + d^y_{yy}\sigma_y + d^y_{zz}\sigma_z + d^y_{yz}\tau_{yz} + d^y_{zx}\tau_{zx} + d^y_{xy}\tau_{xy} \\ P_z &= d^z_{xx}\sigma_x + d^z_{yy}\sigma_y + d^z_{zz}\sigma_z + d^z_{yz}\tau_{yz} + d^z_{zx}\tau_{zx} + d^z_{xy}\tau_{xy} \end{aligned} \tag{4.193}$$

The factors d^k_{ij} are called the *piezoelectric coefficients*. They are expressed in coulomb/newton. They are often written in the form of a matrix with three rows and six columns, which has to be multiplied by the column vector formed by the stresses σ and τ in order to obtain (4.193). Formally, 18 coefficients are therefore necessary to define piezoelectricity. In reality, crystal symmetries considerably reduce this number. In the case of quartz, for example, the matrix of the d^k_{ij} is reduced to

$$\begin{pmatrix} d^x_{xx} & d^x_{yy} & 0 & d^x_{yz} & 0 & 0 \\ 0 & 0 & 0 & 0 & d^y_{zx} & d^y_{xy} \\ 0 & 0 & 0 & 0 & 0 & 0 \end{pmatrix} \tag{4.194}$$

where

$$d^x_{xx} = -d^x_{yy} = -\frac{1}{2}\, d^y_{xy} = -2.3 \cdot 10^{-12}\ \text{C/N} \tag{4.195}$$

$$d^x_{yz} = -d^y_{zx} = -0.7 \cdot 10^{-12}\ \text{C/N} \tag{4.196}$$

Barium titanate has three independent coefficients d^k_{ij}, approximately 100 times higher than those of quartz.

Piezoelectricity is a property used in a large number of devices using the conversion of electric energy to mechanical energy or *vice versa*. Examples are stress gauges, certain microphones, ultrasound transmitters, delay lines, surface wave devices, *et cetera*. As is the case for all elastic bodies, piezoelectric crystals have mechanical resonance frequencies. Because of the polarization that is associated with them, the vibrations may be captured and very easily maintained by means of an amplifying circuit. Very precise frequency standards are obtained in this way.

4.9.6 Electrostriction and Pyroelectricity

Like piezoelectricity, *electrostriction* is a variation of the dimensions of a dielectric under the action of a field E. This variation is also related to the displacement of charges accompanying polarization.

Electrostriction differs from piezoelectricity in that the deformation is not proportional to the applied field E but to the square of the field strength. This is a general property of dielectrics that is found in crystal media and in amorphous, solid, or liquid media. It does not have any reciprocal effect. This means that a mechanical stress does not produce polarization by inverse electrostriction. In fact, electrostriction is almost always a negligible effect.

By way of conclusion, we mention the existence of *pyroelectricity*, a property of certain crystals that modifies their polarization for a change in temperature ΔT. This is a linear phenomenon described by equations of the type:

$$\Delta P_i = \pi_i \, \Delta T, \qquad i = x,y,z, \tag{4.197}$$

where the π_i factors are *pyroelectric coefficients*. Because most often the variation in temperature is relatively slow, polarization by pyroelectricity is nearly always masked by a transport of charges to the surface of the crystal due to imperfect insulation.

4.10 NATURAL INSULATORS AND SYNTHETIC INORGANIC INSULATORS

4.10.1 Introduction

There are a large number and variety of dielectric materials used today. They may be classified according to different criteria such as their state: solid, liquid, or gas; their origin: natural or artificial; *et cetera*. Regardless of the criterion adopted, the very important position of synthetic organic dielectrics makes it necessary to treat them separately. They are the subjects of Sections 4.11 through 4.14. This section therefore deals with all the other dielectrics as a whole.

Because of the breadth of this subject, we can only give a brief survey of a few of the most important dielectrics here.

4.10.2 Ceramics

Ceramics are hard and often brittle substances that usually exhibit both crystalline and amorphous phases. They are manufactured from aluminum silicate, magnesium silicate, quartz, aluminum oxide, titanium dioxide, *et cetera*. The part to be manufactured is first of all produced by molding, with the starting materials in the form of a powder. Suitable binders are used to make the molded part provisionally mechanically resistant enough. The part then passes into a kiln where it is subjected to the thermal treatment called *sintering,* during which the binders are eliminated and the grains are bound together. Sintering has the characteristic of taking place at a temperature below the melting point of the constituents. During this treatment, the grains are progressively joined together by two surface processes occurring at the points of contact as soon as the temperature is high enough. First, the diffusion at the initial points of contact causes a rearrangement of the atoms resulting in the formation of small contact areas increasing in size. Second, the sublimation on adjacent surfaces of the contact areas provides additional atoms cooperating to increase these areas. In fact, the current of particles emitted by sublimation is higher on a convex surface than on a concave surface to which the atoms are more tightly bound (Fig. 4.46).

Fig. 4.46 Diagrammatic Representation of the Successive Stages in the Joining of Two Grains by Sintering

Sintering takes place between 1200°C and 1500°C depending on the components and is accompanied by a shrinkage of 10–20%. It is therefore difficult to conform to type dimensional tolerances. The only possibilities of machining are grinding and polishing using diamond powder or silicon carbide. Holes may be drilled using ultrasound. Often porous, ceramics may be covered with a glass layer in order to seal them and protect them against pollution. A metal layer makes it possible to attach metal parts by soldering.

Ceramics are resistant to surface currents and arcs. They preserve their good electrical properties up to approximately 1000°C. They do not age and are very resistant to chemical agents, with the exception of concentrated phosphoric acid and hydrofluoric acid. They are expensive.

Ceramics are classified as a function of their composition.

With 40–50% aluminum silicate, 30–40% quartz, and 20–30% feldspar (double silicate of aluminum and potassium, or another alkali metal), we obtain a porcelain formed from quartz and aluminum silicate crystals dispersed in an amorphous mass. This ceramic is used for large-sized insulators.

With 70–90% magnesium silicate, 5–10% aluminum silicate, and 5–10% feldspar, we obtain a ceramic formed from crystallized magnesium silicate dispersed in an amorphous mass. It is used at high frequencies and as a support for film or wire-wound resistances.

By adding titanium compounds such as titanium dioxide, magnesium titanate, or barium titanate to the silicates of aluminum and magnesium, we obtain ceramics with very high permittivity used for the manufacture of small capacitors.

Ceramics with a high content of alumina (50–80%) are used as substrates for the manufacture of hybrid integrated circuits with thick layers because the metal deposits conveniently adhere there.

4.10.3 Natural Organic Compounds

In terms of number, the natural organic compounds form an important category. They include paper, mineral oils, paraffin, various waxes such as ozokerite, cellophane, cotton, *et cetera*. The trend is for all of these products to be replaced by synthetic organic compounds. This development is slower so far as paper is concerned, the porosity of which remains a useful original property, allowing impregnation by various agents such as oils. However, its hygroscopic nature is a drawback.

4.10.4 Electronegative Gases

Electronegative gases are those with molecules containing halogen atoms. The particular electronic structure of the halogens (Section 1.2.13) gives these gases a certain affinity for electrons which strongly inhibits the development of discharges.

Among these gases, sulphur hexafluoride SF_6, available on an industrial scale at reasonable cost, is most widely used. Its dielectric strength at 60 Hz under atmospheric pressure is compared to that of dry air and other less common fluorine compounds in Figs. 4.47 and 4.48.

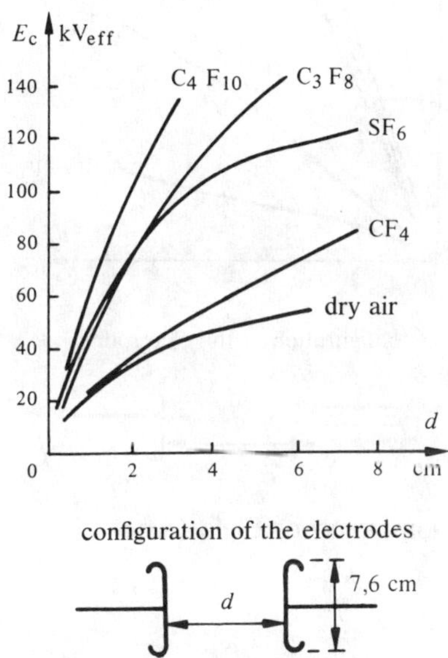

Fig. 4.47 E_c **in a Homogeneous Field (After [63])**

The increase in E_c as a function of pressure, at 60 Hz and for the electrode configuration of Fig. 4.47 with $d = 1.27$ cm, is shown in Fig. 4.49. In an inhomogeneous field, E_c may decrease when the pressure increases for certain electronegative gases and for certain ranges of pressure.

These few characteristics are sufficient to demonstrate the remarkable properties of SF_6. This gas is used pure or diluted in nitrogen in various high-voltage devices and in particular in circuit breakers where it makes it possible to considerably reduce the dimensions.

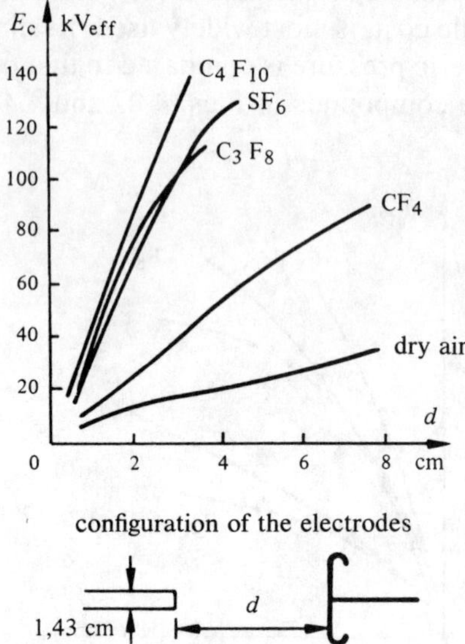

Fig. 4.48 E_c **in an Inhomogeneous Field (After [63])**

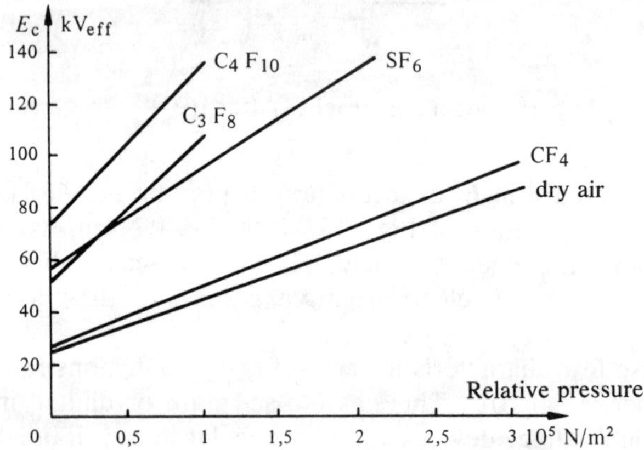

Fig. 4.49 (After [63])

4.11 SYNTHETIC ORGANIC INSULATORS: INTRODUCTION TO POLYMERS

4.11.1 Organic Chemistry: Recap and Definitions

Organic chemistry is the study of compounds formed from carbon and hydrogen possibly with other elements such as nitrogen, oxygen, phosphorous, sulfur, certain metals, and halogens. There is an extremely large number of organic substances which are highly varied. More than one million are known at the present time. Their study and their nomenclature are simplified by the fact that certain arrangements of atoms reoccur in a large number of different organic molecules. These arrangements, the building blocks of organic chemistry are called *groups* or *radicals*. The simplest is the *hydroxyl* group formed by one hydrogen atom and one oxygen atom and designated $-OH$ to indicate that it provides one valence bond.

4.11.2 Groups in Hydrocarbons

Hydrocarbons are compounds formed exclusively from carbon and hydrogen. Groups that can be found in these substances arc *hydrocarbon groups*. Their name is most often derived from that of the corresponding hydrocarbon (Tables 4.50 and 4.51). There are several classes of hydrocarbons, only two of which are mentioned here because of their relationships with the insulators discussed later. These are:

- the *aliphatic* hydrocarbons with a skeleton of an open chain of carbon atoms. *Alkanes* are aliphatic hydrocarbons that do not have double or triple bonds between the carbon atoms;
- the *aromatic* hydrocarbons, characterized by the fact that they contain at least one hexagonal ring with three double bonds.

The groups derived from the alkanes form the *alkyls,* often designated by the general symbol R (Table 4.50).

The groups derived from the aromatic hydrocarbons are the *aryls,* often designated by the general symbol Ar (Table 4.51).

Table 4.50
The Three Simplest Alkanes and the Corresponding Alkyls. Two Groups are Derived from Propane; They are Distinguished by a Prefix

Alkanes	*Alkyl Groups*
Methane	**Methyl**
Ethane	**Ethyl**
	n-propyl
Propane	
	isopropyl

Table 4.51
Two Very Simple Aromatic Hydrocarbons and the Corresponding Aryl Groups

Aromatic Hydrocarbons	*Aryl Groups*
Benzene	**Phenyl**
Toluene	**p-Tolyl**

4.11.3 Polymers: Definitions

Certain simple molecules have the property, under appropriate physicochemical conditions, of joining up together to form larger molecules constituting a repetition of the initial molecule in space. This type of reaction is called *polymerization* and the resulting substance is called a *polymer*. *Monomer* is the term given to the molecule and thus the starting substance.

Several methods of polymerization are used.

- In *mass polymerization,* the monomer is simply placed in a reactor, possibly with catalysts added. Usually, the reaction can only take place at high pressures and high temperatures. Complex apparatus is therefore used. The polymer has a high purity.
- In *solution polymerization,* the monomer and the catalysts, if present, are dissolved in a solvent. This process makes it possible to

operate at lower temperatures and to more efficiently extract the heat of reaction. Compared to mass polymerization, it requires an additional operation: the separation of the polymer from the solvent. Also, solvent impurities are likely to contaminate the polymer.

- *Emulsion polymerization* is characterized by the fact that the substances involved in the reaction form two nonmiscible phases. The polymer is either formed in one of the phases or at the surface separating the phases.

Two types of polymerization reactions are distinguished.

- *Addition polymerization* (or *polyaddition*) is the assembly of the monomers without the formation of a reaction product. The formation of polyethylene (Section 4.12.1) by opening the C=C double bond in ethylene is a typical example.
- In the case where a reaction product is formed, silicone rubber (Section 4.13.2) is an example of this, we speak of *condensation polymerization* or *polycondensation*.

4.11.4 Molecular Structure of Polymers: Definitions

A polymer formed from a single type of monomer is called a *homopolymer*. If it consists of two or more different monomers, it is called a *copolymer*.

A homopolymer may have three different structures (Table 4.52).

There are a larger number of copolymer structures and they are more complex than homopolymer structures because they also depend on the regular or random arrangement of the different monomers along the chains.

The molecular structure of polymers has a decisive effect on their mechanical properties.

Table 4.52
Diagrammatic Representation of the Structures of a Homopolymer

Linear
Branched
Reticulated
M: monomer.

Thermoplastics have linear and weakly branched structures (Section 4.12). As their name indicates, these substances are easily deformed when the temperature increases, which is explained by the weak bonds (of the Van der Waals type (Section 1.3.16)) connecting neighboring chains. *Plasticizers* can act on these bonds by further reducing the bond energy between chains.

Elastomers (Section 4.13), the mechanical behavior of which is comparable to that of rubber, have weakly reticulated structures, i.e., they form a lattice with wide cells.

When the lattice cells become tight, the elasticity disappears. The material is called *thermosetting* (Section 4.14) because the formation of the contracted lattice is accelerated by an increase in temperature.

Locally, linear structures may be aligned parallel to each other forming a zone in which the arrangement of the atoms becomes regular, i.e., crystalline (Fig. 4.53). The *crystallinity of a polymer* is the ratio of the volume of the zones to the total volume of the specimen.

Fig. 4.53

The variation in the modulus of elasticity E_Y of a thermoplastic as a function of the temperature has four stages (or ranges) (Fig. 4.54).

The temperature below which the material becomes brittle is called the vitreous transition temperature T_v (Section 6.5.2). The existence of these stages is explained by the progressive release of the interaction forces between the chains.

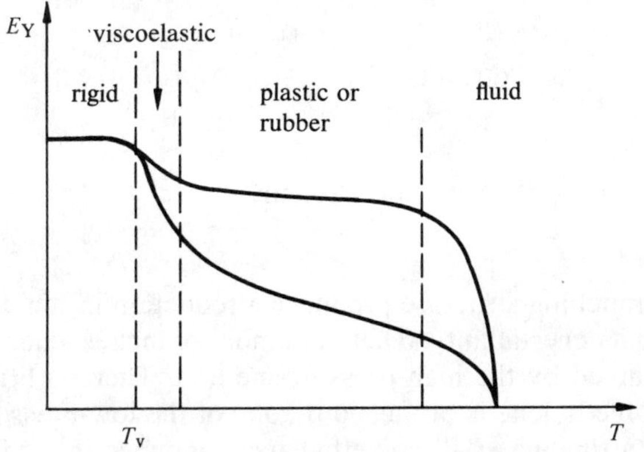

Fig. 4.54

4.12 THERMOPLASTICS

4.12.1 Polyethylene

Formed by the polymerization of ethylene C_2H_4 (Fig. 4.55), the polyethylene molecule has the form of a chain of carbon atoms with free valences occupied by hydrogen atoms (Fig. 4.15). This chain contains 1000 to 20,000 monomers. The molecular weight of the monomer being 28, the weight of the molecule may reach approximately 30,000 to 550,000. The polyethylene molecule may have branching structures (Fig. 4.56).

```
H   H
|   |
C = C
|   |
H   H
```

Fig. 4.55 Ethylene

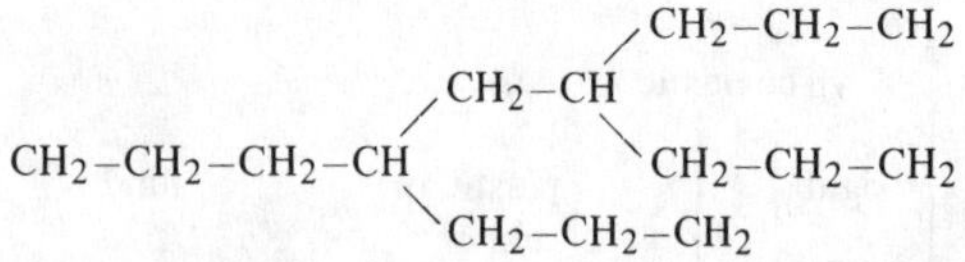

$$
\begin{array}{l}
\qquad\qquad\qquad\qquad\qquad\qquad\quad\;\; /CH_2-CH_2-CH_2 \\
\qquad\qquad\qquad\qquad\;\; /CH_2-CH \\
CH_2-CH_2-CH_2-CH \qquad\qquad\;\; \backslash CH_2-CH_2-CH_2 \\
\qquad\qquad\qquad\qquad\;\; \backslash CH_2-CH_2-CH_2
\end{array}
$$

Fig. 4.56 Branched Polyethylene Molecule

This branching structure produces a reduction in the density of the material and its crystallinity. There are more branches when the polyethylene is obtained by the high-pressure method. There is little branching when the polyethylene is produced by one of the low-pressure methods.

As a starting material, polyethylene is obtained in the form of granules or powder. It is shaped by extrusion, injection, blasting, *et cetera*. It is worked between 180°C and 250°C and presents a contraction of 2–3% when it is cooled to room temperature. Polyethylene powder is used in the techniques of fluidized bed deposition, electrostatic atomization, *et cetera*.

In semifinished form, polyethylene can be supplied in tubes and various sections, sheets, and strips with thickness as slight as 20 microns. In this form, it can replace paper, the hygroscopic characteristic of which is a drawback for certain applications.

Polyethylene may be reticulated by a physical or chemical process. In both cases, the hydrogen atoms have to be removed in order to produce direct bonds between the carbon atoms belonging to adjacent chains. The physical process uses the ejection of H atoms by electron beam irradiation or x-ray irradiation. It is used for small thicknesses because of the difficulty in obtaining homogeneous reticulation when the parts become large. The chemical process involves the addition of peroxides. By increasing the temperature after shaping (by extrusion for example), the peroxide is made to decompose. The free radicals formed at this point in time are responsible for capturing the hydrogen atoms.

Polyethylene lends itself to the manufacture of thermoretractable tubes. The procedure is as follows. A tube is extruded then dilated by inflation while it is in the rubber stage. In this state it is cooled which freezes the tensions induced by the binding forces between the chains. When the temperature is increased again up to the rubber stage, these tensions appear again tending to make the tube go back to its primitive dimensions. Thermoretractable tubes are also obtained by irradiation. This produces new bonds between chains also creating internal tensions which are manifested as soon as the temperature is sufficiently increased.

The properties of polyethylene vary as a function of its density, situated around 0.92 with a crystallinity of 70% for high-pressure quality, and 0.96 with a crystallinity of 93% for the low-pressure quality. The modulus of elasticity, the resistance to traction and to bending, and the hardness increase with density. Low-pressure polyethylene is therefore better mechanically but the high-pressure quality is better electrically because high crystallinity is most often the source of inhomogeneities which can lead to breakdown.

Polyethylene has a high dielectric strength and a good resistance to the spread of surface discharges. The dielectric strength in an alternating field may be increased by aromatic additives which also improve the resistance to partial discharges. The permittivity is independent of the frequency up to the microwave region. The same applies to tan δ because the molecule is nonpolar. The variation in ε_r and tan δ as a function of temperature is only a result of the change in density. Polyethylene is quite chemically inert but it does burn.

Polyethylene is very widely used as an insulator in high-frequency (coaxial) and high-voltage cables up to 220 kV.

4.12.2 Polyvinyl Chloride (PVC)

The molecule of polyvinyl chloride (Fig. 4.15) is obtained by polymerization of the vinyl chloride monomer (Fig. 4.57). The random arrangement of the chlorine atoms (atactic molecule) with diameters larger than those of the hydrogen atoms prevents the formation of crystalline zones. PVC is essentially amorphous. The dipole moment of the C–Cl bond creates dipolar interaction forces between adjacent chains responsible for the mechanical rigidity of PVC.

```
H   H
|   |
C = C
|   |
H   Cl
```

Fig. 4.57 Vinyl Chloride

Several polymerization procedures can be used. Mass polymerization is the most recent. It has the advantage that catalyst residues are the only possible impurities. Polymerization in the presence of water requires the addition of substances promoting the formation of suspensions or emulsions in addition to the catalysts.

Pure PVC is unstable to heat and to light. Lead stearate is a stabilizer frequently used to improve PVC's resistance to these agents. PVC has a high glass-transition temperature T_v lying between 75°C and 80°C which makes it unsuitable for a large number of applications (for example cables) without adding to it plasticizers such as dioctyl phthalate. Plasticizers act by reducing the intensity of the dipole bonds between adjacent chains and make it possible to reduce T_v to between -10°C and -5°C. Neutral substances such as chalk and kaolin may be incorporated in PVC without altering its properties. In this way, a *filled* PVC is obtained at lower cost per unit volume. Above 150°C, PVC decomposes with the liberation of hydrochloric acid. Placed in a flame, it burns but does not itself propagate the flame.

Available in the form of powder or granules, PVC is worked by extrusion, injection, or hot-pressing at temperatures that depend on the additives and the forming method. It is also obtained in the form of profiled sections.

PVC is used for the sheathing and insulation of power cables (up to approximately 10 kV) and low-frequency telecommunication cables. A large number of small insulating parts in PVC are manufactured by injection molding.

4.12.3 Polystyrene

Polystyrene is an aromatic compound; its molecule has the form of an ethylene molecule in which one hydrogen atom is substituted by a benzene ring (Fig. 4.58). Mass polymerization of liquid styrene at room temperature gives a polystyrene with the best electrical properties. Other polymerization methods can also be used: emulsion, precipitation, *et cetera*.

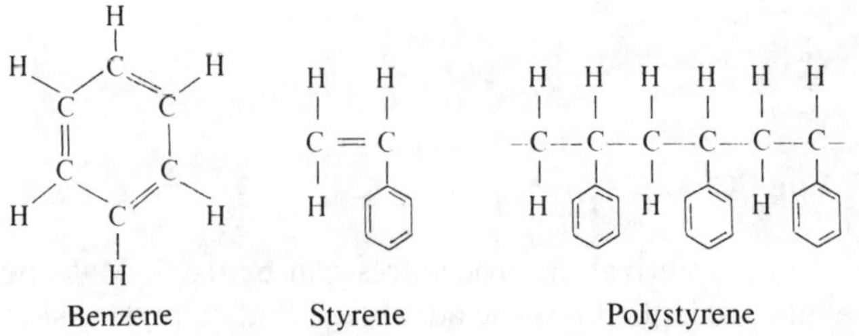

Benzene Styrene Polystyrene

Fig. 4.58

The benzene rings are situated on the carbon atom chain in a totally random fashion (atactic molecule) so that polystyrene is not crystalline. Its molecular weight fluctuates between 150,000 and 400,000 which represents 1500 to 4000 monomers. It would be expected that polystyrene is strongly polar because of the high degree of asymmetry introduced to the molecule by the benzene rings. In reality, the benzene rings are blocked by their very size. This has two consequences.

First, polystyrene is nonpolar because the permanent dipole moments, randomly oriented, are incapable of aligning themselves under the effect of an external field.

Second, polystyrene is rather brittle and sensitive to shocks because the benzene rings provide an obstacle to the relative movement of neighboring chains. This drawback is not important when polystyrene is used in the form of thin sheets as in capacitors where its high resistivity and its low dielectric losses are best exploited. It can be made less brittle by mixing it with an elastomer with a butadiene base but this is detrimental to its electrical properties.

Polystyrene has a higher resistance to ionizing radiation than that of most other thermoplastics. Transparent, it can be easily colored. It burns easily with a characteristic aromatic odor. Polystyrene is already plastic at 80°C and is usually worked at temperatures below those used for other thermoplastics. In the manufacture of capacitors with wound dielectric, it is used in the form of hot-drawn and therefore thermoretractable sheets which solves the problem of compacting the conducting-insulating layers in an elegant manner. It is easy to assemble polystyrene parts by soldering. This material also lends itself very well to the injection molding of parts weighing between 1 g and 10 kg, such as bobbins, all sorts of insulating cases, and supports. A disadvantage of polystyrene is its relatively high price.

4.12.4 Polypropylene

The molecule of polypropylene is a molecule of ethylene in which one hydrogen atom has been replaced by a methyl group $-CH_3$ (Fig. 4.59). Depending on the polymerization process, an atactic or isotactic polypropylene is obtained. Only the isotactic molecule gives a material which is of use for applications. Its crystallinity lies between 50% and 75%. A macromolecule contains between 6000 and 18,000 monomers, which corresponds to an average molecular weight on the order of 500,000.

```
H  H          H   H   H   H   H   H
|  |          |   |   |   |   |   |
C = C        -C---C---C---C---C---C-
|  |          |   |   |   |   |   |
H H-C-H       H  CH3  H  CH3  H  CH3
    |
    H
Propylene          Polypropylene
```

Fig. 4.59

Polypropylene, like polyethylene, is nonpolar and has a low specific gravity. Compared to polyethylene, it has the advantage of a better resistance to tensile stress and a greater hardness. Unfortunately, its glass-transition temperature is relatively high: −12°C. Polypropylene in the form of thin drawn sheets has the highest crystallinity and a tensile strength reaching 120–200 N/mm^2 (rupture). In bulk polypropylene, the mechanical properties may be improved by adding glass fibers or asbestos fibers.

The permittivity of polypropylene is constant *versus* frequency up to 10^9 Hz. The only variation of ε_r *versus* temperature is that produced by a change in the specific gravity. The dielectric strength of thin sheets reaches 300 kV/mm. Polypropylene burns.

Polypropylene is currently formed by injection molding at a temperature between 190°C and 240°C. It has a shrinkage of 1.5%. The thinnest sheets of 6–12 microns are obtained by drawing thicker sheets at a temperature close to the end of crystallization point.

Because of the combination of good electric and mechanical properties, polypropylene is widely used in the following areas in particular: antenna insulators, cable joints, various junction boxes, cooling fins in motors. Polypropylene is useful combined with paper in power capacitors for compensating the reactive component of the current.

4.12.5 Polytetrafluorethylene (PTFE)

If the four atoms of the hydrogen of the ethylene molecule are replaced by four atoms of fluorine, we obtain tetrafluorethylene (TFE), with its polymerization giving polytetrafluoroethylene or PTFE (Fig. 4.60), widely known under the name of Teflon™. Emulsion or suspension polymerization methods can be used and provide the polymer in the form of a very fine powder. The molecular weight lies between 500,000 and 5 million. The symmetry of the macromolecule makes polytetrafluoroethylene a nonpolar material on the one hand and a material with high crystallinity on the other hand.

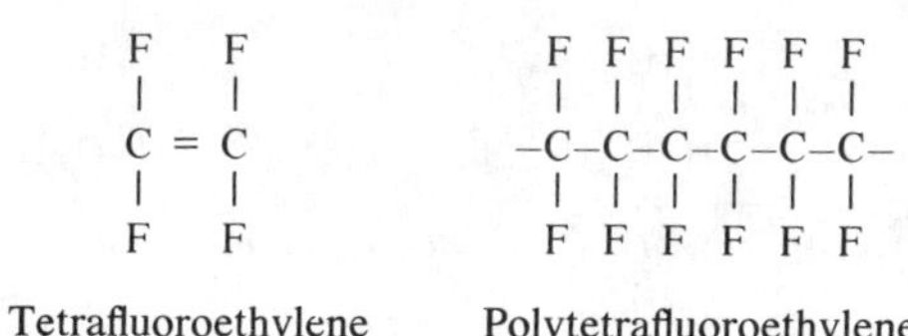

Fig. 4.60

The permittivity does not vary up to 10^{10} Hz. The dielectric losses on the order of 10^{-4} to 10^6 Hz only reach $5 \cdot 10^{-4}$ at 10^9 Hz which is particularly low. The resistivity is high and largely unaffected by the presence of humidity. However, the degradation under the effect of discharges (glow discharges, *et cetera*) limits the permissible electric field.

The range of temperatures that can be used for permanent operation is exceptional: −200°C to +200°C. The material remains flexible at the lowest temperatures. The mechanical properties are very mediocre. Creep is rather high. The coefficient of friction is low.

The forming of polytetrafluoroethylene differs radically from that of the other thermoplastics because of a very high crystallization point, 327°C, and because of the considerable viscosity of the material at this temperature. The procedures used are similar to those used in powder metallurgy. Polytetrafluoroethylene is easily machined by conventional methods. Various fillers can be incorporated in it, in particular titanium dioxide TiO_2, in order to lower the price per unit volume.

PTFE is currently used in high-frequency connectors. Filled with glass fibers, suspension insulators can be made from it that do not allow snow to accumulate on them. It is also used for the manufacture of spacers in cryogenic cables. Generally speaking, its use is indicated under extreme thermal conditions.

4.12.6 Other Thermoplastics

Polycarbonate (Fig. 4.61) has a molecular weight of approximately 150,000. It has low crystallinity.

The importance of polycarbonate in electrical engineering is growing. Polycarbonate resists impact extremely well and withstands relatively high temperatures: 120°C permanently. Its electrical properties are average. Thermoretractable polycarbonate tapes are frequently used to insulate windings. Very thin sheets may be obtained, down to 2 microns, with constant thickness well adapted to metallization by aluminum, from which comes the interest in polycarbonate for capacitors with reduced dimensions.

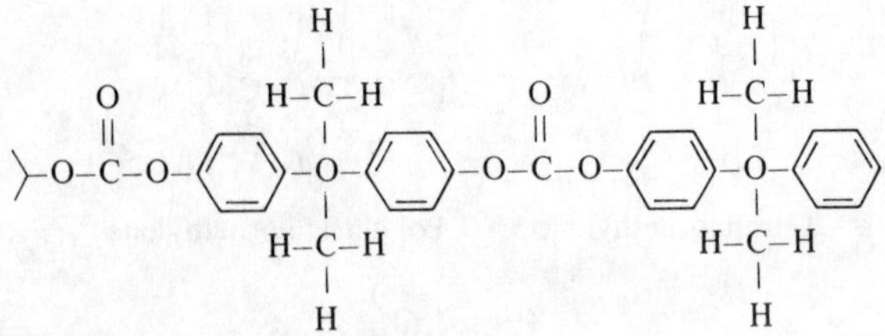

Fig. 4.61 Polycarbonate

The *polyamides* form an important group of thermoplastics, some of which, such as the *nylons,* find applications in electrical engineering, especially when the mechanical properties (hardness, resistance to abrasion) are more important than the electric properties. The electric properties are particularly limited by the hygroscopic nature of these materials.

The above-mentioned thermoplastics should be considered as examples of the materials that are most commonly used. Other thermoplastics are described, for example, in [64].

4.13 ELASTOMERS

4.13.1 Polyisoprene

Polyisoprene (Fig. 4.62) is natural rubber. This polymer can also be obtained by synthesis. Vulcanization is produced by sulfur which establishes bonds between adjacent chains by causing the double bonds to break.

Polyisoprene has a high electrical resistivity but it has a poor resistance to surface currents. The presence of sulfur makes it slightly polar. The maximum permissible temperature under permanent conditions is only 60°C. It ages relatively badly and does not resist ozone or certain mineral oils. Salts, bases, concentrated sulphuric acid, and organic acids, with the exception of nitric acid, do not attack it.

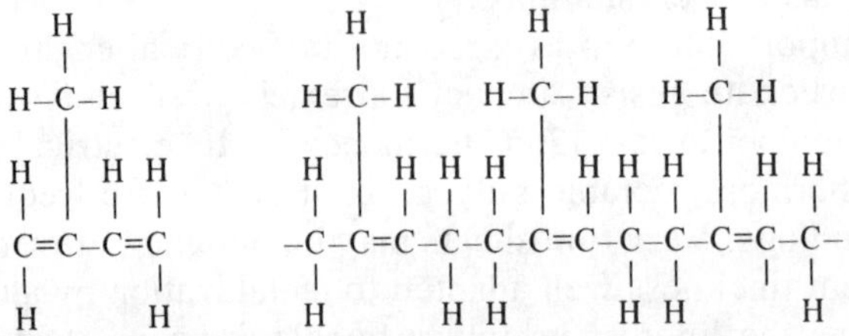

Fig. 4.62

Expensive in terms of its performance, polyisoprene is being progressively replaced by other synthetic rubbers such as the EPRs or the EPDMs (Section 4.13.4).

4.13.2 Silicone Rubbers

Silicone rubbers are polymers with high molecular weight lying between 10,000 and 700,000. They are distinguished by the presence of silicon in the main chain of the molecule. The degree of reticulation may be easily controlled so that these materials can just as easily come in the form of elastomers as thermosetting materials.

Silicone rubbers are essentially formed by extrusion, injection, and hot-pressing. Vulcanization may be produced (by increasing the temperature) in the forming machine itself or in a separate vessel.

The specific characteristic of silicone rubbers is an excellent thermal behavior resulting from the stability of the Si–O bond forming the main chain. They withstand permanent temperatures lying between −50°C and +180°C. Their mechanical properties are inferior to those of the other elastomers. They can be improved by adding quartz powder or titanium dioxide powder. They are nonpolar. They resist surface currents well and their resistivity varies little as a function of temperature (approximately two orders of magnitude up to 200°C). They are incombustible, and even decomposed in the heat of a fire, for example, they still provide a certain insulation which is of value to certain security installations. Gasoline, aliphatic chlorinated hydrocarbons, and aromatic solvents cause them to inflate and lose their mechanical properties. They do not resist concentrated acids and bases, nor steam above 130°C. They are expensive.

Silicone rubbers are used for the insulation of conductors and cables. They are used to make insulator fins to be mounted on tensile parts made of polyepoxides because they preserve good insulating properties in a polluted atmosphere. They are used to isolate various high-voltage components and electronic components. They have the advantage of not exerting any mechanical stress on these components during variations of temperature.

4.13.3 Polyurethanes

The polyurethanes have very mediocre electrical properties, inadequate at high voltages and high frequencies. Their mechanical properties are good, in particular, resistance to wear, to aging, and to oils.

The degree of reticulation may be varied to a large extent with a corresponding variation in the hardness of the product. Certain hard polyurethanes decomposing above 320°C are used to insulate fine wires (enameled wires) with the advantage that the wires are automatically exposed when soldering.

4.13.4 Other Elastomers

The copolymer obtained by the reaction of ethylene and propylene molecules in equal molar concentrations is known as ethylene-propylene rubber or EPR (Fig. 4.63).

```
              H
              |
   H  H  H–C–H  H
   |  |     |    |
 –C–C——C——C–
   |  |     |    |
   H  H     H    H
```

Fig. 4.63 EPR

The formula does not have a double bond. The result is an excellent resistance of the material to oxidation, even in the presence of ozone, and the fact that it is necessary to use a peroxide to produce reticulation. EPR has an excellent resistance to partial discharge and its other electric properties are satisfactory.

By adding a diene (hydrocarbon with a double bond) of the type shown in Fig. 4.64 in a molar concentration of 2–3% to EPR, a MEPD (modified ethylene propylene diene) is obtained. Its electric properties are practically identical to those of EPR when it is reticulated using a peroxide. They are inferior if the reticulation is made using sulfur, which is made possible by the double bond of the diene. The advantage of MEPD over EPR is a more rapid reticulation and a better surface condition. These elastomers are widely used for the manufacture of injection molded accessories used in the high-voltage sector.

```
                    H
                    |
   H    R′–C=C–R″
   |        |
 –C——–C————
   |        |
   H        H
```

Fig. 4.64

4.14 THERMOSETTING MATERIALS

4.14.1 Nonsaturated Polyesters

The most important organic acids are the *carboxylic acids*. They are formed from a carboxyl group (Fig. 4.65) joined to an alkyl or aryl group.

The *esters* are derivatives of the carboxylic acids, in which the hydrogen atom is replaced by an alkyl or aryl group, identical or different to that of the acid. Different families of esters are obtained (Fig. 4.66). These esters are called nonsaturated because the carboxyl group is nonsaturated.

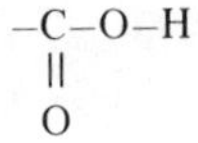

```
–C–O–H
 ||
 O
```

Fig. 4.65

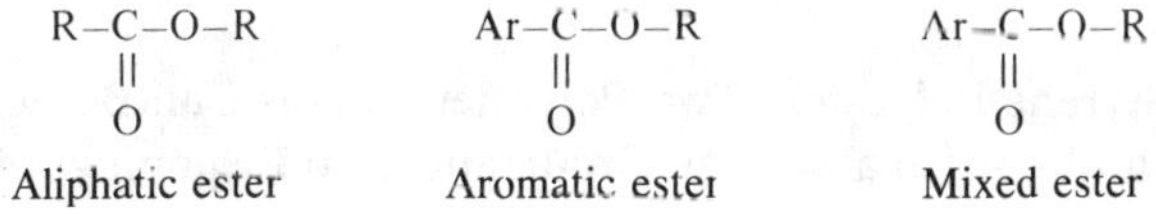

R–C–O–R	Ar–C–O–R	Ar–C–O–R
‖	‖	‖
O	O	O
Aliphatic ester	Aromatic ester	Mixed ester

Fig. 4.66

The polymerization of esters and the copolymerization of esters with other monomers supplies a large variety of products given the general name of polyester. The styrene polyester (Fig. 4.67) is frequently used in electrical engineering. Its resin is formed from a nonsaturated polyester with high molecular weight in suspension in styrene. Because the latter is itself nonsaturated, the valence bonds necessary for its reticulation are provided by both components of the resin.

The hardener is an organic peroxide, benzoyl peroxide, for example. Reticulation normally takes several hours. This time may be reduced to a few minutes by using *accelerators* to increase the speed of decomposition of the peroxide and therefore the liberation of its free radicals. An increase in temperature has the same effect but to a lesser degree. Polyesters are formed by molding, ordinary casting, centrifuge casting for the manufacture of tubes, hardening, and coating.

Various fillers may be added such as chalk, kaolin, lime, quartz powder, glass fibers, *et cetera*.

The permittivity and the dielectric losses strongly depend on the nature of the fillers and the reticulation conditions. At low frequencies,

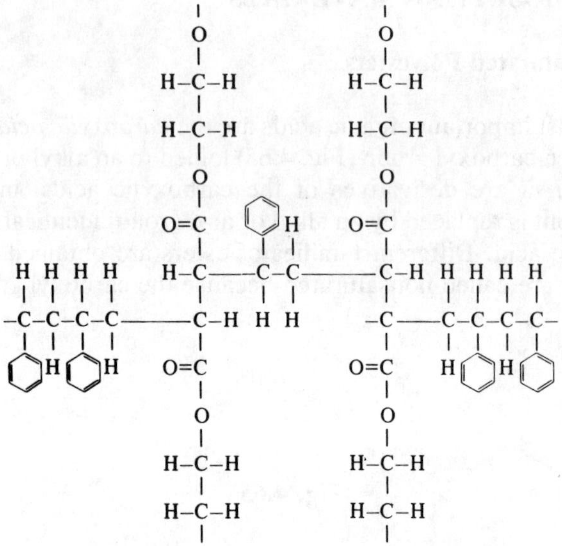

```
                    |                 |
                    O                 O
                    |                 |
                  H-C-H             H-C-H
                    |                 |
                  H-C-H             H-C-H
                    |                 |
                    O                 O
                    |                 |
                   O=C       H       O=C
                    |        |  |     |
 H  H  H  H       H-C-------C-C-------C-H  H  H  H  H
 |  |  |  |         |       |  |      |    |  |  |  |
-C-C-C-C-----------C-H     H  H     -C-----C-C-C-C-
 |  |  |  |         |                 |    |  |  |  |
    H     H        O=C               O=C   H     H
                    |                 |
                    O                 O
                    |                 |
                  H-C-H             H-C-H
                    |                 |
                  H-C-H             H-C-H
                    |                 |
```

Fig. 4.67 Styrene Polyester. Two Polyester Chains Can Be Seen Vertically and Horizontally, Also a Styrene Group and Two Beginnings of Polystyrene Chains

the losses of the polyester filled with glass fibers are greater than those of pure polyester although tan δ is lower in glass than in the polyester. This fact must be attributed to the interfacial polarization appearing at the surfaces separating the glass and the polyester. The resistance to surface currents is good.

Highly reticulated, the polyester is brittle. Flexible grades may be obtained by restricting the reticulation (with a smaller amount of peroxide) and by adding certain fillers. It burns, liberating soot and a sweetish odor characteristic of styrene. It propagates the flame, but impregnated with chlorinated paraffin for example, it may be made self-extinguishable. It is weakly hygroscopic and its chemical stability is good.

Polyester is used in the manufacture of large insulating parts: light antenna masts, antenna protection, switching cabinets, cable junction boxes, various casings.

4.14.2 Polyphenols

Phenols can be considered as derivatives of water in which one of the H atoms is replaced by an aryl group (Fig. 4.68).

Ar–O–H

Fig. 4.68 The Phenols

The simplest of the phenols in which Ar is a phenyl group is called phenol. When Ar is a tolyl group we obtain a cresol. There are three types of cresols (Fig. 4.69) corresponding to the three types of tolyls.

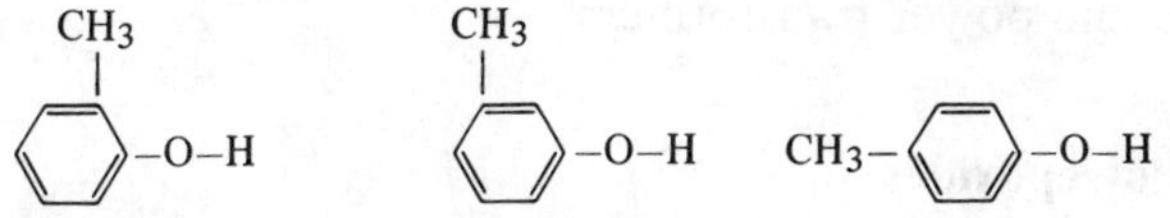

CH_3 (ring)–O–H — o-cresol; CH_3 (ring)–O–H — m-cresol; CH_3–(ring)–O–H — p-cresol

Fig. 4.69

The polyphenols are produced by the reaction of a mixture of phenols with formaldehyde. In the particular case when pure phenol is used, we obtain the oldest and the best known of the polyphenols: bakelite (Fig. 4.70).

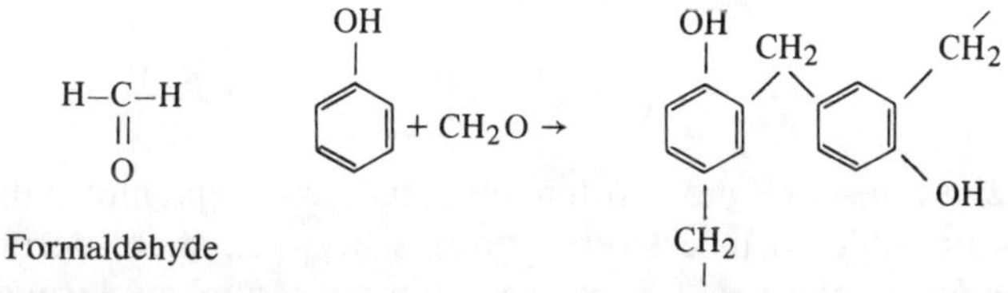

H–C–H (=O) Formaldehyde

OH (ring) + CH_2O → OH, CH_2, CH_2, OH, CH_2

Fig. 4.70 Diagrammatic Representation of the Reaction for the Formation of Bakelite

Mixtures of phenols and cresols are very widely used and give macromolecules similar to that of bakelite. Polyphenols are mainly used for the impregnation of fibrous materials: asbestos, fabrics, papers, wood. This impregnation is carried out using aqueous or alcoholic solutions. Small parts are produced from pure polyphenols.

The electric properties of polyphenols satisfy current requirements but the resistance to surface currents is mediocre. A rather marked hygroscopic sensitivity makes it difficult to use them in a humid atmosphere. Impregnated tissues have good mechanical properties: a high modulus of elasticity, tensile strength, and a good resistance to bending and impact. They can be machined by conventional methods. They are chemically stable and resist alcohol, acetone, petrol, and oils well. They are less resistant to acids and bases, which has to be taken into account in the manufacture of printed circuits. Finally, polyphenols are inexpensive and this is why they are used on such a large scale.

Polyphenols are used for manufacturing bobbins, printed circuits of current quality, insulation parts for electromechanical relays, *et cetera*. In the high-voltage sector, their resistance to oil and to mechanical stresses have made them materials of first choice for insulating parts of circuit breakers and power transformers.

4.14.3 Polyepoxides

Polyepoxide resins are given this name because they are formed from linear molecules with epoxide groups at the ends (Fig. 4.71).

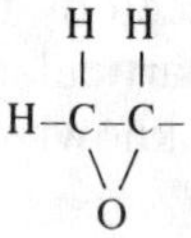

```
  H H
  | |
H–C–C–
   \/
   O
```

Epoxide group

Fig. 4.71

These groups are most often obtained from epichlorhydrin (Fig. 4.72). It is possible to insert other groups between them derived from various phenol derivatives, for example, dihydroxydiphenylpropane (Fig. 4.73), where the hydrogen atoms of the hydroxyl groups combine with the chlorine of the epichlorhydrin forming HCl which is neutralized by the basic medium in which the reaction is carried out.

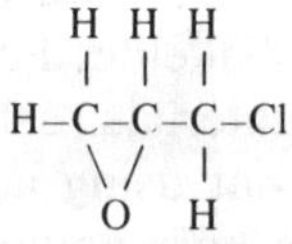

```
  H H H
  | | |
H–C–C–C–Cl
   \/ |
   O  H
```

Epichlorhydrin

Fig. 4.72

The polyepoxide resin molecule then has the form described in Fig. 4.74.

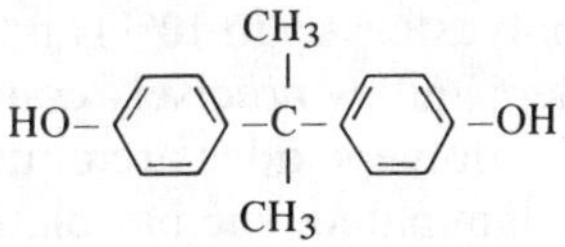

$$HO-C_6H_4-C(CH_3)_2-C_6H_4-OH \qquad HOROH$$

Fig. 4.73 Dihydroxydiphenylpropane Molecule and its Abbreviated Representation

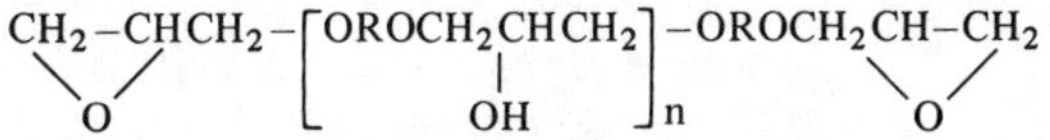

$$\underset{\diagdown O \diagup}{CH_2-CHCH_2}-\left[OROCH_2\underset{|\atop OH}{CH}CH_2\right]_n-OROCH_2\underset{\diagdown O \diagup}{CH-CH_2}$$

Fig. 4.74

Two types of hardeners may be used. The first type comprises the amines. They produce reticulation at room temperature. The second type is the organic acids which only produce reticulation after the resin has been heated to a temperature between 140°C and 200°C depending on the case. Hot-hardened polyepoxides have the best electric properties and the best resistance to heat. The resins should not be stored for longer than a year at room temperature because of the instability of the epoxide groups.

There is a great variety of products resulting from the number of different components that can be inserted between the epoxide groups, and this variety is increased even further by using fillers. Fillers make it possible to modify the mechanical properties in the sense of hardening or, on the contrary, a softening of the material. The mineral fillers (quartz, mica, alumina), used in concentrations representing up to 2–2.5 times the weight of the resin, make it possible to reduce the dilatation coefficient of the hardened material by approximately a factor of two. Reduced in this way, this coefficient is still 2.2 times greater than that of steel, which results in the risk of cracks in metal-polyepoxide assemblies.

Polyepoxides have excellent adherence to metals, ceramics, and mica. Except in the case of modification by fillers, the permittivity and dielectric losses are practically independent of the frequency. The resistance to alcohols, oils, and gasoline, to weak acids and bases, and to water is fairly good. The polyepoxides are not resistant to acetone or to chlorinated hydrocarbons. Some of them are not weather-proof and are not resistant to surface currents.

In the form of the raw material, they are formed by casting or immersion. The shrinkage during reticulation (2–3%) is small compared to other thermosetting plastics (polyester: up to 10%) and may be lowered to less than 0.1% by means of fillers and by precisely controlling the temperature during the hardening. The polyepoxides are obtained in the form of semifinished products, usually laminated, such as plates and tubes. Their area of application includes the protection of antennas, including radomes, tensile parts of insulators made of plastic materials, the insulation of the slots of large machines, certain parts of circuit breakers, the substrates for printed circuits, and the mechanical protection of electronic components.

4.15 TABLES OF DIELECTRIC PROPERTIES

The most important properties of the common polymers are given in Tables 4.75 and 4.76.

Table 4.75
(After [54, 65])

Average Properties	Units	Thermoplastics						
		Low-density Polyethylene	*Plasticized PVC*	*Polystyrene*	*Polypropylene*	*Polytetrafluoroethylene*	*Polycarbonate*	*Polyamide*[3] *+ Glass Fiber*
ε_r at 60 Hz	–	2,25-2,35	5-9	2,45-2,55	2,2-2,3	2	3,17	4-4,6
1 kHz	–	2,25-2,35	4-8	2,4-2,65	2,2-2,3	2	3,02	3,9-4,4
1 MHz	–	2,25-2,35	3,3-4,5	2,4-2,65	2,2-2,3	2	2,96	3,4-3,9
tan δ at 60 Hz	10^{-3}	< 0,5	80-150	0,1-0,3	< 0,5	< 0,3	0,9	18-25
1 kHz	10^{-3}	< 0,5	70-160	0,1-0,3	< 0,5	< 0,3	2,1	20-25
1 MHz	10^{-3}	< 0,5	40-140	0,1-0,4	< 0,5	< 0,3	10	17-22
ρ	Ω·m	$> 10^{14}$	10^{9}-10^{11}	$> 10^{14}$	$> 10^{14}$	$> 10^{13}$	$2 \cdot 10^{14}$	$1{,}5$-$5 \cdot 10^{13}$
E_c	kV/mm	17-28	11-32	16-28	20-26,4	17-24	14-16	16-20
Density umique	10^3 Kg/m^3	0,910-0,925	1,16-1,35	1,04-1,06	0,902-0,910	2,1-2,2	1,2	1,3-1,5
σ stress at failure n	N/mm^2	7-14	10-24	35-63	30-39	11-20	56-67	100-250
Δ1/1 elongation at failure	%	200-550	200-450	1-2,5	200-700	100-200	60-100	1,5-6
E_y Young's modulus	N/mm^2	130-240	– (2)	2 800-3 500	1 100-1 400	400	2 450	600-1 200
Thermal conductivity	10^2 W/m°C	1,9	0,7-0,95	0,57-0,79	0,67	1,4	1,1	0,35-0,40
Linear thermal expansion	10^{-5}/°C	16-18	7-25	6-8	6-8,5	10	6,6	1,2-3,2
Permissible T_{max} under .1) permanent conditions[1]	°C	95	65-80	66-77	121-160	260	120	150-200

(1) This value does not take into account mechanical deformations.
(2) Strongly depends on the plasticizer.
(3) Type 6/6 nylon plus 20–40% of glass fibers.

Table 4.76
(After [64, 65])

Average Properties	*Units*	*Thermosetting plastics*			*Elastomers*		
		Polyester (2)	*Polyphenol (3)*	*Polyepoxide (4)*	*Polyepoxide (5)*	*Silicone Rubber (6)*	*MEPD*
ε_r at 60 Hz	–	4,1-5,5	5-6,5	3,5-5	3,5-5	3	2,2-3,3
1 kHz	–	4,2-6	4,5-6	3,5-4,5	3,5-5	3	–
1 MHz	–	4-5,5	4,5-5	3,3-4	3,5-5	3	–
tan δ at 60 Hz	10^{-3}	10-40	60-100	2-10	10	10	5
1 kHz	10^{-3}	10-60	30-80	2-20	10	10	–
1 MHz	10^{-3}	10-30	15-30	3-50	10	10	–
ρ	$\Omega \cdot m$	10^{12}	10^{9}-10^{11}	10^{10}-10^{15}	$> 10^{12}$	10^{12}	10^{14}
E_c	kV/mm	14-20	10-12	16-20	12-16	20	20
Density umique	10^3 kg/m^3	1,5-2,1	1,25-1,30	1,1-1,4	1,6-2	1	1,1-1,2
σ stress at failure n	N/mm^2	210-350	50-56	30-90	70-210	0,5-3	5-10
$\Delta 1/1$ elongation at failure	%	0,5-2	1-1,5	3-6	4	50-150	300-500
E_y Young's modulus	N/mm^2	10 000-30 000	2 800-7 000	2 450	21 300	–	–
Thermal conductivity	10^{-2} W/m°C	–	0,7-1,4	–	0,95-2,4	0,3	35
Linear thermal expansion	10^{-5}/°C	1,5-3	2,5-6	4,5-6,5	1,1-3,5	36	–
Permissible T_{max} under (1) permanent conditions(1)	°C	150-175	120	120-280	150-260	170	80

(1) This value does not take into account mechanical deformations.
(2) Reinforced by glass fibers.
(3) Pure resin.
(4) Pure resin.
(5) Reinforced by glass fibers.
(6) Cold hardened.

4.16 PROBLEMS

4.16.1 Calculate the electric field of a point dipole in polar coordinates (Fig. 4.77). Determine the equation for the lines of force of the electric fields.

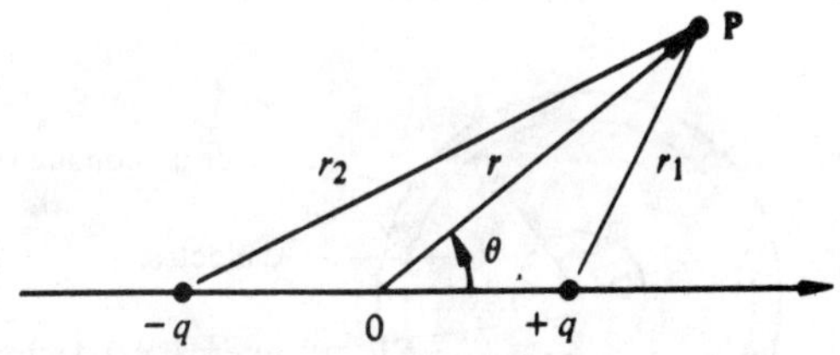

Fig. 4.77

4.16.2 Calculate the local field by Onsager's method. This consists of surrounding the reference point by a sphere with a radius small enough so that it only contains one dipole (Fig. 4.78). The relative permittivity of this sphere is equal to one.

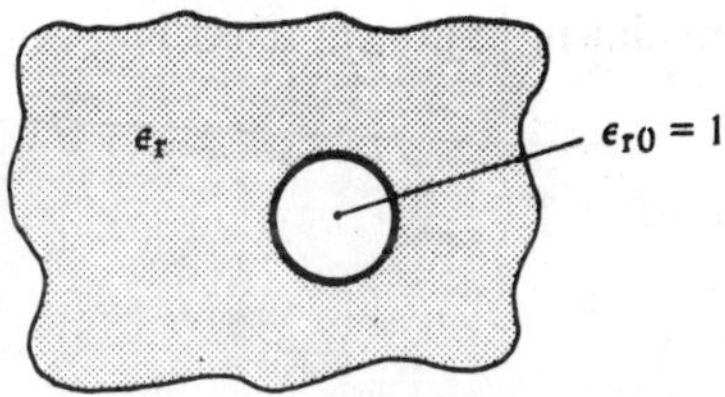

Fig. 4.78

Use the result to calculate ε_r in a dielectric where only the orientation polarization is significant. Compare the numerical values of ε_r based on the Lorentz and Onsager theories.

4.16.3 Argon under normal conditions ($1.0 \cdot 10^5$ Pa, 0°C) has a relative permittivity of 1.00044. Estimate the radius of the argon atom. By how much is the nucleus displaced with respect to the system of electrons in a field of 10^4 V/m?

4.16.4 Magnesium oxide MgO has the NaCl structure with a unit cell with a side of 0.42 nm. It does not have a permanent dipole moment. Its relative permittivity is equal to 9.6 and its relative susceptibility for electronic polarization is equal to 2. Calculate the displacement of a Mg^{++} ion with respect to an O^{--} ion under the action of a field of 10^4 V/m.

4.16.5 Let the coaxial cable shown in Fig. 4.79 be subjected to the direct current voltage U_0. Determine r_i so that E is a minimum at the surface of the internal conductor for a given value of r_e.

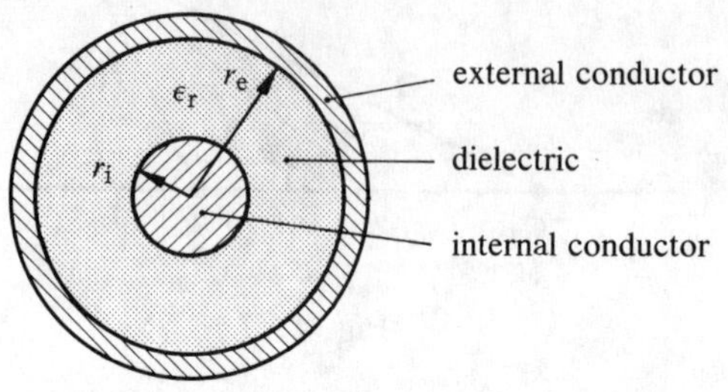

Fig. 4.79

4.16.6 With regard to the cable defined in Section 4.16.5, calculate r_i and r_e so that the electric field is everywhere less than or equal to E_{max} and so that the amount of insulating material used is a minimum.

Chapter 5
Thermodynamic Transformations

5.1 INTRODUCTION

5.1.1 Definition

Thermodynamic transformations are the modifications of matter of a physical and chemical nature, reversible or irreversible, produced by a change in temperature.

5.1.2 Examples

Thermodynamic transformations occupy a central position in the science of materials.

On the one hand, they are involved in the manufacturing processes of almost all materials; the few following examples give an impression.

Semiconductors are purified by the melting zone method (Section 2.7.14). In superconductors, the fluxoids can be pinned (Section 2.8.20) by means of precipitates obtained by thermal treatment. Essentially, the manufacture of magnetic alloys (Sections 3.9 and 3.10) amounts to a sequence of thermal treatments often alternating with mechanical treatments *et cetera*.

On the other hand, thermodynamic transformations may have adverse effects that have to be taken into account. A material that has undergone a modification at a high temperature may be in an unstable state after cooling to the temperature of its normal use. These properties therefore evolve more or less slowly over time. The disaccommodation and magnetic aging of iron (Section 3.9.2) are typical examples of this phenomenon.

5.1.3 General Approach

The study of every thermodynamic transformation is a subject in itself, treated by the appropriate methods of analysis. The results of such studies are most often presented in the form of thermodynamic diagrams such as:

- phase diagrams at equilibrium;
- phase transformation diagrams.

This chapter is limited to a general presentation of these diagrams so that they can be profitably used. One section is devoted to diffusion problems. The diffusion equations are given in Section 1.7.

5.2 THE CONCEPT OF PHASE

5.2.1 Definitions

In the language of thermodynamics, we call a *system* the material specimen under study and the environment with which it interacts taken as a whole.

A *phase* of a system is a part of it in which the *composition* (nature and concentration of the atoms present) and the *structure* (spatial arrangement of these atoms) are constant or vary continuously. Continuous change of the structure is understood, for example, as the modification of the base vectors of a crystal under the action of localized mechanical stresses or under the effect of an inhomogeneous temperature field.

A phase is separated from the system by defined surfaces over which the composition or the structure vary in a discontinuous manner.

5.2.2 Examples of Phases

The three states of matter—gas, liquid, and solid—form the simplest examples of phases in the case of pure substances. However, a single phase does not necessarily correspond to each state, as is shown by Table 5.1 illustrating the polymorphism (Section 1.4.11) of iron.

The situation is much more complex when we are dealing with mixtures of several components. Systems consisting entirely of gases still have a single phase because all gases are miscible in all proportions. The same does not apply to systems that are entirely liquid or solid. Although water and alcohol are miscible in all proportions, as are copper and nickel (solid solution), a mixture of water and oil forms an emulsion, i.e., a fine dispersion of droplets of oil in water corresponding to the existence of two

Table 5.1
Phases of Iron in the Solid State

Temperature °*C*	*Crystal Structure*	*Designation of the Phase*
−273 to 912	CC	α
912 to 1394	FCC	γ
1394 to 1538*	CC	δ

* Melting point

phases in the same liquid. A comparable phenomenon takes place in the solid state when in iron, for example, the concentrations given in Table 3.87 are exceeded. We then say that one phase *precipitates* in another.

Polymers reinforced with glass fiber or with silica powder or other components added to them also provide an example of materials with several phases.

5.3 GIBBS'S PHASE RULE

5.3.1 Statement and Area of Application

Gibbs's phase rule, which can be proven by thermodynamics [66], relates the variance f of the system (Section 5.3.2), the number p of phases simultaneously present, and the number c of components (Section 5.3.3) by the formula:

$$f = c - p + 2 \tag{5.1}$$

It can be applied to systems in equilibrium which can undergo reversible transformations under the action of only three types of variables, namely: pressure, temperature, and relative mean concentration of each component over the entire system. If one of these variables is kept constant, which is often the case for pressure, a large number of systems only being subjected to atmospheric pressure the variations of which often have a negligible effect, the phase rule can be written

$$f = c - p + 1 \tag{5.2}$$

5.3.2 Variance f of a System: Definition

The *variance* f of a system is the number of variables (pressure, temperature, or concentration) that can be modified independently of each other over a certain range without the number of phases present in the system changing. The number of *degrees of freedom* of the system is another name for f.

5.3.3 Number of Components of the System: Definition

The *number of components of the system* is the minimum number of distinct chemical constituents for which the concentration has to be known in order for the composition of each phase to be defined.

This number of components is obtained practically using the relationship:

$$c = CCI - R, \tag{5.3}$$

where CCI represents the number of *independent chemical constituents*. This number is obtained by subtracting the number of conditions restricting the chemical constituents, for example the condition of electric neutrality of the system when certain constituents are ionized, from the total number of chemical constituents.

R is the number of *independent* chemical reactions that effectively occur in the system. Reactions are called independent if they cannot be expressed in terms of a sequence of other reactions already contained in the system.

Consider, for example, a system in which calcium carbonate (solid) is in equilibrium with calcium oxide (solid) and carbon dioxide gas according to the reaction:

$$CaCO_3 \rightleftharpoons CaO + CO_2 \tag{5.4}$$

The three independent chemical constituents are related by an independent chemical reaction (5.4). This system therefore has two components.

In each system, the number of components is perfectly defined. On the other hand, the identity of the components can be chosen with a certain freedom as soon as the number of components is less than the total number of chemical constituents.

5.4 PHASE DIAGRAMS AT EQUILIBRIUM

5.4.1 General Considerations

A phase diagram at equilibrium, or more simply an equilibrium diagram, is a diagram describing the state to which the system tends as a function of the values of the variables under consideration. This final state, which represents the equilibrium state of the system, may be reached very quickly or, on the contrary, after an infinite time. The equilibrium diagram does not give any information on this subject. This is provided by the phase transformation diagram (Section 5.5).

When the system only has one component, and the variables are, for example, pressure and temperature, the equilibrium diagram has two dimensions. With two components, this diagram will have three dimensions. In order to keep to a two-dimensional representation, which is much more practical, pressure is given the role of a parameter. Temperature is then plotted as an ordinate and the ratio of the amounts of the two components is plotted as the abscissa. This diagram, called a binary diagram (two components) has a particular importance. On the one hand, there are a large number of two-component systems that are of interest; on the other hand, systems with more components are often described by a series of binary diagrams in which the variables that are not represented appear as parameters.

The binary diagram provides the following information:

- the number of phases present;
- the composition of each phase;
- the fraction of each component present in each phase.

The binary diagrams shown below (Sections 5.4.3, 5.4.5, and 5.4.7) refer to crystalline chemical constituents, in particular, metals and ceramics.

5.4.2 Cooling Curve: Definition

The cooling curve of a system is the graph of the temperature as a function of time during a cooling process, from the liquid state to the solid state, slowly enough so that at each point in time, the temperature can be considered as homogeneous throughout the system. These curves are used to construct the equilibrium diagrams.

5.4.3 The Simple Binary Phase Diagram

The simple binary phase diagram is the equilibrium diagram corresponding to the case when the two components A and B are miscible in all proportions in the solid state as well as in the liquid state. The form of this diagram can be deduced from the cooling curves of A, B, and the mixture A + B, shown in Fig. 5.2 where T represents the temperature and t represents the time.

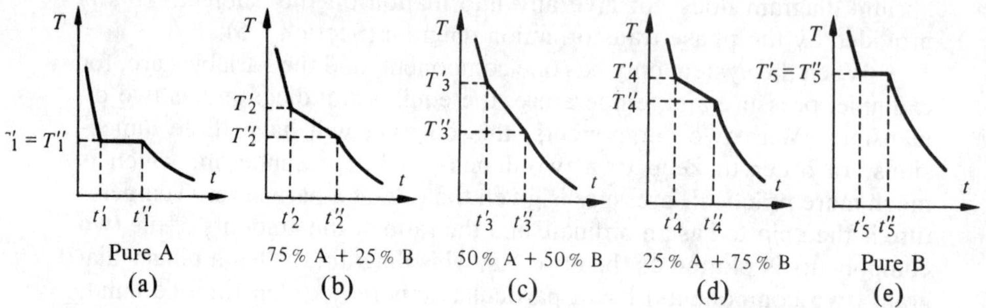

Fig. 5.2 The Points in Time t_i' Correspond to the First Appearance of the Solid Phase, and the Points in Time t_i'' Correspond to the Complete Disappearance of the Liquid Phase

Each of the diagrams of Fig. 5.2 may be considered as a cross section of the three-dimensional diagram shown in Fig. 5.3.

The phase rule gives the following interpretation for Fig. 5.2.

In the case of Fig. 5.2(a), there is only one component ($c = 1$). At the time preceding the time t_1', only one phase, a liquid phase, is present ($p = 1$). Equation (5.2) therefore indicates that there is one degree of freedom ($f = 1$). Consequently, the temperature can vary within certain limits without modifying the number of phases present. In the interval $t_1' < t < t_1''$, $p = 2$ from which we obtain $f = 0$; as long as the liquid and solid phases are present at the same time, the temperature remains constant. During this time, the latent heat of fusion of A leaves the system. As soon as $t > t_1''$, the system is entirely solid ($p = 1$) and the temperature continues to drop. The case of Fig. 5.2(e) is similar.

The case of Figs. 5.2(b) to 5.2(d) is mainly distinguished from Figs. 5.2(a) and 5.2(e) for $t_i' < t < t_i''$. In these intervals, $p = 2$, but since $c = 2$, the equation (5.2) gives $f = 1$. The temperature of the system therefore continues to drop during the solidification phase.

The simple binary phase diagram is obtained by plotting the points T_i' and T_i'' of a sufficiently large number of cooling curves as a function of

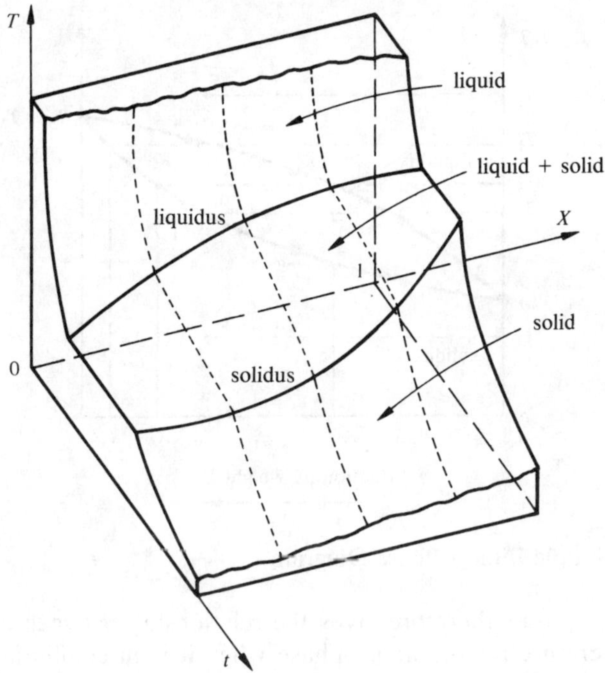

Fig. 5.3 *X* Represents the Fraction by Weight of B in the System

the overall composition of the system expressed in terms of the fraction by weight *X* of B (Fig. 5.4).

The geometric loci of the points T_i' and T_i'' are called the *liquidus* and the *solidus,* respectively. Fig. 5.4 corresponds to the picture one would have of the three-dimensional diagram of Fig. 5.3, observed from a point situated on the *t*-axis at infinity.

For a substance with overall composition X_0, the simple binary diagram (Fig. 5.4) gives the following information:

- at temperature T_a, the substance is entirely liquid;
- at T_b, a solid phase of composition X_{si} enriched in B relative to the liquid phase appears in the liquid phase with composition X_0;
- at T_c the liquid phase with composition X_ℓ and a solid phase with composition X_s coexist;
- at T_d, the last trace of the liquid phase with composition $X_{\ell f}$ disappears in the solid phase with composition X_0;
- at T_e, the substance is entirely solid.

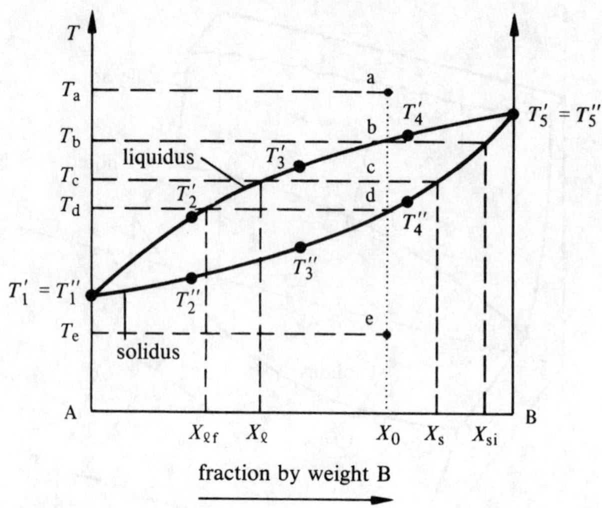

Fig. 5.4 Simple Binary Phase Diagram

The *liquidus* therefore gives the relationship between composition and temperature for the liquid phase when it is in equilibrium with the solid phase. Similarly, the *solidus* gives the relationship between the composition and temperature for the solid phase when it is in equilibrium with the liquid phase.

5.4.4 Lever Rule

Apart from the composition of each phase, the diagram makes it possible to calculate the amount of matter present in each phase using the lever rule as follows:

M = total mass of the system

M_ℓ = mass of the liquid phase

M_s = mass of the solid phase

The mass of B in the system is written as

$$X_0 M = X_\ell M_\ell + X_s M_s \tag{5.5}$$

Now,

$$M = M_\ell + M_s \tag{5.6}$$

from which we have

$$\frac{M_s}{M} = \frac{X_0 - X_\ell}{X_s - X_\ell} \tag{5.7}$$

and

$$\frac{M_\ell}{M} = \frac{X_s - X_0}{X_s - X_\ell} \tag{5.8}$$

or also

$$\frac{M_\ell}{M_s} = \frac{X_s - X_0}{X_0 - X_\ell} \tag{5.9}$$

This equation expresses the *lever rule*.

5.4.5 Binary Eutectic Diagram

It frequently happens that two components that are entirely miscible in the liquid state only have a limited solubility in each other in the solid state. In this case, and if there are only two solid phases, the diagram is either of the eutectic type (Fig. 5.5) or of the peritectic type (Section 5.4.7).

Let α be the solid phase of substance A which is pure or contains an amount of substance B less than the solubility limit of B in A. Similarly, let β be the solid phase of B which is pure or contains a certain amount of A in solid solution. The diagram contains

- two liquidus arcs intersecting at the eutectic point defining the eutectic temperature T_e and the eutectic composition X_e;
- two solidus arcs situated between the melting points of pure A and pure B on the one hand, and T_e on the other hand;
- two solvus arcs.

Solvus is the name given to the relationship between the composition and temperature of a solid phase in contact with a different solid phase. When a mixture with overall composition X_0 is cooled starting from a temperature T_a, the following stages occur in the following order:

- at T_b, the phase α appears with composition $X_{\alpha 1}$;
- at T_c, the phase α with composition X_α coexists with the liquid phase

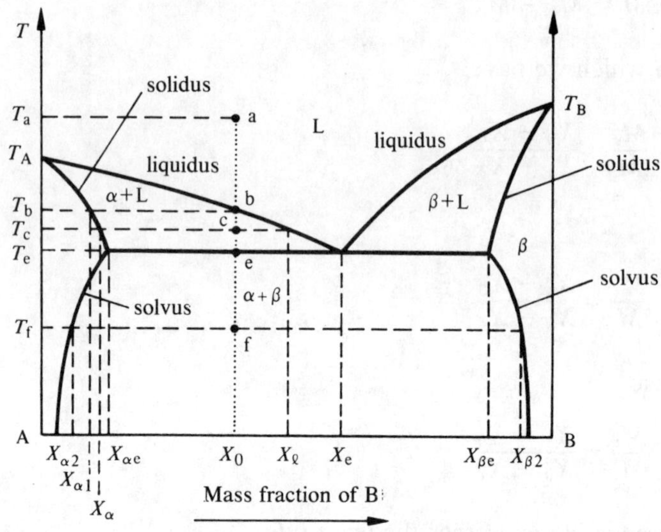

Fig. 5.5 Binary Eutectic Diagram

with composition X_ℓ. The system keeps one degree of freedom, the temperature;

- at T_e, the β phase appears. The simultaneous presence of the three phases, liquid, α, and β brings the variance of the system to zero (5.2). The solidification of the remaining liquid phase therefore takes place at constant temperature. During this transformation, α and β are deposited at the same time forming a heterogeneous mixture of small particles of each of the phases. This mixture is called the eutectic mixture. The α and β phases have the compositions $X_{\alpha e}$ and $X_{\beta e}$;
- at T_f, only the phases α and β remain with compositions $X_{\alpha 2}$ and $X_{\beta 2}$;
- when the temperature drops still further, X_α decreases, whereas X_β increases because the solubilities of B in α and of A in β decrease.

On the right in Fig. 5.5, we can see the solvus giving the mass fraction of B in β in equilibrium with α. The solvus on the left of the figure gives the fraction of B in α in equilibrium with β.

The amounts of each substance in each phase are given in the binary eutectic diagram, as in the simple binary diagram, by the lever rule (5.7) and (5.8).

The structure of the material in the solid state depends on the overall composition X_0. If $X_0 < X_e$, the first solid phase appearing during the cooling is the α phase. When T_e is reached, the growth of the grains of α that have already been formed is interrupted and at $T < T_e$, these are encased in the eutectic mixture. If $X_0 > X_e$, the same phenomenon occurs but with the β phase. Finally, if $X_0 = X_e$, the solid phase only consists of the eutectic mixture. It is formed at a constant temperature as in the case of the pure substance (Fig. 5.6).

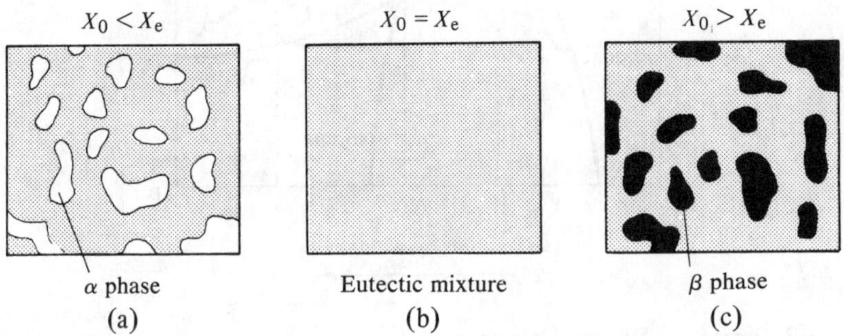

Fig. 5.6 Diagrammatic Representation of the α and β Phases in the Eutectic Mixture

5.4.6 Eutectic Transformations and Eutectoids

A eutectic transformation is an isothermal transformation during which a liquid phase is simultaneously transformed into two different solid phases, finely interspersed in each other (Fig. 5.6(b)).

By analogy, the isothermal transformation by means of which one solid phase gives rise simultaneously to two solid phases different from the first, finely interspersed in each other, is called a *eutectoid* transformation. The corresponding binary diagram has the same form as the eutectic diagram, the liquidus and solidus arcs being replaced by solvus arcs.

5.4.7 Peritectic Binary Diagram

The peritectic binary diagram corresponds to the situation in which the temperature of the isothermal transformation lies between the melting points of the two constituents. The two liquidus and solidus arcs are then placed on each side of the horizontal line corresponding to the isothermal transformation temperature (Fig. 5.7).

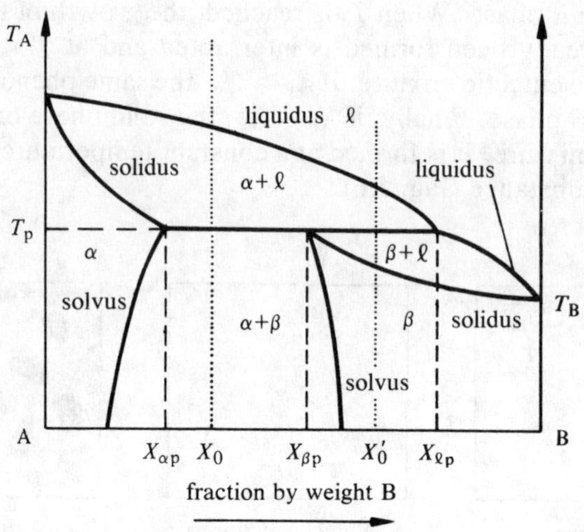

Fig. 5.7 Peritectic Binary Diagram

The interpretation of the peritectic diagram for $X < X_{\alpha p}$ and $X > X_{\ell p}$ involves only the concepts already encountered in the simple binary and eutectic binary diagrams. On the other hand, in the region $X_{\alpha p} < X < X_{\ell p}$, a new type of transformation appears which will be illustrated by studying the cooling of two mixtures with overall composition X_0 and X_0'. Let M be the total mass of the system.

At the moment when the mixture X_0 reaches T_p, three phases exist at the same time: α with composition $X_{\alpha p}$, β with composition $X_{\beta p}$, and a liquid phase ℓ with composition $X_{\ell p}$. The variance of the system is therefore reduced to zero. The liquid phase therefore disappears during an isothermal transformation.

At the beginning of this transformation, the mass of phase α, $M_{\alpha i}$, and the mass of the liquid $M_{\ell i}$ are given by the lever rule.

$$M_{\alpha i} = \frac{X_{\ell p} - X_0}{X_{\ell p} - X_{\alpha p}} M \tag{5.10}$$

$$M_{\ell i} = \frac{X_0 - X_{\alpha p}}{X_{\ell p} - X_{\alpha p}} M \tag{5.11}$$

At the end of this reaction, we have

$$M_{\alpha f} = \frac{X_{\beta p} - X_0}{X_{\beta p} - X_{\alpha p}} M \tag{5.12}$$

$$M_{\beta f} = \frac{X_0 - X_{\alpha p}}{X_{\beta p} - X_{\alpha p}} M \tag{5.13}$$

It can be seen that $M_{\alpha i}$ is greater than $M_{\alpha f}$($\sim$0.85 M and $\sim$0.7 M respectively in Fig. 5.7) which shows that the new solid phase β is formed from ℓ and α. It cannot be otherwise: because ℓ is richer in B than β, during the solidification it is necessary to make up a deficit in A which can only be done by reducing the amount of phase α.

Such a transformation, during which a solid phase is formed from another solid phase and a liquid phase, is called a *peritectic transformation*.

The cooling of a mixture with composition $X_{\beta p} < X'_0 < X_{\ell p}$ takes place somewhat differently. When the temperature T_p is reached, the variance of the system is still zero but it is the α phase that disappears to the advantage of the nascent β phase and the liquid phase already present. This is also a peritectic transformation. The balance of the masses present in each phase at the beginning and at the end of the transformation can also be written as

$$M_{\alpha i} = \frac{X_{\ell p} - X'_0}{X_{\ell p} - X_{\alpha p}} \tag{5.14}$$

$$M_{\ell i} = \frac{X'_0 - X_{\alpha p}}{X_{\ell p} - X_{\alpha p}} M \tag{5.15}$$

$$M_{\alpha f} = 0 \tag{5.16}$$

$$M_{\beta f} = \frac{X_{\ell p} - X'_0}{X_{\ell p} - X_{\beta p}} M \tag{5.17}$$

$$M_{\ell f} = \frac{X'_0 - X_{\beta p}}{X_{\ell p} - X_{\beta p}} M \tag{5.18}$$

The mass in the liquid phase has decreased during the transformation (approximately from 0.88 M to 0.75 M in Fig. 5.7). In conclusion, it should be noted that if X'_0 is only very slightly larger than $X_{\beta p}$, α can reappear in β after the disappearance of the liquid phase.

5.4.8 Peritectic and Peritectoid Transformations

During the peritectic transformation, two initial phases, one solid and the other liquid, give birth to a new solid phase. At the end of the transformation, one or the other of the initial phases has disappeared.

When the two initial phases are solid, the transformation is called, by analogy, a *peritectoid transformation.*

5.4.9 Binary Phase Diagrams in General

In reality, binary systems are often characterized by several isothermal transformations. A binary diagram therefore frequently occurs as a collection of diagrams belonging to the categories described above (Sections 5.4.3, and 5.4.5 to 5.4.7).

5.4.10 Example of the Iron-Carbon System

In the Fe-C system, carbon is in the form of graphite at equilibrium. However, under normal cooling conditions (natural cooling in the open air, for example), it is not graphite that is formed but iron carbide Fe_3C. Since this metastable compound takes several millions of years to be transformed in its turn to iron and graphite, it makes sense to speak of the Fe-Fe_3C equilibrium diagram (Fig. 5.8).

An immediate precipitation of graphite, which is useful in certain cases, can be obtained using catalysts such as silicon (Section 3.9.3).

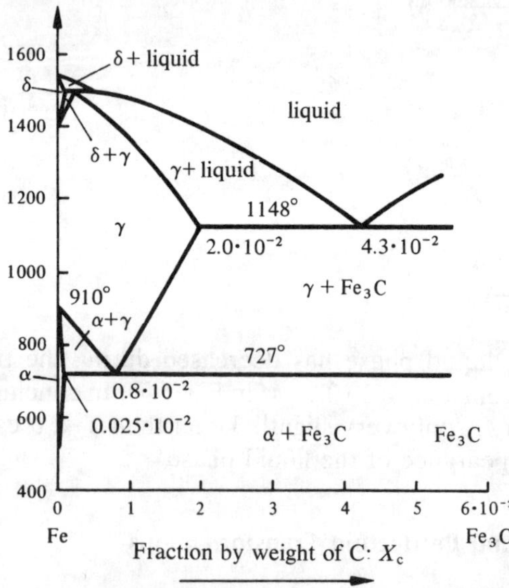

Fig. 5.8 Fe-Fe_3C Equilibrium Diagram

In addition to the liquid phase ℓ, the Fe-Fe_3C diagram involves the three phases of iron: α, γ, and δ (Table 5.1) and iron carbide. At 1495°C and over a small interval of concentration X_c, a peritectic transformation takes place corresponding to one of the following reactions:

$$\ell + Fe_\delta \rightarrow Fe_\gamma + Fe_\delta \qquad \text{if } X_c < 0.2 \cdot 10^{-2} \qquad (5.19)$$

$$\ell + Fe_\delta \rightarrow \ell + Fe_\gamma \qquad \text{if } X_c > 0.2 \cdot 10^{-2} \qquad (5.20)$$

$$\ell + Fe_\delta \rightarrow Fe_\gamma \qquad \text{if } X_c = 0.2 \cdot 10^{-2} \qquad (5.21)$$

At 1148°C and for $X_c > 2 \cdot 10^{-2}$, a eutectic transformation takes place, the composition of the eutectic mixture corresponding to $X_c = 4.3 \cdot 10^{-2}$. At 727°C, a eutectoid transformation takes place involving one of the three reactions:

$$Fe_\alpha + Fe_\gamma \rightarrow Fe_\alpha + Fe_3C \qquad \text{if } X_c < 0.8 \cdot 10^{-2} \qquad (5.22)$$

$$Fe_\gamma + Fe_3C \rightarrow Fe_\alpha + Fe_3C \qquad \text{if } X_c > 0.8 \cdot 10^{-2} \qquad (5.23)$$

$$Fe_\gamma \rightarrow Fe_\alpha + Fe_3C \qquad \text{if } X_c = 0.8 \cdot 10^{-2} \qquad (5.24)$$

These transformations are of key importance in the thermal treatment of steels. The composition $X_c = 0.8 \cdot 10^{-2}$ corresponds to the eutectoid mixture.

5.5 NONEQUILIBRIUM TRANSFORMATIONS

5.5.1 Minor Segregation

The conditions under which materials are manufactured and used always depart more or less from the succession of equilibrium states infinitely close to each other that form the strict area of validity of the equilibrium diagrams. It is therefore important to know to what extent these diagrams can still be used in the case of rapid cooling, for example.

Unfortunately, it is not possible to give an answer with general application. Sometimes, a simple deformation of the equilibrium diagram occurs; at other times, new phases and even compounds emerge such as iron carbide in the Fe-C system.

During rapid cooling, one often observes the formation of a concentration gradient inside a phase itself. This phenomenon, called *minor segregation,* can be represented on an equilibrium diagram. The cases of simple binary diagrams and peritectic binary diagrams are shown below.

When a simple binary system (Fig. 5.9) with overall composition X_0 is cooled starting from the liquid state, the grains of α, at the moment of their appearance at temperature T_1, have composition $X_{\alpha 1}$. At T_2, the dimension of these grains has increased and the external layer that is deposited at this temperature has the composition $X_{\alpha 2}$. Since the growth of the grains is too rapid for the concentration at the center to take the equilibrium value $X_{\alpha 2}$ by diffusion, a concentration gradient is established. The average concentration of B inside the grain therefore remains less than the equilibrium concentration corresponding to the solidus. It is given by the arc $y' - z'$. The system completely passes to the solid state when this arc intersects the vertical passing through X_0. The last liquid volume called upon to disappear is therefore more enriched in B for a rapid cooling ($X_{\ell 4} < X_{\ell 3}$), and on the other hand, the temperature range over which the solidification takes place is broader ($T_3 < T_4$).

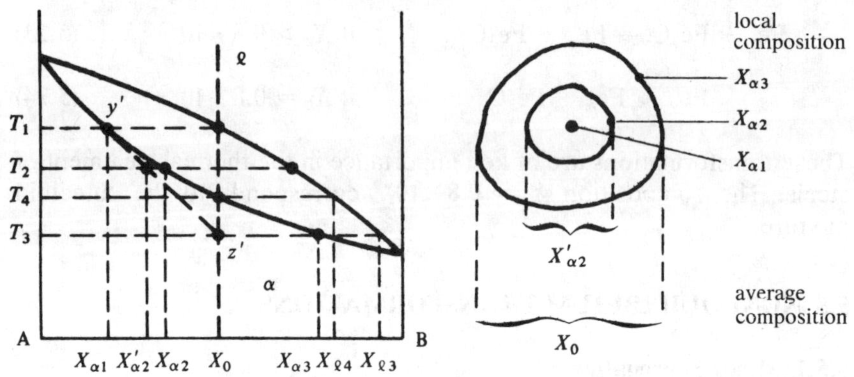

Fig. 5.9 Minor Segregation in a Simple Binary System. Equilibrium Diagram and Structure of a Grain

The behavior of a peritectic binary system is somewhat different (Fig. 5.10).

Below T_1, grains of α are formed presenting minor segregation with the average concentration of B being given by the arc $y' - z'$. When the temperature reaches T_p, the peritectic transformation starts. The new β phase forming from α and ℓ appears at the surface of the α grains. This surface layer of β then slows down the exchange between α and ℓ necessary to achieve the peritectic transformation, which because of this usually remains incomplete. The concentration of B in β is given by the arc $y'' - z''$.

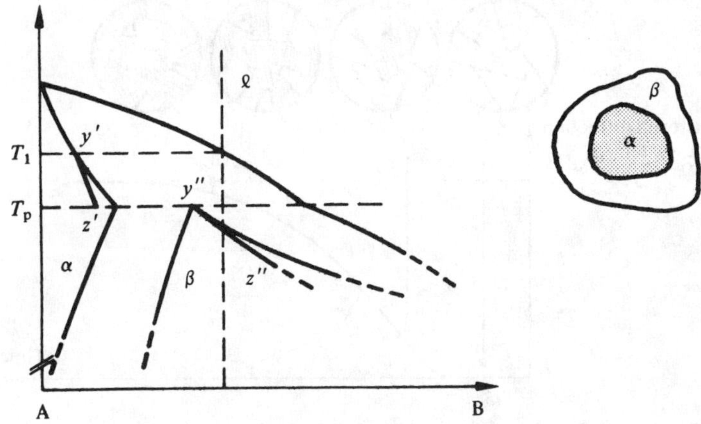

Fig. 5.10 Minor Segregation in a Binary Peritectic System. Equilibrium Diagram and Structure of a Grain

5.5.2 Isothermal Transformation Diagrams

The position of the arcs $y' - z'$ in Fig. 5.9, and of $y' - z'$ and $y'' - z''$ in Fig. 5.10, obviously depends on the speed of the cooling.

In other words, the time has to appear in the quantitative description of a nonequilibrium phase transformation. There are several ways of introducing it. The simplest corresponds to the isothermal transformation diagram. This is obtained for a given transformation at a fixed temperature by plotting the fraction of the new phase formed as a function of time (Fig. 5.11).

In practice, this diagram is established by taking a sufficient number of specimens which are simultaneously brought to the selected temperature. During the transformation, specimens are withdrawn successively at specified time intervals in which the transformation is blocked, most often by sudden cooling. It then remains to observe the extent of transformation obtained.

5.5.3 TTT Diagrams

The TTT (time, temperature, transformation) diagram combines the information of a family of isothermal transformation diagrams taken at different temperatures, the degree of transformation appearing as a parameter (Fig. 5.12).

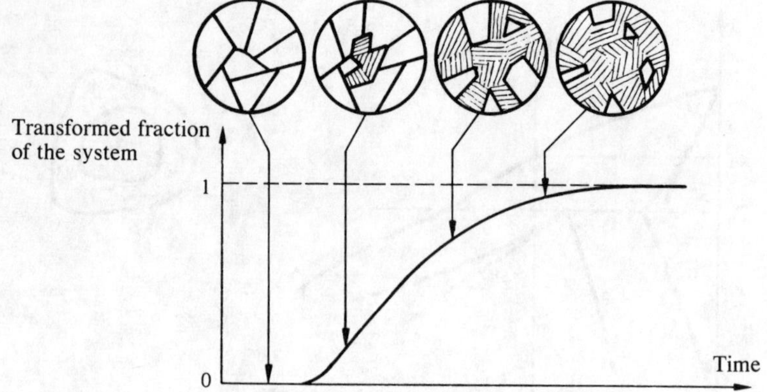

Fig. 5.11 Diagram of Isothermal Transformation and Diagrammatic Representation of the Corresponding Structure of the Material

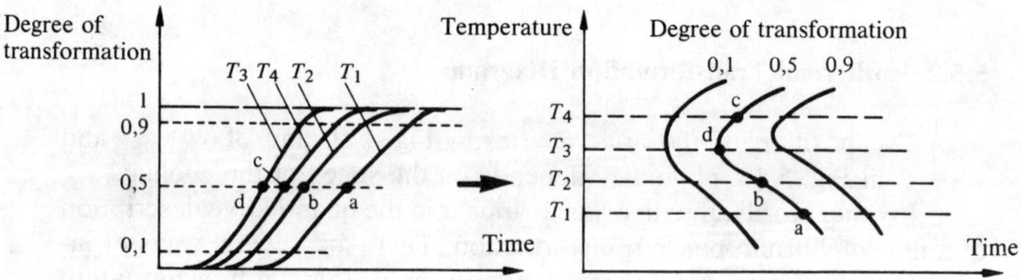

Fig. 5.12 Construction of the TTT Diagram

Chapter 6
Mechanical Properties

6.1 GENERAL CONSIDERATIONS

6.1.1 Definition

The *mechanical properties* are the properties of a material that govern its deformation when it is subjected to the action of forces.

6.1.2 Reference Mechanical Behaviors

Mechanical properties depend on a large number of characteristics, such as,

- the nature of the valence bonds (Section 1.3);
- the cohesion energy (Section 1.3.3);
- the state: crystalline or amorphous;
- the presence of imperfections in the crystal structure (Section 1.5);
- the simultaneous presence of several phases (Ch. 5);
- the chemical nature, *et cetera*.

To this list should be added the mode of application of the forces. The mechanisms of rupture (Section 6.4), for example, can differ totally depending on whether a part is subjected to a constant force or a variable force over time (vibrations).

The description of the general mechanical behavior of materials is simplified by the fact that three extremes are used as references. These are elastic behavior (Section 6.2), plastic behavior (Section 6.3), and viscoelastic behavior (Section 6.5).

6.2 ELASTIC BEHAVIOR

6.2.1 Definitions

A material has an *elastic behavior* if it returns to its initial dimensions as soon as the action of the applied forces is removed. In other words, elastic deformation is, by definition, *reversible*.

Four parameters are used to describe elastic behavior. These are: the elastic modulus (or Young's modulus) E_Y, the shear modulus G, the modulus of compressibility κ, and Poisson's ratio ν.

6.2.2 Elasticity Modulus E_Y

The elastic modulus E_Y is defined by Hooke's law (6.1) relating the axial stress σ (it acts along only one direction) to ε, the resulting relative elongation (Fig. 6.1).

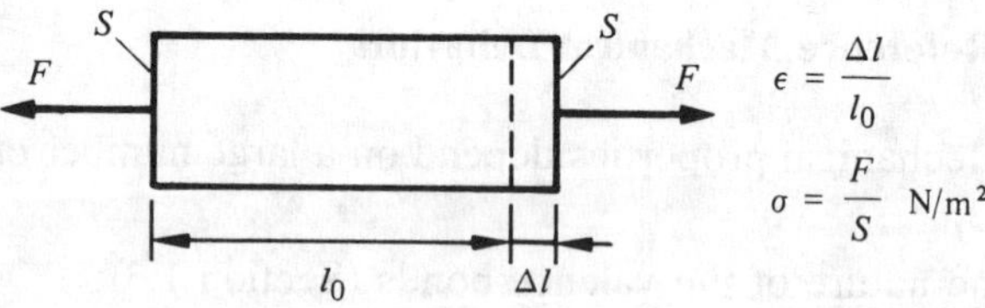

Fig. 6.1

$$E_Y = \frac{\sigma}{\varepsilon} \tag{6.1}$$

The elastic modulus is a measure of the rigidity of a material. It is directly related to the valence forces and increases with the intensity of these. For small deformations (ε = 1 to 2%), E_Y is a constant. It does not even depend on whether σ is acting as a compression or as a traction because the cohesive energy passes through a minimum for a zero deformation (Fig. 1.14).

6.2.3 Shear Modulus G

The shear modulus G is defined as the ratio of the shear stress τ to the shear angle γ that results (Fig. 6.2).

$$G = \frac{\tau}{\gamma} \tag{6.2}$$

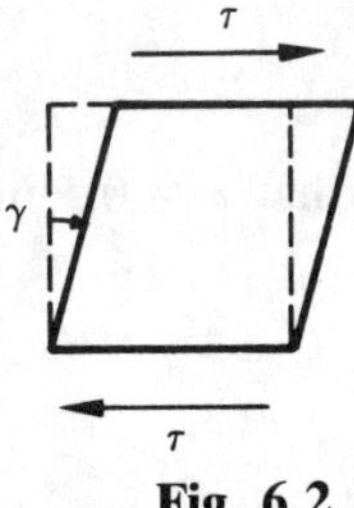

Fig. 6.2

Similar to E_Y, the shear modulus is a constant for small deformations.

6.2.4 Modulus of Compressibility κ

The modulus of compressibility κ is defined by the relationship:

$$\kappa = -\frac{1}{\Omega}\frac{\Delta\Omega}{\Delta P} \quad \text{N/m}^2 \tag{6.3}$$

where $\Delta\Omega = \Omega - \Omega_0$, Ω_0 being the initial volume and Ω being the volume after the application of a hydrostatic pressure ΔP. The compressibility modulus is directly related to the cohesive energy (1.41).

6.2.5 Poisson's Ratio ν

Poisson's ratio ν is defined as the ratio of the lateral contraction to the longitudinal strain produced by an axial force (Fig. 6.3).

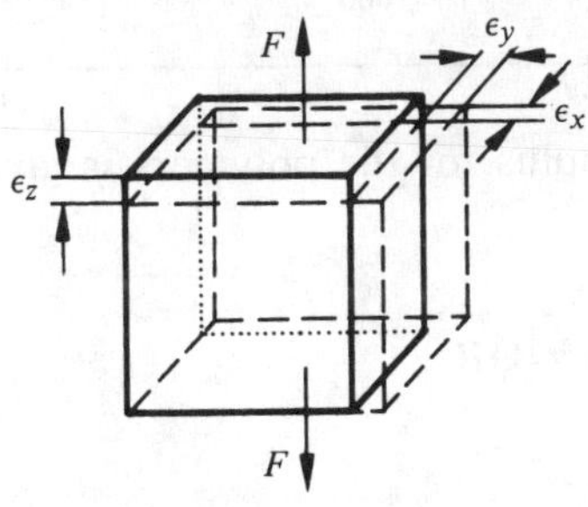

Fig. 6.3

In an isostropic body $\varepsilon_x = \varepsilon_y$. By giving ε a plus sign for an elongation and a minus sign for a contraction, we have

$$\nu = \frac{-\varepsilon_x}{\varepsilon_z} \tag{6.4}$$

It can easily be shown that $\nu = 0.5$ in the theoretical case of an incompressible material.

6.2.6 Remark

Only two of the parameters out of the four that were just defined have to be considered as fundamental because the other two can be deduced from them by relationships that can be simply expressed in the case of isotropic materials.

6.2.7 Numerical Values of E_Y, G, and ν

Table 6.4

Material	E_Y $10^9\ N/m^2$	G $10^9\ N/m^2$	ν
Aluminum	70	25	0.33
Copper	120	45	0.36
Iron	150	53	0.28
Soft steel	205	82	0.26
Lead	16	6	0.40
Nickel	205	74	0.30
Tungsten	355	148	—
Silicon carbide	344	—	—
Sintered aluminum	320	—	—
Glass	69	22	0.23
Diamond	1000	—	—

The elastic modulus for the polymers is given in Tables 4.75 and 4.76.

6.3 PLASTIC BEHAVIOR

6.3.1 Definition

A crystalline material has a *plastic behavior* when, acted on by a stress exceeding a certain minimum threshold, it has a permanent residual deformation called the *plastic deformation* after the stress has been removed.

Seen from the outside, a similar behavior can be observed with amorphous materials. Based on different mechanisms, it has a different name (viscous behavior, Section 6.5).

6.3.2 Mechanism of Plastic Deformation

Plastic deformation corresponds to a displacement of atoms. This is mainly the result of slipping of the reticular planes caused by movement of dislocations (Table 1.33). In accordance with the results obtained in Section 1.5.7, these slipping movements take place in directions for which the atoms are at a minimum distance from each other (minimum Burgers vector).

6.3.3 Tensile Test

The tensile test is the simplest experimental method for studying plastic behavior. It is carried out using the apparatus shown in Fig. 6.5.

A test piece of the material to be studied is held between two clamps. A mechanism makes it possible to move these apart while the elongation of the test piece and the force to which it is subjected are recorded. This recording, called the stress-strain curve, is shown diagrammatically in Fig. 6.6.

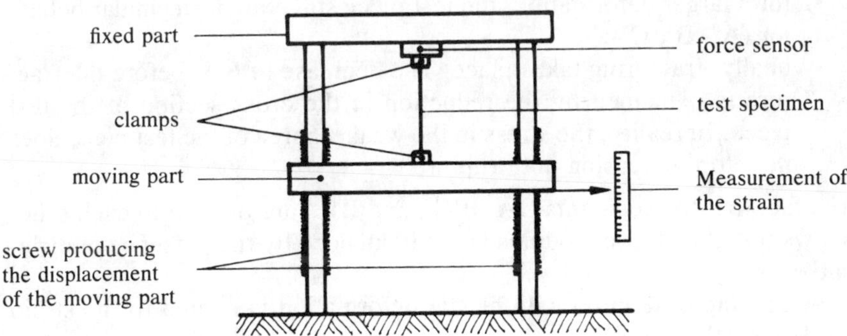

Fig. 6.5 Schematic of a Tensile Test Apparatus

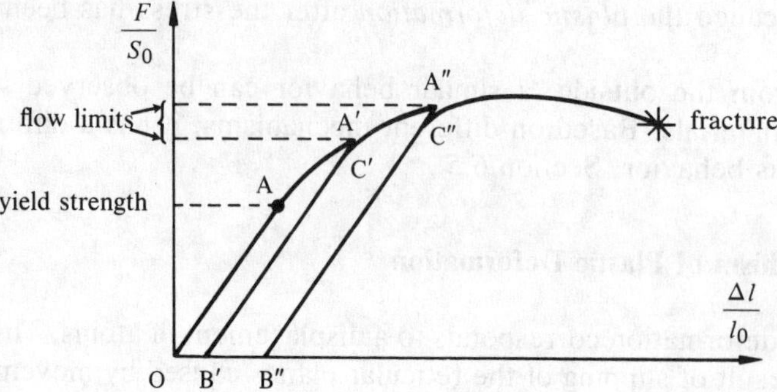

Fig. 6.6 Typical Stress-Strain Curve: *F*, Applied Force; S_0, Initial Cross Section of the Specimen; l_0, Initial Length of the Specimen; Δl, Absolute Elongation

This diagram can be interpreted in the following manner:

- up to the point A, the specimen behaves in an elastic manner following Hooke's law (6.1); the stress corresponding to the point A is called the *yield strength* σ_{el};
- beyond A, the force is no longer a linear function of the strain and if having arrived at the point A′, the force is released, the test piece has a permanent strain $\Delta l/l_0 = OB'$;
- by stressing the elongated test piece even more, it again behaves in an elastic manner up to a point C′ close to A′, i.e., over a wider range than for the first traction. The stress corresponding to C′ is called the *flow limit* σ_{fl}. This depends on the elongation;
- for a larger deformation, the test piece still exhibits a similar behavior (A″, B″, C″);
- finally, fracturing takes place. The decrease in F/S_0 before the fracture is explained by the reduction in the cross section of the test piece. In reality, the stress in the weakest area of the test piece does not stop increasing until rupture.

The fact that the segments OA, B′C′, and B″C″ are parallel to each other shows that the elastic modulus is not influenced by the presence of dislocations.

The increase in σ_{fl} with plastic deformation is related to the strain hardening (Section 6.3.6).

The stress-strain curve is very sensitive to the purity of the materials and the size of the grains. It is a very widely used control method.

6.3.4 Conventional Yield Strength and Flow Limits

It is difficult to determine σ_{el} and σ_{fl} from the stress-strain curve because the place where the function of Fig. 6.6 departs from a straight line is poorly defined. This is why in practice so-called *conventional* values of σ_{el} and σ_{fl} are used. By definition, they correspond to an agreed $\Delta l/l_0$, for example 0.2% or 0.5%.

6.3.5 Rational Stress-Strain Curve

The *rational stress-strain curve* is the representation of the actual stress σ *versus* ε.

This curve gives a better reflection of reality than the ordinary stress-strain curve (Fig. 6.6) but it cannot be directly obtained from standard tensile test apparatus which limits its use in practice.

Let l and S be the length and the cross section (variable) of the test piece during the test. If the test piece is regularly deformed and its density does not vary, we have

$$S_0 l_0 = Sl \tag{6.5}$$

By setting

$$e = \frac{l - l_0}{l_0} = \frac{\Delta l}{l_0} \tag{6.6}$$

the stress σ can be expressed by

$$\sigma = \frac{F}{S} = \frac{Fl}{S_0 l_0} = \frac{F}{S_0}(1 + e) \tag{6.7}$$

On the other hand,

$$\varepsilon = \int_{l_0}^{l_0+\Delta l} \frac{\mathrm{d}l}{l} = \log \frac{l_0 + \Delta l}{l_0} = \log(1 + e) \tag{6.8}$$

The relationships (6.7) and (6.8) make it possible to plot the rational stress-strain curve (Fig. 6.7) from the ordinary stress-strain curve in the case when the assumptions leading to (6.6) are satisfied.

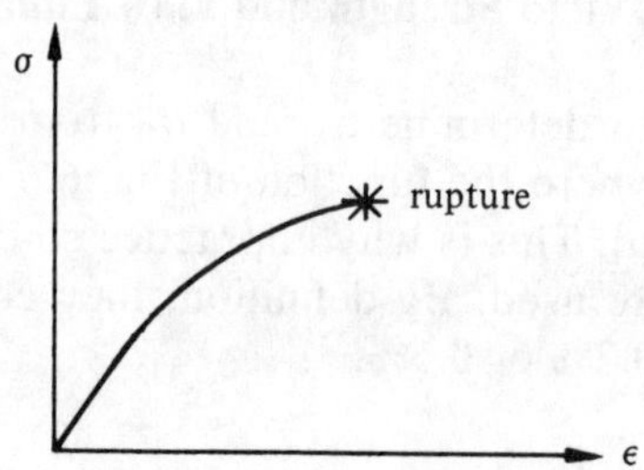

Fig. 6.7 Typical Rational Stress-Strain Curve

The slope $d\sigma/d\varepsilon$ is called the *consolidation ratio*. It is a measure of the capacity for strain hardening (Section 6.3.6).

6.3.6 Increasing the Yield Strength

In many applications plastic deformation cannot be tolerated, which explains the interest in procedures capable of increasing the yield strength of materials. In order to achieve this end, it is necessary to hinder the displacement of the dislocations (Section 1.5.4 *et seq.*) as much as possible. This is done in several ways:

- by *strain hardening,* i.e., by preliminary plastic deformation. During this process, the number of dislocations increases considerably; they eventually form a kind of jumbled structure within which they block each other by interaction of their compression and depression zones. Figure 6.8 shows these zones for the case of an edge dislocation and the diagrammatic representation of this dislocation by a T. The vertical bar of the T corresponds to the additional reticular plane and therefore to the compression zone. Fig. 6.9 shows the interaction of two parallel edge dislocations;
- by producing a *solid solution* with a foreign element. Whether this is placed in an interstitial or substitution position, it produces a local strain of the crystal lattice and therefore a local stress field acting on the dislocations. If the temperature is high enough, the foreign atoms can diffuse and accumulate in the vicinity of the dislocations creating what is called the Cottrell cloud;
- by *precipitation of a new phase*. The precipitates are either coherent, i.e., their crystal axes are parallel to the axes of the phase in which they are formed (Fig. 6.10), or incoherent (Fig. 6.11).

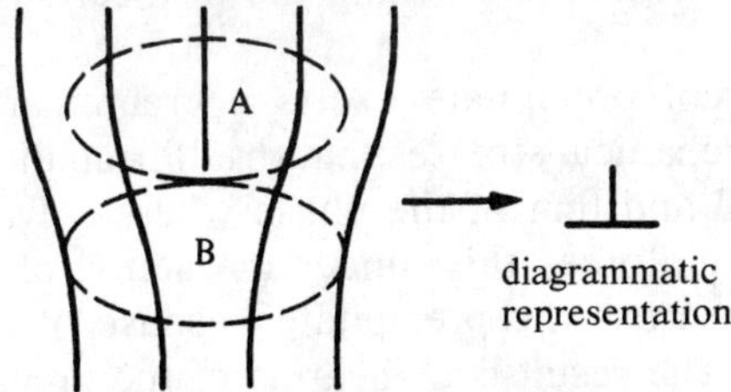

Fig. 6.8 A, Compression Zone; B, Depression Zone

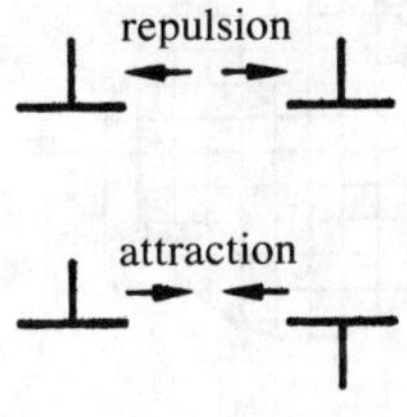

Fig. 6.9

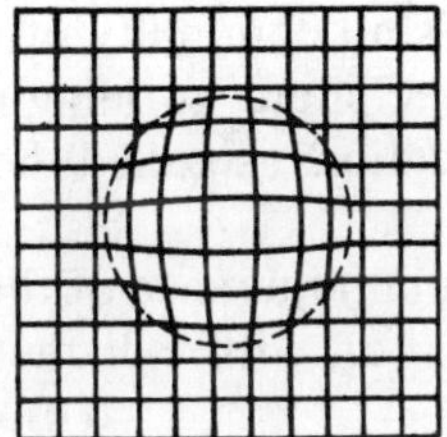

Fig. 6.10 Coherent Precipitate

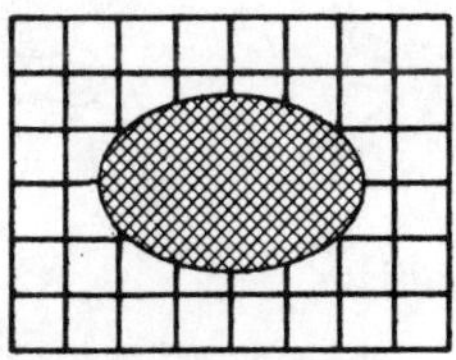

Fig. 6.11 Incoherent Precipitate

The mechanisms for blocking the dislocations are different in the two cases.

The coherent precipitate exerts a preliminary braking action by means of the mechanical stresses that result from the differences between its own unit cell and that of the phase where it was formed. A second braking action reinforces this: under the action of sufficient stress, the dislocations can cross the precipitate because of the continuity of the crystal lattices. The result is a shearing of the precipitate increasing the mechanical stresses around it (Fig. 6.12).

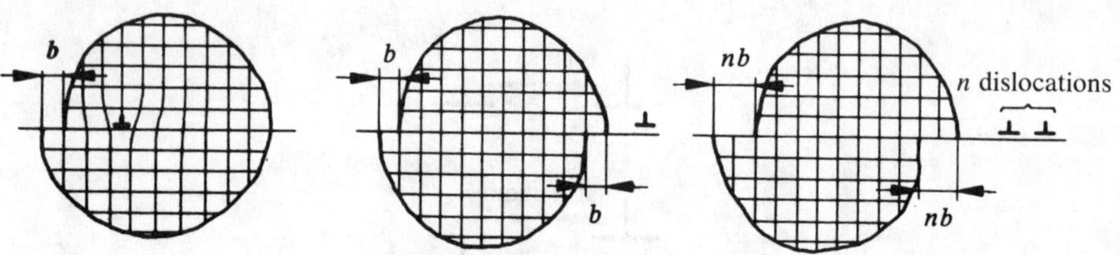

Fig. 6.12 Shearing of a Coherent Precipitate

On the other hand, an incoherent precipitate cannot be crossed by a dislocation because there is no continuity of the slip plane. The dislocation therefore skirts the precipitate, forming new dislocations around it. This behavior of the dislocations is called the *Orowan mechanism* (Fig. 6.13).

In practice, the maximum increase of the yield strength is obtained with very fine precipitates, i.e., with a diameter less than 10 nm.

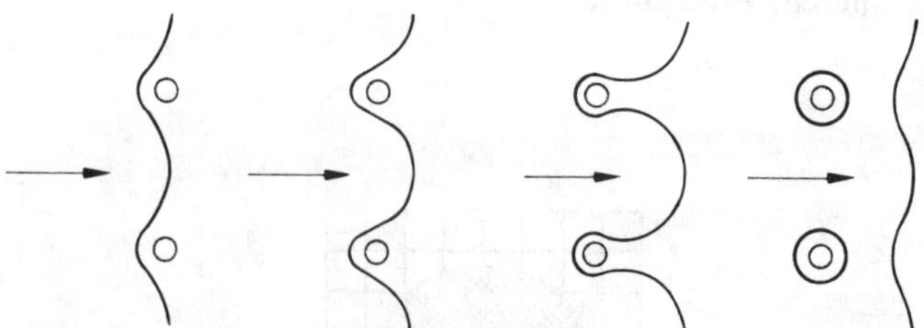

Fig. 6.13 The Orowan Mechanism

6.4 FRACTURE

6.4.1 Introduction

Only the fracture of metals is considered here. To a certain extent, it can serve as a reference for the study of fracture in other materials. A distinction is made between fragile fracture, ductile fracture, and fatigue fracture. The type of fracture observed depends not only on the metal considered and the shape of the part but also on how the forces are applied.

6.4.2 Brittle Fracture

The characteristics of brittle fracture are as follows:

- it results from the extremely rapid propagation of a crack;
- it is produced at the maximum of the applied force;
- the part under stress does not exhibit any (or very little) plastic deformation just before the rupture;
- the rupture zone has a granular, often shiny, appearance.

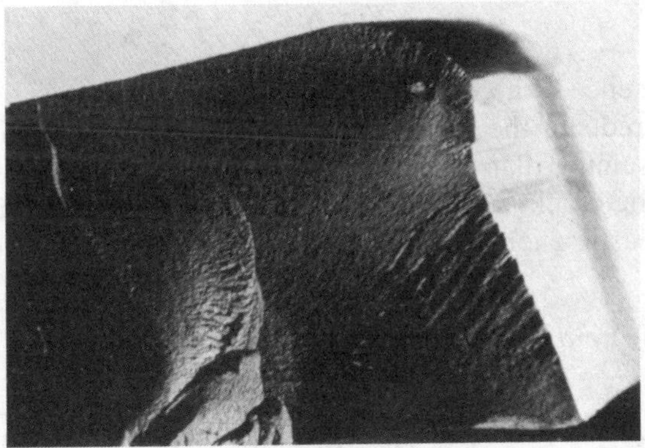

Fig. 6.14 Appearance of a Brittle Fracture (Steel)

Two mechanisms are involved in brittle fracture: the slipping of entire crystal planes (cleavage planes), and slipping along the grain boundaries where resistance has been reduced by the accumulation of impurities.

6.4.3 Griffith's Model

This simple model of brittle fracture uses the assumption of an isotropic material.

In a brittle material, small cracks are permanently present; they are deformed under the action of a force so that they take an elliptical shape (Fig. 6.15).

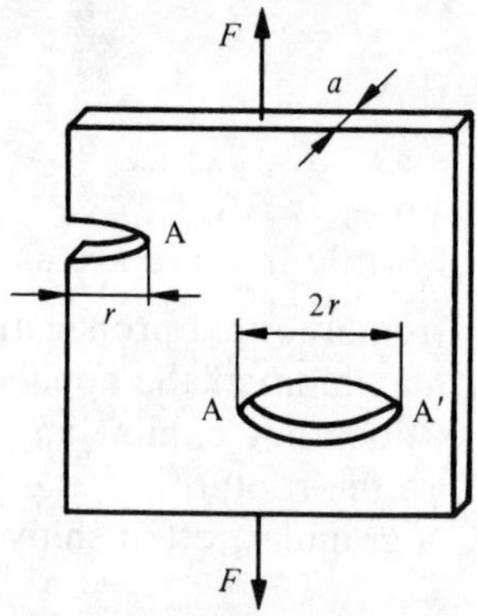

Fig. 6.15

The presence of a crack has two effects.

First, it reduces the elastic energy stored in the neighborhood of the crack. By assuming that this energy vanishes inside a circle of radius $\boldsymbol{r}$ centered on the crack, whereas it is not modified outside this circle, the variation in elastic energy in the plate of Fig. 6.15 is equal to

$$\Delta W_{\mathrm{el}} = -\pi \boldsymbol{r}^2 a \frac{\sigma^2}{E_{\mathrm{Y}}} \tag{6.9}$$

Second, it increases the surface energy of the plate by increasing its total area by a value ΔW_{s} given by

$$\Delta W_{\mathrm{s}} = 4a\boldsymbol{r}\gamma, \tag{6.10}$$

where γ(N/m) is the energy per unit area.

The evolution of the crack depends on the variation of the sum:

$$\Delta W = \Delta W_{\mathrm{el}} + \Delta W_{\mathrm{s}} \tag{6.11}$$

versus $\boldsymbol{r}$ (Fig. 6.16).

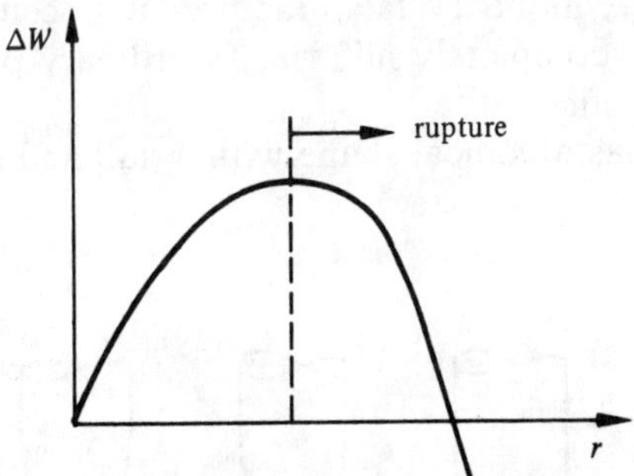

Fig. 6.16 Graph of ΔW According to (6.11)

As long as an increase in r produces an increase in ΔW, the crack has no tendency to widen. On the other hand, as soon as an increase in r reduces the energy of the plate, the crack increases leading to rupture in a very short time. This is produced for a stress σ_r given by

$$\frac{\partial \Delta W}{\partial r} = 0 = 2\pi ra \frac{\sigma^2}{E_Y} + 4a\gamma \qquad (6.12)$$

from which we have

$$\sigma_r = \sqrt{\frac{2\gamma E_Y}{\pi r}} \qquad (6.13)$$

During this calculation, it has been implicitly assumed that there was no plastic deformation preceding the fracture. In crystalline media, such a deformation is always produced in the vicinity of the points A and A′. For a given stress, the result is an increase in the maximum width of the crack that does not cause a fracture.

6.4.4 Ductile Fracture

Ductile fracture is characterized by a significant plastic deformation of the test piece, mainly localized in a necking zone. The mechanism leading to fracture is shown diagrammatically in Fig. 6.17.

Small cavities are formed in the necking zone as soon as it appears. With the deformation, the initial axial stress is progressively transformed into a three-dimensional stress state. The initial cavities are deformed and merge, creating larger cavities. These cavities reduce the cross section of the metal until fracture takes place.

In high-purity monocrystals, fracture only occurs when the necking zone has become completely filiform. In ordinary polycrystalline specimens, fracture occurs earlier.

The break has a conical shape with a dull and fibrous surface (Fig. 6.18).

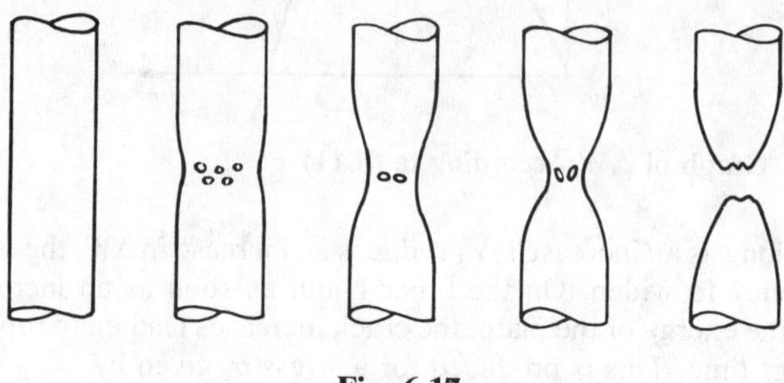

Fig. 6.17

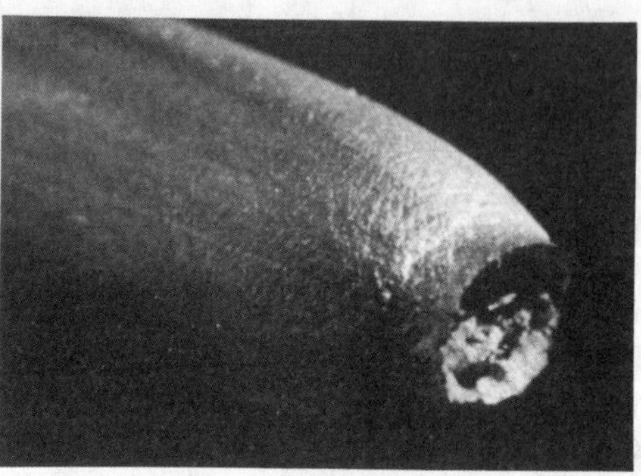

Fig. 6.18 Appearance of a Ductile Fracture (Annealed Copper)

6.4.5 Fatigue Fracture

Fatigue fracture that occurs following the application of a certain number of alternating stresses. It occurs at a value of the stress less than that which causes fracture under a static load. The *endurance limit* σ_f is the maximum stress that does not cause fracture regardless of the number of alternating stresses (Fig. 6.19).

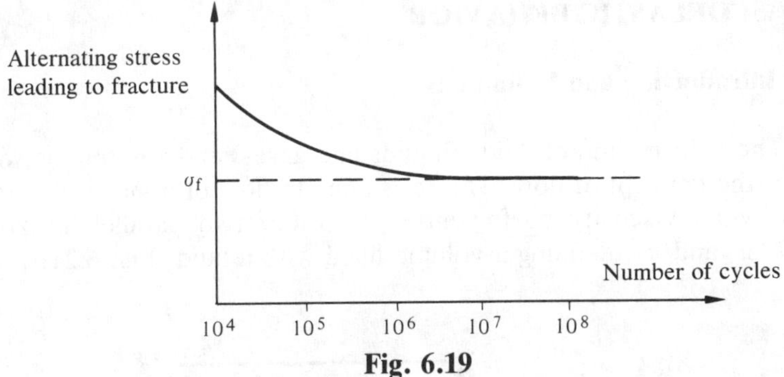

Fig. 6.19

In iron alloys, we have approximately

$$\sigma_f \sim \frac{\sigma_{el}}{2} \tag{6.14}$$

Fatigue fracture occurs in two stages. In the first stage, the initial crack grows slowly over successive periods separated by pauses. This stage corresponds to zone a of the break (Fig. 6.20) which has the form of a shiny shell. The veins of this shell indicate a pause in the evolution of the crack.

When the intact cross section of the specimen becomes too small to withstand the applied forces by itself, the crack suddenly develops very rapidly: in fact the last intact section undergoes a brittle fracture. This second stage corresponds to zone b of the break which reveals the characteristic granular appearance of brittle fracture.

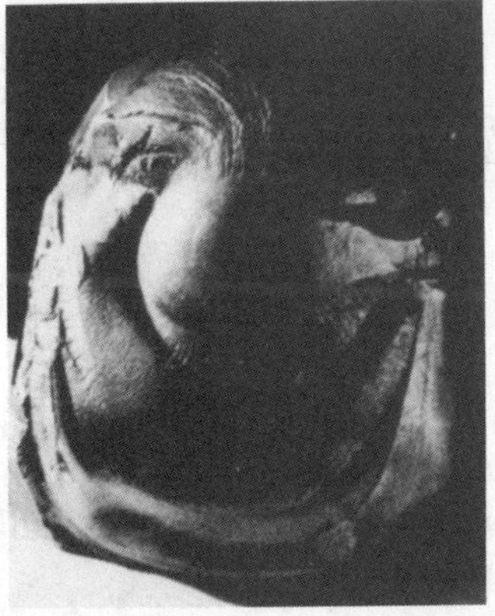

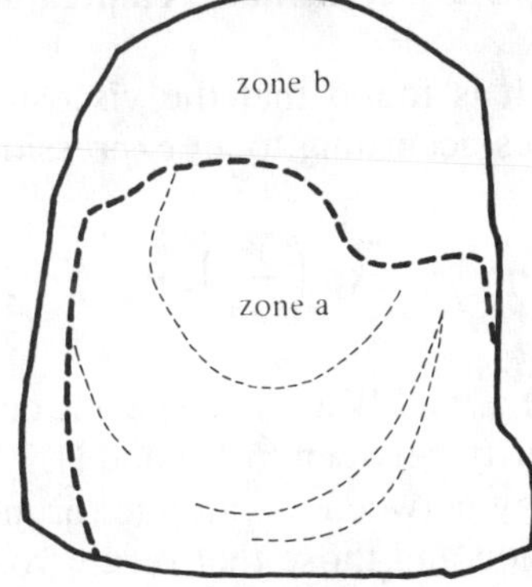

Fig. 6.20

6.5 VISCOELASTIC BEHAVIOR

6.5.1 Introduction and Definitions

The deformation of fluids (liquids and gases) and amorphous solids implies the concept of flow. The resistance to flow of a material is measured by its viscosity coefficient η. Consider two parallel horizontal planes, π_1 and π_2, defining a volume filled with a fluid (Fig. 6.21).

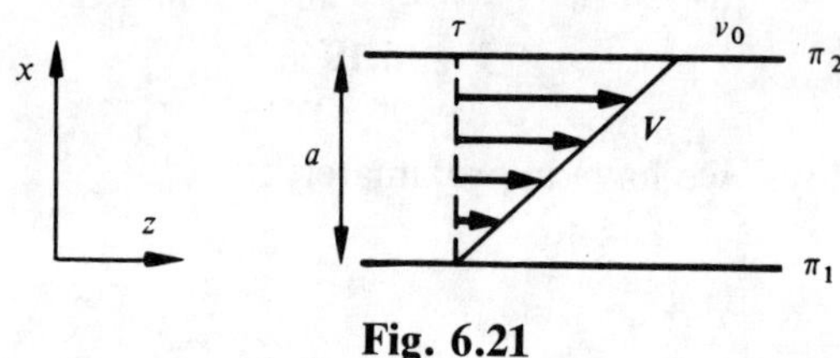

Fig. 6.21

The plane π_1 is stationary whereas π_2 moves with a velocity v_0 under the action of a tangential stress τ. The flow is assumed to be laminar.

The viscosity coefficient η is the ratio of the stress producing the flow to the velocity gradient appearing in the fluid:

$$\eta = \frac{\tau}{\mathrm{d}v/\mathrm{d}x} \tag{6.15}$$

A fluid in which η is independent of τ and time is called *newtonian*.

6.5.2 Glass-Transition Temperature

It is found that the viscosity increases when the temperature decreases according to an expression of the type:

$$\eta = \eta_0 \exp\left(\frac{W}{k_B T}\right), \tag{6.16}$$

where η_0 and W are two specific constants of the material considered. The similarity between (6.16) and (1.79) is not fortuitous. There is an obvious analogy between the atomic (or molecular) movements corresponding to diffusion and those that result from a mechanical stress.

Below a certain temperature T_v, the atoms lose their capacity to move because the thermal energy is not sufficient to weaken the valence bonds to the point of allowing these movements. The elastic properties of these bonds therefore reemerge at the macroscopic level: the materials recover their elastic behavior and become brittle.

T_v is called the *glass-transition temperature*.

6.5.3 Nonnewtonian Materials

A large number of materials, including polymers, are not newtonian. Their mechanical behavior may be described by phenomenological models in which the viscous and elastic behaviors are superimposed (viscoelasticity). These models as well as a study of the mechanical characteristics of amorphous polymers based on the properties of the macromolecule chains themselves, can be found in [67]. Figure 6.22 is an example of the mechanical behavior of a viscoelastic material.

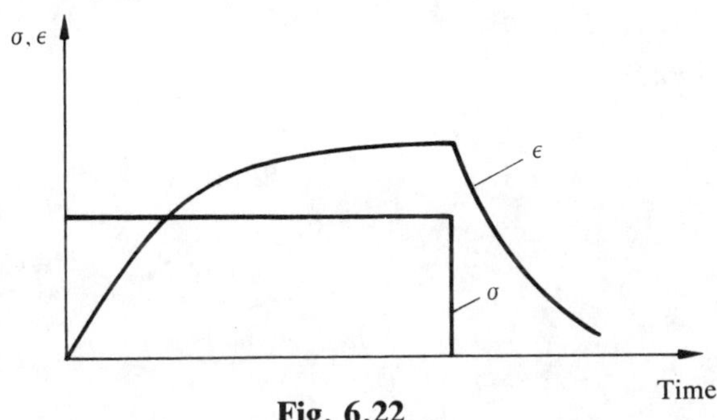

Fig. 6.22

Chapter 7
Appendices

7.1 THE LAGUERRE POLYNOMIALS

The Laguerre polynomials $L_n(X)$ with degree n are given by the expression:

$$L_n(x) = \exp(x) \frac{d^n[x^n \exp(-x)]}{dx^n} \tag{7.1}$$

The associated Laguerre polynomials $L_n^s(X)$, of order s and degree $n - s$, can be deduced from the Laguerre polynomials by

$$L_n^s(x) = \frac{d^s L_n(x)}{dx^s} \tag{7.2}$$

7.2 MAXWELL'S DISTRIBUTION

7.2.1 Area of Application

Maxwell's distribution governs the behavior of a system of classical point particles

- in thermodynamic equilibrium,
- not interacting with each other except during collisions assumed to be elastic, and
- not subjected to the action of an external potential.

Such a system of particles is called a perfect gas. The atoms of a perfect gas are often considered as particles *without mutual interaction*. It should be understood by this that they do not create any potential responsible for interaction forces with a certain range, such as the valence forces, for example. However, these particles certainly have some form of interaction at the time of collision! Otherwise, it would be impossible to establish any thermodynamic equilibrium.

7.2.2 Derivation of Maxwell's Distribution

Maxwell's distribution answers the following question in a precise manner. What is the number dN of particles possessing, at a given time, a velocity with components that lie between v_x and $v_x + dv_x$; v_y and $v_y + dv_y$; v_z and $v_z + dv_z$. The answer is in the form:

$$dN = Nf(v_x, v_y, v_z)\, dv_x\, dv_y\, dv_z, \tag{7.3}$$

where N is the total number of particles and f is the distribution to be found.

The probability that a particle has a certain velocity component along one axis is totally independent of the components that it possesses along the two other axes. The validity of this postulate of Maxwell can be verified by the Boltzmann distribution (Section 7.3) established by other methods, of which Maxwell's distribution is a particular case. We therefore set

$$f(v_x, v_y, v_z) = g(v_x)\, g(v_y)\, g(v_z) \tag{7.4}$$

The g functions represent the velocity distributions along each axis. Let

$$v = \sqrt{v_x^2 + v_y^2 + v_z^2} \tag{7.5}$$

be the scalar speed of a particle. The probability for a particle that this speed lies within any interval (v, $v + dv$) must be independent of the direction, otherwise we are dealing with a gas flowing spontaneously in certain directions, which is contrary to the proposed condition of equilibrium. Consequently,

$$f(v_x, v_y, v_z) = f(v) = g(v_x)\, g(v_y)\, g(v_z) \tag{7.6}$$

This equation, apparently very general, is sufficient to establish Maxwell's distribution, the use of additional arguments being necessary only to determine the value of two parameters.

The functions f and g satisfying (7.6) are obtained by the following procedure: we take the natural logarithm of both sides of (7.6):

$$\ln f(v_x, v_y, v_z) = \ln f(v) = \ln g(v_x) + \ln g(v_y) + \ln g(v_z), \tag{7.7}$$

and then the partial derivative of (7.7) with respect to v_x:

$$\left.\frac{\partial \ln f(v)}{\partial v_x}\right|_{v_y, v_z} = \frac{\mathrm{d} \ln f(v)}{\mathrm{d}v} \frac{\partial v}{\partial v_x} = \frac{v_x}{v} \frac{\mathrm{d} \ln f(v)}{\mathrm{d}v} = \frac{\mathrm{d} \ln g(v_x)}{\mathrm{d}v_x} \tag{7.8}$$

This equation has its variables separated. It can be written in the form:

$$\frac{1}{v} \frac{\mathrm{d} \ln f(v)}{\mathrm{d}v} = \frac{1}{v_x} \frac{\mathrm{d} \ln g(v_x)}{\mathrm{d}v_x} \tag{7.9}$$

By successively taking the derivatives of both sides of (7.7) with respect to v_x, v_y and v_z, we similarly obtain

$$\frac{1}{v} \frac{\mathrm{d} \ln f(v)}{\mathrm{d}v} = \frac{1}{v_x} \frac{\mathrm{d} \ln g(v_x)}{\mathrm{d}v_x} = \frac{1}{v_y} \frac{\mathrm{d} \ln g(v_y)}{\mathrm{d}v_y} = \frac{1}{v_z} \frac{\mathrm{d} \ln g(v_z)}{\mathrm{d}v_z} \tag{7.10}$$

Each term of (7.10) depends on a different variable. All these terms are therefore equal to one and the same constant which will be called $-\beta$. By integrating (7.9), we thus obtain

$$g(v_x) = C \exp(-\beta v_x^2) \tag{7.11}$$

The integration constant C is determined by the normalization condition:

$$\int_{-\infty}^{+\infty} g(v_x)\, \mathrm{d}v_x = 1 \tag{7.12}$$

This integral is calculated by mathematical relation (7.8.1). We have

$$C = \sqrt{\frac{\beta}{\pi}} \tag{7.13}$$

It remains to determine the constant β. The theory of perfect gases shows that the energy of the particles, which are reduced to their kinetic energy of translation, is proportional to the absolute temperature and is equal to $\tfrac{1}{2}k_{\mathrm{B}}T$ per degree of freedom. Consequently,

$$\frac{1}{2} m\overline{v_x^2} = \frac{1}{2} k_{\mathrm{B}} T = \int_{-\infty}^{+\infty} \frac{1}{2} m v_x^2 g(v_x)\, \mathrm{d}v_x \tag{7.14}$$

By substituting (7.11) and (7.13) into (7.14) and performing the integration, we obtain

$$\beta = \frac{m}{2k_B T}, \tag{7.15}$$

from which we have

$$g(v_x) = \sqrt{\frac{m}{2\pi k_B T}} \exp\left(-\frac{mv_x^2}{2k_B T}\right) \tag{7.16}$$

The expressions for $g(v_y)$ and $g(v_z)$ are similar to (7.16). The distribution $f(v_x, v_y, v_z)$ is therefore known from (7.6):

$$f(v_x, v_y, v_z) = \left(\frac{m}{2\pi k_B T}\right)^{3/2} \exp\left(-\frac{m}{2k_B T}(v_x^2 + v_y^2 + v_z^2)\right) \tag{7.17}$$

In velocity space, all representative points of the particles with the same scalar speed v are situated on a sphere with radius v. The number of particles with scalar speeds between v and $v + dv$ is proportional to the volume $4\pi v^2 dv$ situated between the spheres with radius v and radius $v + dv$. This property of the velocity space makes it possible to express (7.3) only in terms of v. We have

$$dN = N 4\pi \left(\frac{m}{2\pi k_B T}\right)^{3/2} v^2 \exp\left(-\frac{mv^2}{2k_B T}\right) dv \tag{7.18}$$

This is Maxwell's distribution.

The meaning of dN differs slightly in (7.3) and in (7.18) where dN represents the number of particles with a scalar speed that lies between v and $v + dv$. It can easily be shown that (7.18) satisfies the normalization relationship:

$$\frac{1}{N}\int_0^\infty dN = 1 \tag{7.19}$$

7.2.3 Graphic Representation

Fig. 7.1 represents Maxwell's distribution for nitrogen at different temperatures.

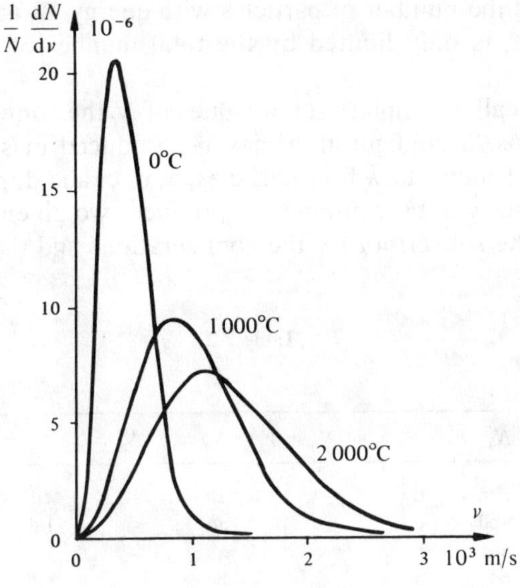

Fig. 7.1

7.3 BOLTZMANN'S DISTRIBUTION

7.3.1 Area of Application

Boltzmann's distribution governs the behavior of a system of point particles with classical characteristics,

- in thermodynamic equilibrium,
- without mutual interaction in the sense of Section 7.2.1, and
- capable of being subjected to the action of an external potential.

The last point gives a greater generality to Boltzmann's distribution compared to Maxwell's distribution.

7.3.2 Derivative of the Boltzmann Distribution

Let N be the total number of particles and W_s be the sum of the total energies of these particles. The quantities N and W_s are constants because the system is closed and isolated.

Let W_j, with $j = 1, 2, \ldots, m$, where m can tend to infinity, be the set of possible energies for the particles. These energies are either discrete values or intervals W, $W + dW$ in the case of a continuous distribution.

Let N_j be the number of particles with energy W_j at the same time. The number N_j is only limited by the total number of particles in the system.

We shall call a complete set of values of N_j the *configuration*. Apart from exceptions, a configuration may be produced in several different ways. Table 7.2 shows how four particles, a, b, c, and d, possessing three possible energies can be combined to produce two given configurations. We shall use the subscript i for the configurations and the subscript j for the energies.

Table 7.2

$N_1 = 3$	$N_2 = 1$	$N_3 = 0$	$N_1 = 2$	$N_2 = 0$	$N_3 = 2$
abc	d	–	ab	–	cd
abd	c	–	ac	–	bd
acd	b	–	ad	–	bc
bcd	a	–	bc	–	ad
			bd	–	ac
			cd	–	ab

Statistical thermodynamics shows that for N sufficiently large, the average occupation of an energy level W_j is equal to the most probable occupation of this level. The determination of the set of average N_j therefore reduces to finding the configuration that can be produced in the largest number of possible ways.

The only characteristic that makes it possible to distinguish the particles from each other is their energy. A configuration i may therefore be reduced in ν_i different ways, where

$$\nu_i = \frac{N!}{N_1!N_2! \ . \ . \ . \ N_m!} \tag{7.20}$$

Therefore, the problem amounts to finding the set of N_j making (7.20) a maximum, under the conditions:

$$N = \sum_j N_j \tag{7.21}$$

and

$$W = \sum_j W_j N_j \tag{7.22}$$

This is a problem which is solved by the Lagrange method for constrained maxima. For practical reasons, a search is not made for the maximum of ν_i but for the maximum of ln ν_i which amounts to the same thing. Let Φ be the Lagrange function. By definition,

$$\Phi = \ln \nu_i + \lambda_1 \left(N - \sum_j N_j\right) + \lambda_2 \left(W - \sum_j N_j W_j\right) \tag{7.23}$$

where λ_1 and λ_2 are the Lagrange multipliers. From (7.20), we have

$$\Phi = \ln N! - \sum_j \ln N_j! + \lambda_1 \left(N - \sum_j N_j\right) + \lambda_2 \left(W - \sum_j N_j W_j\right) \tag{7.24}$$

The set of required N_j is obtained by solving the system of equations (7.21), (7.22), and (7.25).

$$\frac{\partial \Phi}{\partial N_j} = 0 = -\frac{\partial}{\partial N_j} \ln N_j! - \lambda_1 - \lambda_2 W_j \tag{7.25}$$

For N_j sufficiently large, it is possible to use Stirling's formula in the form:

$$\ln N! \cong N \ln N, \tag{7.26}$$

which, introduced in (7.25) gives

$$N_j = \exp\left[-(\lambda_1 + \lambda_2 W_j)\right] \tag{7.27}$$

λ_1 is easily eliminated with the help of condition (7.21). We obtain

$$\frac{N_j}{N} = \frac{\exp(-\lambda_2 W_j)}{\sum_j \exp(-\lambda_2 W_j)} \tag{7.28}$$

It remains to find the value of λ_2. For this it is necessary to resort to physical arguments. According to Section 7.2, equation (7.28) is valid in the case of a perfect gas. In one-dimensional space, the average energy $\bar{W}$ of the particles of such a gas is equal to $\frac{1}{2} k_B T$ so that

$$\bar{W} = \frac{1}{2} k_B T = \frac{\sum_j N_j W_j}{N} \tag{7.29}$$

If we take the momentum p_j of the particles as a variable instead of their energy W_j and if we replace the sum in (7.29) by an integral, the distribution of the velocities of the atoms of a perfect gas being continuous, we obtain

$$\frac{1}{2} k_B T = \frac{\frac{1}{2m} \int_{-\infty}^{+\infty} p^2 \exp\left(-\lambda_2 \frac{1}{2m} p^2\right) \mathrm{d}p}{\int_{-\infty}^{+\infty} \exp\left(-\lambda_2 \frac{1}{2m} p^2\right) \mathrm{d}p} \tag{7.30}$$

By using relations (7.8.1) and (7.8.2), we have

$$\lambda_2 = \frac{1}{k_B T} \tag{7.31}$$

We can now obtain a total explicit expression for (7.28):

$$\frac{N_j}{N} = \frac{\exp(-W_j/k_B T)}{\Sigma \exp(-W_j/k_B T)} \tag{7.32}$$

and obtain the Boltzmann distribution (7.32) shown in Fig. 7.3.

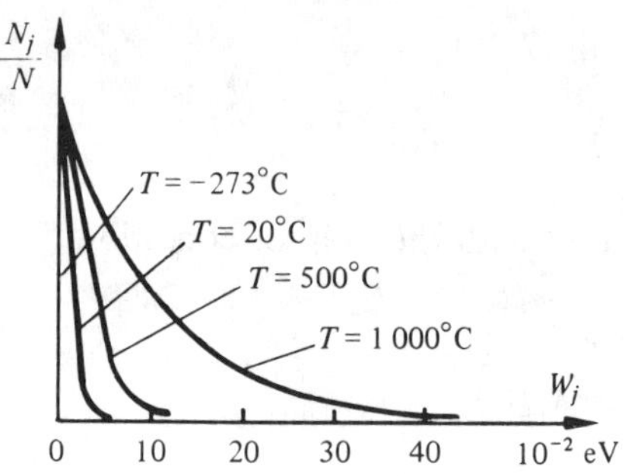

Fig. 7.3

7.4 THE FERMI-DIRAC DISTRIBUTION

7.4.1 Area of Application

The Fermi-Dirac distribution governs the behavior of a set of point particles with quantum characteristics,

- in thermodynamic equilibrium,
- without mutual interaction in the sense of Section 7.2.1, and
- capable of being subjected to an external potential.

Because the particles have quantum characteristics, they are subject to the Pauli exclusion principle. The Fermi-Dirac distribution therefore appears as a Boltzmann distribution modified by the extra condition of obeying the Pauli principle. However, this condition has a considerable effect and the two distributions only resemble each other in the high-energy region.

7.4.2 Derivation of the Fermi-Dirac Distribution

Consider a system in which each energy level may either be unoccupied, or occupied by a single particle (nondegenerate system).

Let us consider two particles in this system with initial energies W_1 and W_2, and energies W_3 and W_4 after having collided with each other.

The conservation energy theorem dictates that

$$W_1 + W_2 = W_3 + W_4 = W_1 - \delta W + W_2 + \delta W \tag{7.33}$$

because the collisions are assumed to be elastic. The term δW represents the transfer of energy from one particle to another.

The chances of observing the collision described by (7.33) depend on

- the probabilities that the two particles have energies W_1 and W_2;
- the probabilities that the two states W_3 and W_4 are not occupied. Otherwise, the Pauli principle would forbid the collision.

Let $f(W)$ be the probability that one particle in the system has energy W. The term $1 - f(W)$ represents the probability that the level with energy W is unoccupied. Because all the probabilities concerned are independent of each other, the frequency ν of collisions of the type (7.33) is

$$\nu = \alpha f(W_1)f(W_2)[1 - f(W_3)][1 - f(W_4)] \tag{7.34}$$

where α is a proportionality coefficient depending on the total number of particles and their number per unit volume.

Because the system is in equilibrium, the detailed balance principle of [68] imposes that the frequency of collisions of type (7.33) is equal to the frequency of the inverse collisions given by

$$\nu = \alpha f(W_3)f(W_4)[1 - f(W_1)][1 - f(W_2)] \tag{7.35}$$

By comparing (7.34) and (7.35), we can write

$$\frac{1 - f(W_1)}{f(W_1)} \frac{1 - f(W_2)}{f(W_2)} = \frac{1 - f(W_3)}{f(W_3)} \frac{1 - f(W_4)}{f(W_4)}, \qquad (7.36)$$

or also, by introducing the amount of energy transferred during the collision δW,

$$\frac{1 - f(W_1)}{f(W_1)} \frac{1 - f(W_2)}{f(W_2)} = \frac{1 - f(W_1 - \delta W)}{f(W_1 - \delta W)} \frac{1 - f(W_2 - \delta W)}{f(W_2 - \delta W)} \qquad (7.37)$$

This equation can only be satisfied if

$$\frac{1 - f(W_i)}{f(W_i)} = C \exp(\beta W_i), \qquad (7.38)$$

where i = 1, 2, 3, or 4; C and β are constants. In the general case, i denotes any allowed energy level. From (7.38), we obtain

$$f(W_i) = (C \exp(\beta W_i) + 1)^{-1} \qquad (7.39)$$

Mathematically, C and β can have any values provided that $f(W_i) \leq 1$ in accordance with the definition of a probability. These two constants must be determined on the basis of physical arguments.

Let us consider a range of sufficiently high energies for the probability of a level being occupied to be a very small. By assuming that the Pauli principle does not exist, the probability of a level being occupied by *two* or more particles would be infinitesimal. Consequently, the distribution is not affected over this range by whether the particles are subjected to the Pauli principle or not. In other words, the Fermi-Dirac distribution tends to the Boltzmann distribution over this range. At high energies, (7.39) reduces to

$$f(W_i) = C^{-1} \exp(-\beta W_i) \qquad (7.40)$$

By comparing this expression with the Boltzmann distribution (7.32), it is found that

$$\beta = 1/k_B T \qquad (7.41)$$

The constant C depends on the system of particles studied and their environment. By definition of W_F, we set

$$C = \exp(-W_F/k_B T) \tag{7.42}$$

The constant W_F is called the Fermi energy. By substituting (7.41) and (7.42) into (7.39), we obtain the Fermi-Dirac distribution $F(W_i)$, shown in Fig. 7.4:

$$F(W_i) = \frac{1}{\exp\left(\frac{W_i - W_F}{k_B T}\right) + 1} \tag{7.43}$$

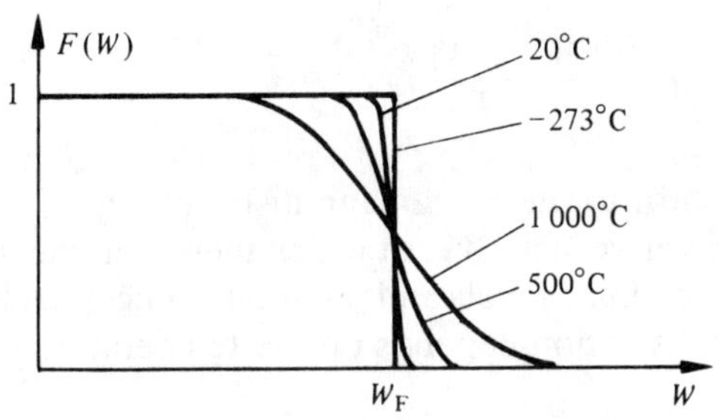

Fig. 7.4

7.4.3 Properties of the Fermi-Dirac Distribution

The Fermi-Dirac distribution depends on one independent variable, the energy W_i, and two parameters, the Fermi energy W_F and the absolute temperature T.

Examination of (7.43) makes it possible to deduce the following properties of the distribution:

- whatever the temperature, the probability of occupation of the states at $W = W_F$ is equal to ½;
- at the temperature of 0 K,

$$F(W) = 1 \text{ for } W < W_F \tag{7.44}$$

$$F(W) = 0 \text{ for } W > W_F \tag{7.45}$$

At absolute zero, all the levels corresponding to energies less than the Fermi energy are occupied, whereas all the levels corresponding to higher energies are empty. At any temperature different from 0 K, three regions can be distinguished:

- the low-energy region where $W \ll W_F$. Then, by truncated expansion of (7.43), we find

$$F(W_i) \cong 1 - \exp\left(\frac{W_i - W_F}{k_B T}\right) \tag{7.46}$$

The probability of nonoccupation of a state is equal to $\exp[(W_i - W_F)/k_B T]$;
- the high-energy region where $W \gg W_F$. In (7.43), 1 becomes negligible compared to the exponential in the denominator, so that

$$F(W_i) \cong \exp\left(-\frac{W_i - W_F}{k_B T}\right) \tag{7.47}$$

and we return to the Boltzmann distribution;
- the region where $W \cong W_F$, in which there is a transition between the intensely occupied levels and the weakly occupied levels. The rapidity of this transition depends on the temperature (Table 7.5).

Table 7.5

	Corresponding Energy	
Occupation Level	k_B T *Units*	*eV, at 20°C*
95%	$W_F - 2.95$	$W_F - 0.075$
90%	$W_F - 2.20$	$W_F - 0.056$
10%	$W_F + 2.20$	$W_F + 0.056$
5%	$W_F + 2.95$	$W_F + 0.075$

7.5 RECIPROCAL LATTICE AND BRILLOUIN ZONES

7.5.1 Introduction

Crystals are represented by means of the concepts of unit cells, lattice point patterns, and base vectors (Section 1.4) with respect to ordinary space.

Certain properties of crystals (Sections 2.6.16 and 7.5.5) are worth describing not in ordinary space but in the space of the wave numbers k_x, k_y, k_z, also called the k space. The reference frame used in this space is formed from the three base vectors of the reciprocal lattice.

7.5.2 Definition of the Base Vectors of the Reciprocal Lattice

Let $\boldsymbol{a}_1$, $\boldsymbol{a}_2$, and $\boldsymbol{a}_3$ be the base vectors of a crystal lattice. The *base vectors* $\boldsymbol{b}_1$, $\boldsymbol{b}_2$, and $\boldsymbol{b}_3$ *of the reciprocal lattice* (of this crystal lattice) are defined by the following relationships:

$$\boldsymbol{b}_1 = 2\pi \frac{\boldsymbol{a}_2 \times \boldsymbol{a}_3}{\boldsymbol{a}_1 \cdot (\boldsymbol{a}_2 \times \boldsymbol{a}_3)}; \; \boldsymbol{b}_2 = 2\pi \frac{\boldsymbol{a}_3 \times \boldsymbol{a}_1}{\boldsymbol{a}_1 \cdot (\boldsymbol{a}_2 \times \boldsymbol{a}_3)}; \; \boldsymbol{b}_3 = 2\pi \frac{\boldsymbol{a}_1 \times \boldsymbol{a}_2}{\boldsymbol{a}_1 \cdot (\boldsymbol{a}_2 \times \boldsymbol{a}_3)} \tag{7.48}$$

Because the $\boldsymbol{a}_i$ vectors have the dimension of a length, the above equations show that the $\boldsymbol{b}_i$ vectors have, as the wave number, the inverse of a length as a dimension.

7.5.3 Definition of the Reciprocal Lattice

The *reciprocal lattice* is the lattice formed from points situated at the end of vectors $\boldsymbol{R}$ such that

$$\boldsymbol{R} = n_1 \boldsymbol{b}_1 + n_2 \boldsymbol{b}_2 + n_3 \boldsymbol{b}_3, \tag{7.49}$$

where n_1, n_2 and n_3 are integers.

7.5.4 Examples of Reciprocal Lattices

In order to construct the reciprocal lattice of a given lattice, it is worth noting that equations (7.48) satisfy the following conditions in which δ_{ij} is the Kronecker symbol:

$$\boldsymbol{a}_i \cdot \boldsymbol{b}_j = 2\pi\delta_{ij} \qquad i, j = 1, 2, 3 \tag{7.50}$$

Any base vector of the reciprocal lattice is therefore orthogonal to the base vectors of the crystal lattice with different indices (Table 7.6). If

Table 7.6

Crystal lattice	Reciprocal lattice
1. Linear	
a_1	 b_1
2. Planar: square	
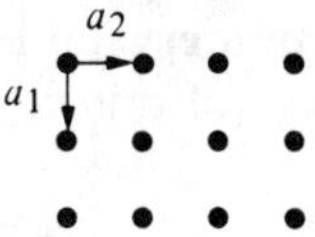 a_2, a_1	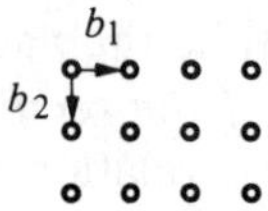 b_1, b_2
3. Planar: centered rectangular	
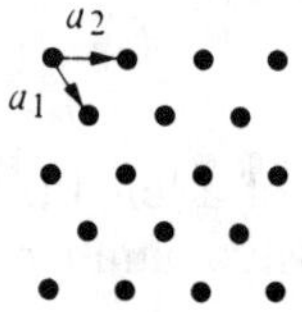 a_2, a_1	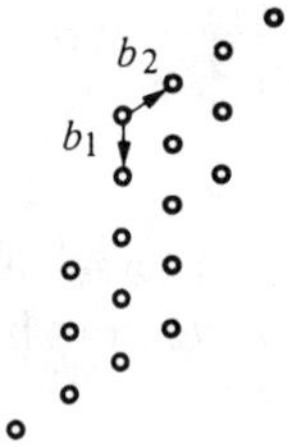 b_2, b_1

●, lattice points
○, reciprocal lattice points

the base for these vectors is known, their direction and modulus are determined by (7.50).

7.5.5 Reciprocal Lattice and Diffraction

The reciprocal lattice has a key role in the study of crystal structures using x-rays (Section 1.6). In fact, the diffraction pattern of a crystal is an image of its reciprocal lattice, so that a major problem of diffraction methods is the reconstitution of the crystal lattice from images of its reciprocal lattice [69, 70]. Here, we shall limit ourselves to showing how the Laue equation, which is a generalization of the Bragg condition (1.76) to three-dimensional space, relates the crystal lattice to the reciprocal lattice. Let

$$\boldsymbol{C} = l_1\boldsymbol{a}_1 + l_2\boldsymbol{a}_2 + l_3\boldsymbol{a}_3, \tag{7.51}$$

with l_1, l_2, and l_3 integers, be a vector of the crystal lattice, the end of which coincides with a lattice point.

We take the scalar product of this vector with a vector of the reciprocal lattice $\boldsymbol{R}$ such as (7.49). Taking into account (7.50), we have

$$\boldsymbol{R} \cdot \boldsymbol{C} = 2\pi(n_1 l_1 + n_2 l_2 + n_3 l_3) \tag{7.52}$$

The expression in the parentheses is still an integer, which makes it possible to write (7.52) in the form:

$$\exp(\mathrm{j}\boldsymbol{R} \cdot \boldsymbol{C}) = 1, \tag{7.53}$$

which is the Laue equation [5]:

$$\exp(\mathrm{j}\Delta\boldsymbol{k} \cdot \boldsymbol{C}) = 1, \tag{7.54}$$

where

$$\Delta\boldsymbol{k} = \boldsymbol{k}' - \boldsymbol{k}, \tag{7.55}$$

and $\boldsymbol{k}$ and $\boldsymbol{k}'$ are the wave vectors of the incident wave and the diffracted wave, respectively.

We easily return to Bragg's equation (1.76) starting from the Laue equation (7.54). For this, it is assumed that the crystal has a linear response. The frequency of the emergent wave is then equal to the frequency of the incident wave. It follows that

$$k = k' \tag{7.56}$$

because the two waves are propagating in the vacuum surrounding the specimen.

A possible configuration for the wave vectors satisfying the condition for Bragg diffraction, is shown in Fig. 7.7.

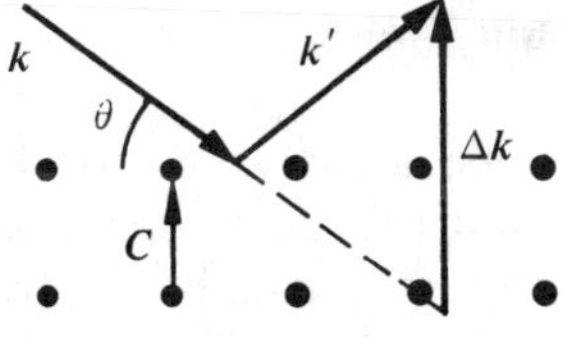

Fig. 7.7

It immediately follows that

$$\boldsymbol{\Delta k} = 2k \sin \theta \tag{7.57}$$

and

$$\boldsymbol{\Delta k} \cdot \boldsymbol{C} = 2kC \sin \theta \tag{7.58}$$

From (7.54):

$$\boldsymbol{\Delta k} \cdot \boldsymbol{C} = 2n\pi \qquad n \text{ integer} \tag{7.59}$$

from which we have

$$2C \sin \theta = n \frac{2\pi}{k} = n\lambda \tag{7.60}$$

This is the Bragg diffraction law.

7.5.6 Definition of the First Brillouin Zone

The first Brillouin zone is a volume in k space determined by the following procedure:

- a reciprocal lattice point is chosen as the origin O;
- O is joined to each of the n_1 lattice points nearest to O by a vector $\boldsymbol{R}_i$, i varying from 1 to n_1;
- the π_i planes ($i = 1$ to n_1) are constructed perpendicular to these vectors passing halfway between O and the lattice points concerned.

The *first Brillouin zone* is the smallest closed volume bounded by the planes π_i.

7.5.7 Higher-order Brillouin Zones

In the order of increasing distances to O, apart from the lattice points considered for the first Brillouin zone, a succession of families of lattice points appears, each family equidistant from the origin.

A Brillouin zone may be defined for each of these families by a similar procedure to that used for the first zone. It also has to be taken into account that the Brillouin zones never interpenetrate and that they all have the same volume. By way of example, the first three zones of a few simple reciprocal lattices are shown in Fig. 7.8.

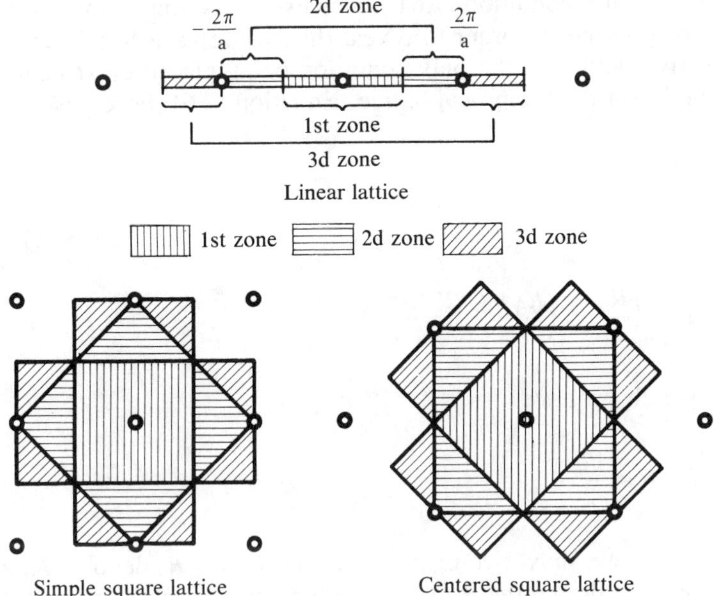

Fig. 7.8

7.5.8 Brillouin Zones and Diffraction

The examination of (7.53) and (7.54) shows that the Laue diffraction condition is satisfied if $\mathbf{\Delta k}$ is a vector $\boldsymbol{R}$ joining two reciprocal lattice points:

$$\mathbf{\Delta k} = \boldsymbol{R}, \tag{7.61}$$

or from (7.55):

$$\boldsymbol{k}' = \boldsymbol{k} + \boldsymbol{R} \tag{7.62}$$

By squaring both sides of (7.62), we have

$$k'^2 = k^2 + R^2 + 2\boldsymbol{k} \cdot \boldsymbol{R} \tag{7.63}$$

By taking into account (7.56), we obtain

$$2\boldsymbol{k} \cdot \boldsymbol{R} + R^2 = 0 \tag{7.64}$$

The diffraction condition (7.64) receives a very simple interpretation in reciprocal space. In order to reveal this, we replace $\boldsymbol{R}$ by $-\boldsymbol{R}$, which is perfectly legitimate, the only condition being placed on $\boldsymbol{R}$ being to join two nodes of the reciprocal lattice. Equation (7.64) becomes

$$2\boldsymbol{k} \cdot \boldsymbol{R} = R^2 \tag{7.65}$$

or

$$2\boldsymbol{k} \cdot \left(\frac{\boldsymbol{R}}{2}\right) = \left(\frac{R}{2}\right)^2, \tag{7.66}$$

from which we have

$$\boldsymbol{k} \cdot \boldsymbol{R} = \frac{R}{2} \tag{7.67}$$

In Fig. 7.9, we show two examples of the vector $\boldsymbol{R}$, denoted $\boldsymbol{R}_1$ and $\boldsymbol{R}_2$, and we have drawn the π planes perpendicular to these vectors passing through their midpoint. These π planes bound the Brillouin zones. Any vector $\boldsymbol{k}$ with origin O whose end lies in a plane π satisfies (7.67). The diffraction condition therefore takes the form of Fig. 7.9.

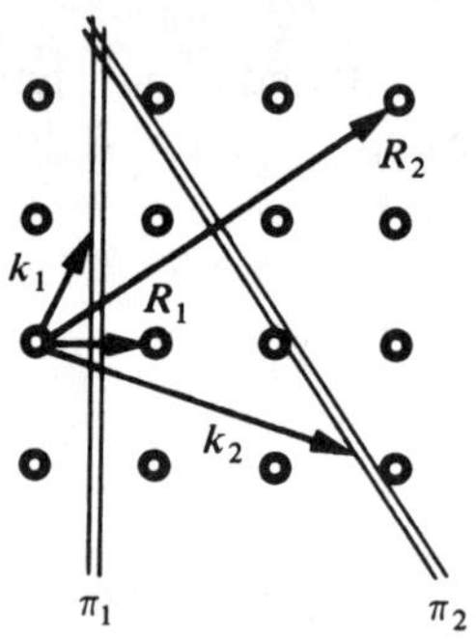

Fig. 7.9

For a wave to be diffracted by a crystal lattice, it is necessary that its wave vector, referred to the origin in the reciprocal lattice, has its end in a plane defining a Brillouin zone.

7.6 MAGNETIC MOMENTS OF ELECTRONS AND ATOMS

7.6.1 System With One Electron

Magnetic moments involved in the various types of magnetism described in Section 3.3 are produced by electrons. As we mentioned in Section 3.2.2, there are two elementary forms of magnetic moment:

- the orbital magnetic moment $\boldsymbol{m}_\mathrm{L}$;
- the spin magnetic moment $\boldsymbol{m}_\mathrm{S}$.

The origin of these moments is briefly described below within the framework of the amperian current model (Section 3.2.3). According to the Bohr model, the electron rotates with an angular velocity ω_L at distance $\boldsymbol{r}$ from the nucleus. This movement corresponds to the elementary current i,

$$i = -e\,\frac{\omega_\mathrm{L}}{2\pi}, \tag{7.68}$$

creating an orbital magnetic moment $\boldsymbol{m}_\mathrm{L}$:

$$\boldsymbol{m}_\mathrm{L} = -e\,\frac{\omega_\mathrm{L}}{2}\,\boldsymbol{r}^2 \tag{7.69}$$

In this movement, the angular momentum of the electron, called the orbital angular momentum $\boldsymbol{L}$ is given by

$$\boldsymbol{L} = \boldsymbol{r} \times m_\mathrm{n}(\omega_\mathrm{L} \times \boldsymbol{r}) = m_\mathrm{n}\,\omega_\mathrm{L}\boldsymbol{r}^2 \tag{7.70}$$

It follows from (7.69) and (7.70) that the orbital magnetic moment is related to the orbital angular momentum by

$$\boldsymbol{m}_\mathrm{L} = -\,\frac{e\hbar}{2m_\mathrm{n}}\,\mathbf{L} \tag{7.71}$$

Because the kinetic moment is always, according to quantum mechanics, a multiple of $\hbar$, it follows from (7.71) that the orbital magnetic moment is a multiple of the Bohr magneton m_B defined by

$$m_\mathrm{B} = \frac{e\hbar}{2m_\mathrm{n}} = 9.273 \cdot 10^{-24}\ \mathrm{A} \cdot \mathrm{m}^2 \tag{7.72}$$

Apart from $\boldsymbol{L}$, the electron has an intrinsic angular momentum $\boldsymbol{S}$ (Section 1.2.2), also producing a magnetic moment, $\boldsymbol{m}_S$, called the spin magnetic moment. The quantities $\boldsymbol{S}$ and $\boldsymbol{m}_S$ are not related in the same manner as $\boldsymbol{L}$ and $\boldsymbol{m}_L$, as would be expected, but by the expression:

$$\boldsymbol{m}_S = -\frac{e\hbar}{m_n}\boldsymbol{S}, \tag{7.73}$$

reflecting what is known as the magnetomechanical anomaly of the electron. Equation (7.73) agrees with the experimental results obtained by spectroscopy (Zeemann effect) on the one hand, and with relativistic quantum theory on the other hand. According to this theory, the modulus of the angular momentum $\boldsymbol{X}$ (orbital or spin) is related to the corresponding quantum number X by

$$|\boldsymbol{X}| = \hbar\sqrt{X(X+1)}, \tag{7.74}$$

from which we have, from (7.71) and (7.73):

$$|\boldsymbol{m}_L| = \frac{e\hbar}{2m_n}\sqrt{l(l+1)} = m_B\sqrt{l(l+1)} \tag{7.75}$$

$$|\boldsymbol{m}_S| = \frac{e\hbar}{m_n}\sqrt{s(s+1)} = 2m_B\sqrt{s(s+1)} \tag{7.76}$$

In these expressions, l is the azimuthal quantum number and s is the spin quantum number (Sections 1.1.5 and 1.1.2).

Apart from the quantization of the *moduli* of the angular momenta and the magnetic moments, there is a quantization of the *projection* of these quantities on to the direction of the applied field $\boldsymbol{H}$. For the angular momenta, we have

$$L_H = l_p\hbar, \tag{7.77}$$

$$S_H = s_p\hbar, \tag{7.78}$$

where L_H and S_H are the projections of $\boldsymbol{L}$ and $\boldsymbol{S}$ on $\boldsymbol{H}$. The quantity l_p is a quantum number that can take all integer values from $-l$ to $+l$; whereas $s_p = \pm 1/2$. The projected magnetic moments m_{LH} and m_{SH}, using (7.71) and (7.73), are given by

$$m_{LH} = l_p m_B \tag{7.79}$$

$$m_{SH} = 2s_p m_B \tag{7.80}$$

According to conventional usage, we can give these results the following diagrammatic representation. We consider the case of the orbital moment, but the image can be transposed immediately for the spin. The orbital movement of the electron is similar to the rotation of a top with the same angular momentum. Let γ be the angle formed by $\boldsymbol{H}$ and the axis of rotation of the top. The magnetic moment $\boldsymbol{m}$ carried by the axis of the top is subjected to a torque $\boldsymbol{C} = \mu_0 \boldsymbol{m}_L \times \boldsymbol{H}$ (Fig. 7.10) in $\boldsymbol{H}$.

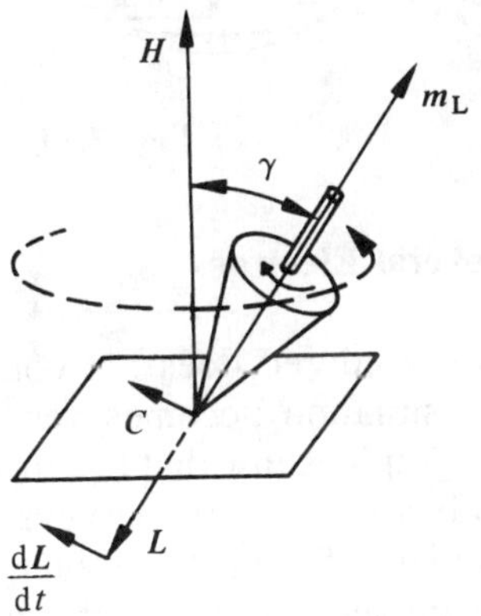

Fig. 7.10

This torque acting on the top makes it precess around $\boldsymbol{H}$ because, according to the angular momentum theorem:

$$\frac{d\boldsymbol{L}}{dt} = \boldsymbol{C} \tag{7.81}$$

The precession angular velocity ω_{pr} can be deduced from this expression:

$$\omega_{pr} = \frac{C}{\sin \gamma} \tag{7.82}$$

Figure 7.11 illustrates the significance of equations (7.77) and (7.79), namely that only certain precession angles γ are allowed. For $l = 2$, there are five of them.

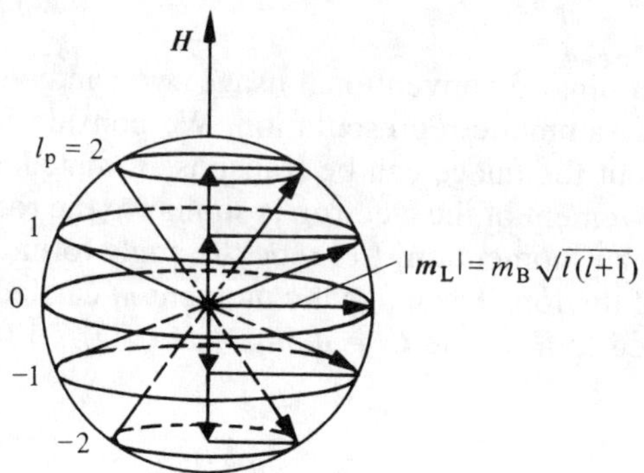

Fig. 7.11

7.6.2 System With Several Electrons

When the atom considered does not contain a single electron but several electrons, the situation becomes very complex. The results of experiments in spectrography show that the angular momenta (orbital and spin) of the electrons belonging to the same atom are added vectorially. Their resultant is quantized. When all the energy states corresponding to an electron shell are occupied, all the possible orientations for the angular momenta of the electrons in this shell are also occupied and the resultant angular momentum is always zero. In the study of the magnetic moment of an atom, it is therefore sufficient to consider electron shells that are incompletely filled. Finally, the modulus of the total atomic magnetic moment, $|\boldsymbol{m}_A|$, resulting from the combination of orbital *and* spin moments of all the electrons can be expressed by a relationship of the same form as that obtained for a single electron:

$$|\boldsymbol{m}_A| = g m_B \sqrt{J(J+1)}, \tag{7.83}$$

in which J is a quantum number, integer, or half-integer, and g is the Landé factor. The latter is a measure of the relative magnitude of the orbital and spin moments. It always lies between 1 and 2. If $\boldsymbol{m}_A$ results from spins only, $g = 2$; if, on the contrary, $\boldsymbol{m}_A$ results from orbital moments only, $g = 1$.

Under the effect of an applied field $\boldsymbol{H}$, $\boldsymbol{m}_A$ describes a precession movement comparable to that which has just been studied in the case of the single electron. Finally, the component of $\boldsymbol{m}_A$ parallel to $\boldsymbol{H}$, m_{AH} is expressed in the form:

$$m_{\mathrm{AH}} = \boldsymbol{M} g m_{\mathrm{B}} \tag{7.84}$$

$\boldsymbol{M}$ is a quantum number either taking integer values:

$$\boldsymbol{M} = 0, \pm 1, \pm 2, \ldots, \pm J \tag{7.85}$$

or half-integer values:

$$\boldsymbol{M} = \pm\frac{1}{2}, \pm\frac{3}{2}, \pm\frac{5}{2}, \ldots, \pm J, \tag{7.86}$$

depending on whether J is an integer of half-integer.

7.7 FUNDAMENTAL PHYSICAL CONSTANTS

Table 7.12

Charge of the electron	$-e$	$-1.60219 \cdot 10^{-19}$	C
Mass of the electron	m_{n}	$9.10956 \cdot 10^{-31}$	Kg
Bohr magneton	m_{B}	$1.16542 \cdot 10^{-29}$	Wb·m
		$9.27410 \cdot 10^{-24}$	A·m²
Boltzmann's constant	k_{B}	$1.38062 \cdot 10^{-23}$	J/K
		$8.61708 \cdot 10^{-5}$	eV/K
	$k_{\mathrm{B}}T$ at 20°C	$2.526 \cdot 10^{-2}$	eV
Planck's constant	h	$6.62620 \cdot 10^{-34}$	J·s
		$4.13571 \cdot 10^{-15}$	eV·s
The Avogadro number	N_{a}	$6.02217 \cdot 10^{26}$	kmole^{-1}
Magnetic permeability of vacuum	μ_0	$4\pi \cdot 10^{-7}$	H/m
Permittivity of vacuum	ε_0	$8.85419 \cdot 10^{-12}$	F/m

7.8 MATHEMATICAL RELATIONSHIPS

7.8.1

$$\int_{-\infty}^{+\infty} \exp(-ax^2)\, \mathrm{d}x = \sqrt{\frac{\pi}{a}}$$

7.8.2

$$\int_{-\infty}^{+\infty} x^2 \exp(-ax^2)\, dx = \frac{1}{2}\sqrt{\frac{\pi}{a^3}}$$

7.8.3

$$\int_0^{\infty} \sqrt{x} \exp(-x)\, dx = \frac{\sqrt{\pi}}{2}$$

7.9 DIMENSIONS AND UNITS

Table 7.13 gives the dimensions, units, and usual symbols for a number of important quantities. The dimensions are indicated by the powers of the following basic units (Vol. I, Table 1.1), the mass m, the length l, the time t, the current i, and the temperature T.

Table 7.13

Quantity	Dimension m	l	t	i	T	Unit	Usual Symbol
Acceleration	0	1	−2	0	0	meter/second^2	a
Admittance	−1	−2	3	2	0	siemens	Y
Magnetization	0	−1	0	1	0	ampere/meter	M
Capacitance	−1	−2	4	2	0	farad	C
Electric field	1	1	−3	−1	0	volt/meter	E
Magnetic field	0	−1	0	1	0	ampere/meter	H
Charge	0	0	1	1	0	coulomb	q
Conductivity	−1	−3	3	2	0	siemens/meter	σ
Stress	1	−1	−2	0	0	newton/meter^2	σ
Current	0	0	0	1	0	ampere	i,I
Current density	0	−2	0	1	0	ampere/meter^2	J
Electric displacement	0	−2	1	1	0	coulomb/meter^2	D
Energy	1	2	−2	0	0	joule	W
Entropy	1	2	−2	0	−1	joule/degree	S
Magnetic flux	1	2	−2	−1	0	weber	ϕ
Force	1	1	−2	0	0	newton	F,f
Frequency	0	0	−1	0	0	hertz	f
Impedance	1	2	−3	−2	0	ohm	Z
Inductance	1	2	−2	−2	0	henry	L
Magnetic induction	1	0	−2	−1	0	tesla	B
Density	1	−3	0	0	0	kilogram/meter^3	ρ
Elasticity modulus	1	−1	−2	0	0	newton/meter^2	E_Y
Shear modulus	1	−1	−2	0	0	newton/meter^2-radian	G
Momentum	1	2	−1	0	0	newton-meter-second	
Electric dipole moment	0	1	1	1	0	coulomb-meter	p
Amperian magnetic moment	0	2	0	1	0	ampere-meter^2	m_A
Magnetic dipole moment	1	3	−2	−1	0	weber-meter	m_d
Number per unit volume	0	−3	0	0	0	meter^{-3}	N
Absolute magnetic permeability	1	1	−2	−2	0	henry/meter	μ
Permittivity	−1	−3	4	2	0	farad/meter	ε
Electric polarization	0	−2	1	1	0	coulomb/meter^2	P
Magnetic polarization	1	0	−2	−1	0	tesla	I
Power	1	2	−3	0	0	watt	P
Resistance	1	2	−3	−2	0	ohm	R
Resistivity	1	3	−3	−2	0	ohm-meter	ρ
Voltage	1	2	−3	−1	0	volt	U
Work	1	2	−2	0	0	joule	W

Solutions to the Problems

CHAPTER 1

1.8.1

$$\Psi_{200} = \frac{1}{8}\frac{1}{\sqrt{2\pi}}\left(\frac{1}{r_0}\right)^{3/2}\left(-\frac{2r}{r_0} + 4\right)\exp(-r/2r_0)$$

$$r_1 = (3 - \sqrt{5})r_0;\ r_2 = (3 + \sqrt{5})r_0$$

1.8.2 $r = 3.72 \cdot 10^{-10}$ m

1.8.3 $\pi/6$; $\pi\sqrt{3}/8$; $\pi\sqrt{2}/6$; $\pi\sqrt{2}/6$.

1.8.4 $1.53 \cdot 10^{19}$ m^{-2}; $1.76 \cdot 10^{19}$ m^{-2}

1.8.5 $2.33 \cdot 10^{3}$ kg m^{-3}

1.8.6 At 500°C: $1.04 \cdot 10^{5}$ m^{-2}; $4.02 \cdot 10^{13}$ m^{-2}. At 1000°C: $1.73 \cdot 10^{19}$ m^{-2}; $6.67 \cdot 10^{17}$ m^{-2}

1.8.7 From (1.85), we obtain:

$$N(x,t) = \frac{1}{2\sqrt{\pi D}}\,t^{1/2}\int_0^\infty f(x')\left[\exp\left(-\frac{(x - x')^2}{4Dt}\right) - \exp\left(-\frac{(x + x')^2}{4Dt}\right)\right]dx'$$

and

$$N(x,t) = \frac{2N_0}{\sqrt{\pi}}\,\mathrm{erf}\left(\frac{x}{2\sqrt{Dt}}\right)$$

CHAPTER 2

2.9.1

$$\bar{v} = \sqrt{\frac{8k_BT}{\pi m_n}}$$

2.9.2

$$R = \frac{-N_n\mu_n^2 + N_p\mu_p^2}{e(N_n\mu_n + N_p\mu_p)^2}$$

Negative R does not necessarily imply that there are more negative charge carriers, because these are the $N\mu^2$ products that appear in the formula.

2.9.3 Let $\rho(x) = \alpha x$ be the charge per unit volume. We find

$$E(x) = \frac{\alpha}{2\varepsilon}\,(x^2 - a^2)$$

It follows from this that on both sides of a plate 2 m thick, for example, a field of 1 V/m, which is a considerable value (Section 2.3.2), only produces a variation of N_n equal to $0.22 \cdot 10^9$ m^{-3}. This variation is negligible in terms of relative value (Section 2.3.2).

2.9.4 N_n = **1.45** $\cdot$ **10^{29}** m^{-3}; μ_n = **1.54** $\cdot$ **10^{-3}** m^2/V $\cdot$ s; **2.41** electrons.

2.9.5 μ_n = **1.23** $\cdot$ **10^{-1}** m^2/V $\cdot$ s; μ_p = **4.11** $\cdot$ **10^{-2}** m^2/V $\cdot$ s.

2.9.6 n = **4.99** $\cdot$ **10^{20}** m^{-3}; p = **6.07** $\cdot$ **10^{10}** m^{-3}.

CHAPTER 3

3.12.1 **353°C; 0.494** m_B.

3.12.2 We have chosen to measure the relative quantities of matter present in the normal structure and in the inverse structure by means of the proportion λ of Fe^{+++} in the A sites. Let m_1 and m_2 be the magnetic moments of Fe^{+++} and M^{++}, respectively. The molecular magnetic moment m is equal to

m_1 = $5m_B$; m_2 = $5m_B$ for Mn^{++}; m_2 = $2m_B$ for Ni^{++}; from which we have Fig. 3.110.

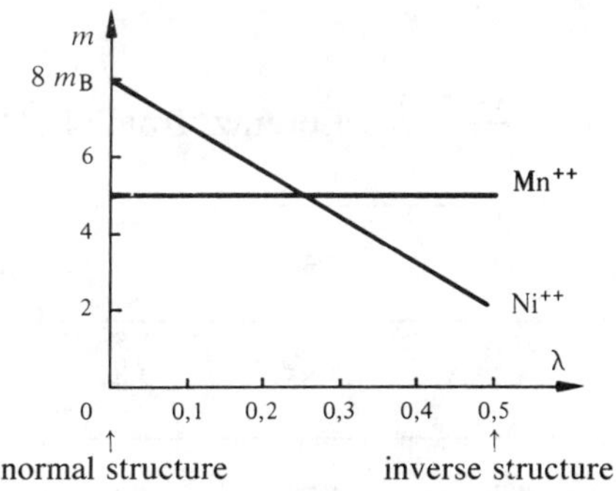

Fig. 3.110

3.12.3

$l = \sqrt{\dfrac{W_{sB} \cdot a}{1.1 \cdot 10^5 \, I_s^2}} = 8.3\ \mu\text{m}$. Ratio of the magnetostatic energies: 2180.

3.12.4

$$\bar{\mu}_{ri} = \mu_{ri} \left(\frac{1}{2} \frac{x+1}{x-1} \ln x \right) \qquad \bar{\nu} = \nu \frac{(x+1)^2}{4x}$$

3.12.5 Equation (3.117).

3.12.6

f Hz	800	1 600	2 400	3 200
R Ω	6.13	8.09	10.87	14.46

CHAPTER 4

4.16.1 Equations (4.29) and (4.30). Equation for the lines of force of $\boldsymbol{E}$: $\sin^2 \theta / \boldsymbol{r} =$ constant.

4.16.2

$$\boldsymbol{E}_{\mathrm{L}} = \boldsymbol{E} \frac{3\varepsilon_r}{2\varepsilon_r + 1} \qquad \varepsilon_r = \frac{1 + 3x + 3\sqrt{x^2 + \frac{2}{3}x + 1}}{4} \quad \text{(Onsager)}$$

where

$$x = \frac{N_{\text{or}} p_{\text{m}}^2}{3k_{\text{B}} T \varepsilon_0} \qquad \varepsilon_{\text{r}} = \frac{2x + 3}{3 - x} \quad \text{(Lorentz, from (4.56))}$$

x	0	1/3	2/3	1	1.5	2
ε_r Lorentz	1	1.37	1.85	2.49	4	7
ε_r Onsager	1	1.36	1.78	2.22	2.92	3.64

4.16.3 Diameter: $1.09 \cdot 10^{-10}$ m. Displacement: $0.5 \cdot 10^{-18}$ m.

4.16.4 Displacement: $3.38 \cdot 10^{-17}$ m.

4.16.5 $r_{\text{i}} = r_{\text{e}}/\text{e}$, e = Euler constant

4.16.6 The ratio $x = r_{\text{e}}/r_{\text{i}}$ is a solution of the equation: $\ln x = 1 - 1/x^2$. We find $x = 2.22$ from which $r_{\text{i}} = U_0/(E_{\text{max}} \ln 2.22) = 1.26 U_0/E_{\text{max}}$ and $r_{\text{e}} = 2.78\ U_0/E_{\text{max}}$.

Bibliography

[1] A. MESSIAH, *Mécanique quantique,* vol. 1, Dunod, Paris, 1959.

[2] E. MERZBACHER, *Quantum Mechanics,* John Wiley, New York, 1970.

[3] D. HARTREE, *The Calculation of Atomic Structures,* John Wiley, New York, 1957.

[4] W. MOORE, *Physical Chemistry,* Prentice-Hall, Englewood Cliffs, NJ, 1972.

[5] C. KITTEL, *Introduction to Solid State Physics,* John Wiley, New York, 1971.

[6] H.S. CARSLAW, J.C. JEAGER, *Conduction of Heat in Solids,* Oxford University Press, London, 1971.

[7] G.T. MEADEN, *Electrical Resistance of Metals,* Heywood Books, London, 1965.

[8] M. ROSE, L. SHEPARD, J. WULFF, *The Structure of Properties of Materials,* vol. 4, John Wiley, New York, 1966.

[9] L. AZAROFF, J. BROPHY, *Electronic Processes in Materials,* McGraw-Hill, New York, 1963.

[10] C. HODGMAN, *Handbook of Chemistry and Physics,* Chemical Rubber Publ., Cleveland, Ohio, 1959.

[11] A.S. GROVE, *Physics and Technology of Semiconductor Devices,* John Wiley, New York, 1967.

[12] K.B. WOLFSTIRN, Holes and Electron Mobilities in Doped Silicon from Radiochemical and Conductivity Measurements, *J. Phys. Chem. Solids,* vol. 16, 1960, p. 279.

[13] S.M. SZE, *Physics of Semiconductor Devices,* John Wiley, New York, 1969.

[14] M. HADLOW, Superconductivity and its Applications to Power Engineering, *Proc. IEE,* vol. 119, no. 8, August 1972.

[15] W.H. KEESOM, J.A. KOL, On the Change of the Specific Heat of Tin when becoming Supraconductive, *Commun. Phy. Lab. Univ. Leiden,* no. 221e, 1932.

[16] C.J. GORTER, H.B.G. CASIMIR, On Supraconductivity, *Physica,* vol. 1, 1934, p. 306.

[17] J. BARDEEN, L.N. COOPER, J.R. SCHRIEFFER, Microscopic Theory of Superconductivity, *Phys. Rev.,* vol. 106, 1957, p. 162 and Theory of Superconductivity, *Phys. Rev.,* vol. 108, 1957, p. 1175.

[18] M. TINKHAM, *Superconductivity,* in *Documents on Modern Physics,* Gordon and Breach, New York, 1965.

[19] I. GIAVER, K. MEGERLE, Study of Superconductors by Electron Tunneling, *Phys. Rev.,* vol. 122, no. 4, 1961, p. 1101.

[20] V.L. GINSBURG, L.D. LANDAU, The Theory of Superconductivity, *Eksp. & Theor. Fiz.,* vol. 20, 1950.

[21] A.A. ABRIKOSOV, The Magnetic Properties of Superconducting Alloys, *J. Phys. Chem. of Solids,* vol. 2, 1957.

[22] Document from l'Institut de physique expérimentale, Université de Lausanne, 1978.

[23] *Rev. Brown Boveri,* no. 2.77, Baden, 1977.

[24] J.K. HOFFER, et al., Stabilizing Superconductors for Power Engineering Applications, *IEEE Trans. P.A.S.,* vol. 94, no. 6, Nov.-Dec. 1975.

[25] B.T. MATTHIAS, *The Empirical Approach to Superconductivity,* Advances in Cryoengineering, no. 13, Plenum Press, 1969.

[26] B.D. JOSEPHSON, Possible New Effects in Superconductive Tunneling, *Phys. Lett.,* no. 1, 1962, p. 251.

[27] W.E. HENRY, Spin Paramagnetism in Cr^{+++}, Fe^{+++} and Gd^{+++} at Liquid Helium Temperatures and Strong Magnetic Fields, *Phys. Rev.,* no. 88, 1952, p. 559.

[28] R. BECKER, W. DÖRING, *Ferromagnetismus,* Springer, Berlin, 1939.

[29] R.M. BOZORTH, *Ferromagnetismus,* Van Nostrand, New York, 1951.

[30] C.G. SHULL, et al., Neutron Diffraction Studies and Antiferromagnetism in Manganese and Nickel Oxides, *Phys. Rev.,* no. 83, 1951.

[31] L. NEEL, Preuves expérimentales du ferrimagnétisme et de l'antiferromagnétisme, *Ann. Inst. Fourier,* no. 1, 1949, p. 163.

[32] E.W. GORTER, Saturation Magnetization and Crystal Chemistry of Ferrimagnetic Oxides, *Philips Res. Report,* no. 9, 1954.

[33] L. NEEL, Propriétés magnétiques des ferrites; ferrimagnétisme et antiferromagnétisme, *Ann. de Phys.,* 12th series t, March-April 1948, pp. 137–198.

[34] J. SMIT, P.J. WIJN, *Les ferrites,* Bibliothèque technique Philips, Eindhoven, 1962.

[35] A. DESCHAMPS, Les ferrites en hyperfréquence, *Câbles et Transmissions,* no. 4, Oct. 1970.

[36] A. DESCHAMPS, Ferrites grenats pour dispositifs microélectroniques de puissance en hyperfréquences, *Câbles et Transmissions,* no. 2, April 1976.

[37] G.F. DIONNE, A Review of Ferrites for Microwave Applications, *Proc. of IEEE,* vol. 63, no. 5, May 1975.

[38] R.S. TEBBLE, D.J. CRAIK, *Magnetic Materials,* Wiley-Interscience, London, 1969.

[39] S. CHIKAZUMI, *Physics of Magnetism,* John Wiley, New York, 1964.

[40] R. WHITE, *Quantum Theory of Magnetism,* McGraw-Hill, New York, 1970.

[41] R. GOLDSCHMIDT, *Courants faibles 2* (Weak Currents 2), Course notes, EPFL, Lausanne, 1971.

[42] G.L. HOUSE, JR., Domain Wall Motion in Grain-Oriented Silicon Steel in Cyclic Magnetic Fields, *J. Appl. Phys.,* no. 38, 1967, pp. 1089–1096.

[43] M. JUFER, A. APOSTOLIDES, An Analysis of Eddy Currents and Hysteresis Losses in Solid Iron Based upon Simulation of Saturation and Hysteresis Characteristics, *IEEE Trans. P.A.S.,* vol. 95, no. 6, Nov.-Dec. 1976.

[44] H. EDELMANN, Model for Calculating Magnetic Hysteresis Loops, *Siemens Forsch. u. Entwickl. Ber.,* vol. 5, Springer Verlag, 1976.

[45] C. HECK, *Magnetic Materials and their Applications,* Butterworth, London, 1974.

[46] *Documentation Imphysil,* Métalimphy, Paris.

[47] *Documentation Philips,* Zürich.

[48] J.E. GOULD, Progress in Permanent Magnet Materials, *Proc. IEE,* no. 166A, 1959.

[49] R. BOLL, *Weichmagnetische Werkstoffe,* Siemens A.G., Berlin, 1977.

[50] E.C. SNELLING, *Soft ferrites,* Iliffe Books, London, 1969.

[51] B.D. CULLITY, *Introduction to Magnetic Materials,* Addison-Wesley, Philippines, 1972.

[52] E. DURAND, *Electrostatique I. Les distributions,* Masson, Paris, 1964.

[53] P. BAUDOUX, *Précis d'électricité fondamentale,* Presses universitaires de Bruxelles-Eyrolles, Paris, 1970.

[54] C.J.F. BÖTTCHER, *Theory of Electric Polarization,* 2nd ed., Elsevier, Amsterdam, 1973.

[55] K.S. COLE, R.H. COLE, Dispersion and Absorption in Dielectrics, *J. Chem. Phys.* no. 9, 1941, pp. 341–351.

[56] O. ZINKE, *Wiederstände, Kondensatoren, Spulen und ihre Werkstoffe,* Springer Verlag, Berlin, 1965.

[57] F.H. KREUGER, *Discharge Detection in High Voltage Equipment,* Temple Press Books, London, 1964.

[58] J.J. O'DWYER, *The Theory of Electrical Conduction and Breakdown in Solid Dielectrics,* Clarendon Press, Oxford, 1973.

[59] *IEEE Transactions on Electrical Insulation,* published bimonthly by IEEE New York.

[60] A.H. BECK, *Handbook of Vacuum Physics,* The Macmillan Company, Pergamon Press, New York, 1965.

[61] W.J. MERZ, The Electric and Optical Behaviour of $BaTiO_3$ Single Domain Crystals, *Phys. Rev.,* no. 76, 1949, p. 1221.

[62] J.F. NYE, *Propriétés physiques des cristaux,* Dunod, Paris, 1961.

[63] G. CAMILLI, et al., Dielectric Behavior of some Fluorogases and their Mixtures with Nitrogen, *Electrical Engineering,* July 1955.

[64] *Plastiques modernes et élastomères,* July-August, CFE, Paris, 1972.

[65] C. BRINKMANN, *Die Isolierstoffe der Elektrotechnik,* Springer Verlag, Berlin, 1975.

[66] J.S. HSICH, *Principles of Thermodynamics,* Scripta Book Company, Washington D.C., 1975.

[67] J. SCHULTZ, *Polymer Materials Science,* Prentice-Hall, Englewood Cliffs, 1974.

[68] C. KITTEL, *Eléments de physique statistique,* Dunod, Paris, 1961.

[69] C.S. BARRETT, T.B. MUSSALSKI, *Structure of Metals: Cristallographic Methods, Principles, Data,* McGraw-Hill, New York, 1966.

[70] A.H. COMPTON, S.K. ALLISON, *X-rays in Theory and Experiment,* Van Nostrand, London, 1967.

Select Bibliography

The Traite d' Electricité, listed below by volume number, is published by the Presses Polytechniques Romandes (Lausanne, Switzerland) in collaboration with the Ecole Polytechnique Fédérale de Lausanne. The title of each volume is given with the year of publication in parentheses. English translations by Artech House are denoted by an asterisk with the year of publication in parentheses.

Vol.	Author	Title
I	Frédéric de Coulon & Marcel Jufer	Introduction à l'électrotechnique (1981).
II	Philippe Robert	Matériaux de l'électrotechnique (1979). *Electrical and Magnetic Properties of Materials (1988).
III	Fred Gardiol	Electromagnétisme (1979).
IV	René Boite & Jacques Neirynck	Theorie des reseaux de Kirchhoff (1983).
V	Daniel Mange	Analyse et synthèse des systèmes logiques (1979). *Analysis and Synthesis of Logic Systems (1986).
VI	Frédéric de Coulon	Theorie et traitement des signaux (1984). *Signal Theory and Processing (1986).
VII	Jean-Daniel Chatelain	Dispositifs à semiconducteur (1979).
VII	Jean-Daniel Chatelain & Roger Dessoulavy	Electronique (1982).
IX	Marcel Jufer	Transducteurs électromécaniques (1979).
X	Jean Chatelain	Machines électriques (1983).
XI	Jacques Zahnd	Machines séquentielles (1980).
XII	Michel Aguet & Jean-Jacques Morf	Energie électrique (1981).
XIII	Fred Gardiol	Hyperfréquences (1981). *Introduction to Microwaves (1984).
XIV	Jean-Daniel Nicoud	Calculatrices (1983).
XV	Hansruedi Bühler	Electronique de puissance (1981).
XVI	Hansruedi Bühler	Electronique de réglage et de commande (1979).
XVII	Philippe Robert	Mesures (1985).
XVIII	Pierre-Gérard Fontolliet	Systèmes de télécommunications (1983). *Telecommunication Systems (1986).

XIX	Martin Hasler & Jacques Neirynck	Filtres électriques (1981). *Electric Filters (1986).
XX	Murat Kunt	Traitement numérique des signaux (1980). *Digital Signal Processing (1986).
XXI	Mario Rossi	Electroacoustique (1984). *Acoustics and Electroacoustics (1988).
XXII	Michel Aguet & Mircea Ianovici	Haute tension (1982).

Index

UNIVERSITY OF BRISTOL LIBRARY
ENGINEERING
88 03353